Ecological Risk Assessment of Contaminants in Soil

Edited by

Nico M. van Straalen

Department of Ecology and Ecotoxicology
Vrije Universiteit
The Netherlands

and

Hans Løkke

Department of Terrestrial Ecology
National Environmental Research Institute
Denmark

An Initiative of the SERAS network

Supported by
the Netherlands Integrated Soil Research Programme

CHAPMAN & HALL

London • Weinheim • New York • Tokyo • Melbourne • Madras

Published by Chapman & Hall,
2–6 Boundary Row, London SE1 8HN, UK

Chapman & Hall, 2–6 Boundary Row, London SE1 8HN, UK

Chapman & Hall GmbH, Pappelallee 3, 69469 Weinheim, Germany

Chapman & Hall USA, 115 Fifth Avenue, New York, NY 10003, USA

Chapman & Hall Japan, ITP-Japan, Kyowa Building, 3F, 2-2-1 Hirakawacho, Chiyoda-ku, Tokyo 102, Japan

Chapman & Hall Australia, 102 Dodds Street, South Melbourne, Victoria 3205, Australia

Chapman & Hall India, R. Seshadri, 32 Second Main Road, CIT East, Madras 600 035, India

First edition 1997

© 1997 Chapman & Hall

Typeset in 10/12pt Times by Saxon Graphics Ltd, Derby

Printed in Great Britain by T.J. International, Padstow, Cornwall

ISBN   0 412 75900 4

A catalogue record for this book is available from the British Library

Library of Congress Catalog Card Number: 96–72372

∞ Printed on permanent acid-free text paper, manufactured in accordance with ANSI/NISO Z39.48-1992 and ANSI/NISO Z39.48-1984 (Permanence of Paper).

Contents

PART THREE   THE SOIL AS AN ECOSYSTEM

PART FOUR   THE SPATIAL COMPONENT OF SOIL COMMUNITIES

Participants of the SERAS workshop

Dr Jørgen A. Axelsen
National Environmental Research
Institute
Department of Terrestrial Ecology
PO Box 314 - Vejlsøvej 25
DK-8600 Silkeborg
Denmark

Dr Göran Bengtsson
Department of Ecology
University of Lund
Sölvegatan 37
S-22362 Lund
Sweden

Dr Ruslan O. Butovsky
All-Russian Research Institute for
Nature Protection
Sadki-Znamenskoje
113628 Moscow
Russian Federation

Dr Trudie Crommentuijn
RIVM-CSR
PO Box 1
3720 BA Bilthoven
The Netherlands

Dr John Dighton
(Division of Pinelands Research
Department of Biology
Rutgers University
Camden
New Jersey 08102
USA

Dr Marianne H. Donker
Department of Ecology and
Ecotoxicology
Vrije Universiteit
De Boelelaan 1087
1081 HV Amsterdam
The Netherlands

Dr Herman Eijsackers
RIVM-Eco
PO Box 1
3720 BA Bilthoven
The Netherlands

Dr. Thomas L. Forbes
National Environmental Research
Institute
Department of Marine Ecology and
Microbiology
Frederiksborgvej 399 - PO Box 358
DK-4000 Roskilde
Denmark

Dr Stephen P. Hopkin
Ecotoxicology Group
School of Animal and Microbial
Sciences
University of Reading
PO Box 228
Reading RG6 2AJ
United Kingdom

Professor Paul C. Jepson
Department of Entomology
Oregon State University

Cordley Hall 2046
Corvallis
Oregon 97331
USA

Dr Jan E. Kammenga
Department of Nematology
Agricultural University Wageningen
Binnenhaven 10
6709 PD Wageningen
The Netherlands

Dr Heinz Köhler
Zoological Institute
Eberhard-Karls-University
Tübingen
Auf der Morgenstelle 28
D-72076 Tübingen
Germany

Dr Werner Kratz
Fachbereich Biologie,
WE 5Freie Universität Berlin
Grunewaldstrasse 34
D12165 Berlin
Germany

Dr Paul H. Krogh
National Environmental Research
Institute
Department of Terrestrial Ecology
PO Box 314 - Vejlsøvej 25
DK-8600 Silkeborg
Denmark

Dr Cristine Kula
Federal Biological Research Centre
for Agriculture and Chemistry
Biology Division
Messeweg 11-12
D38104 Braunschweig
Germany

Professor Otto Larink
Zoologisches Institut

Technische Universität
Braunschweig
Pockelsstrasse 10A
D38092 Braunschweig
Germany

Dr Ryszard Laskowski
Institute of Environmental Biology
Department of Ecosystem Studies
Jagiellonian University
Ingardena 6
PL30-060 Krakow
Poland

Dr Hans Løkke
National Environmental Research
Institute
Department of Terrestrial Ecology
PO Box 314 - Vejlsøvej 25
DK-8600 Silkeborg
Denmark

Professor John C. Moore
Department of Biological Sciences
University of Northern Colorado
Greeley
Colorado 80639
USA

Dr J. Notenboom
RIVM-Eco
PO Box 1
3720 BA Bilthoven
The Netherlands

Dr Leo Posthuma
RIVM-Eco
PO Box 1
3720 BA Bilthoven
The Netherlands

Dr D. Rossel
IGE-Écotoxicologie
École Polytechnique Federale de
Lausanne

Lausanne
Switzerland

Dr Peter C. de Ruiter
Research Institute for Agrobiology
and Soil Fertility (AB-DLO)
PO Box 129
9750 AC Haren
The Netherlands

Dr S. Rundgren
Department of Ecology
University of Lund
Sölvegatan 37
S-223 62 Lund
Sweden

Professor R. Schulin
ETH - Institut für terrestrische
Ökologie
Fachbereich Bodenschutz
Grabenstrasse 3
8952 Schlieren
Switzerland

Dr David J. Spurgeon
Ecotoxicology Group
School of Animal and Microbial
Sciences
University of Reading
PO Box 228
Reading RG6 2AJ
United Kingdom

Dr Cornelis A.M. van Gestel

Department of Ecology and
Ecotoxicology
Vrije Universiteit
De Boelelaan 1087
1081 HV Amsterdam
The Netherlands

Professor Nico M. van Straalen
Department of Ecology and
Ecotoxicology
Vrije Universiteit
De Boelelaan 1087
1081 HV Amsterdam
The Netherlands

Dr Joke van Wensem
Technical Soil Protection
Committee
PO Box 30947
2500 GX Den Haag
The Netherlands

Dr Jason M. Weeks
Institute of Terrestrial Ecology
Monks Wood
Abbots Ripton
Huntingdon, Cambs PE17 2LS
United Kingdom

Dr John A. Wiles
Huntingdon Research Centre Ltd
Department of Ecotoxicology
PO Box 2
Huntingdon, Cambs PE18 6ES
United Kingdom

Preface

The Netherlands Integrated Soil Research Programme (NISRP) – which was started in 1986 – has been aimed at the promotion and programming of research in a large variety of aspects of soil science. These aspects concerned both basic research in, for example soil chemistry, soil physics and soil biology, and applied research. The applied research was focused primarily on remediation technology and risk assessment of polluted soils. The main incentive for initiating and funding of NISRP was the growing awareness during the past decades of continued – and in many cases irreversible – soil deterioration and pollution.

One of the first major problems emphasized within the NISRP framework was to provide a scientific basis for soil quality evalution. It soon became evident that progress could be made only by an interdisciplinary effort. It was clear that soil quality assessment should involve a judgement of both quantity and quality of produced crop, of air and water quality, and of human health in general. Gradually, however, it became more apparent also that the intrinsic values and the role of the soil ecosystem are important in soil quality assessment. Hence, in soil protection regulations and measures, a specific approach was developed to protect soil life in particular.

Despite the development of this approach, insufficient scientific understanding and data are currently available to form a proper basis for soil ecosystem protection. At present, the understanding, the database – and sometimes even the experimental techniques – are found to be insufficient for a practical implementation in soil protection measures. In October 1994, a workshop was organized concerning these research areas, which was partly sponsored by NISRP. This book contains the scientific contributions of that workshop and provides some of the much needed understanding, both at the population and the ecosystem levels. Even more important for setting and implementing regulations, the contributions reveal the necessity of the chosen interdisciplinary approach and highlight the issues that must be resolved at different levels of complexity. From the scope of protecting the soil ecosystem, it may prove to be a milestone in implementing these regulations.

F.A.M. de Haan
Chairman Committee Basic Research
Netherlands Integrated Soil Research Programme

Acknowledgements

The editors are grateful to the many people who have helped in the completion of this volume. Firstly, to the participants of the SERAS workshop on Ecological Principles for Risk Assessment of Contaminants in Soil, in Papendal, Arnhem, The Netherlands, 2–5 October 1994, who set the stage for the scientific contents of this book through their contributions to the discussion, criticism of the introductory presentations and exchanges of information. Written contributions to this volume were reviewed by a number of colleagues, whom we would like to thank for their comments, suggestions and criticism, namely T. Aldenberg, J.P.E. Anderson, J.A. Axelsen, T. Bolger, L. Brussaard, J.P. Cancela da Fonseca, R. Dallinger, M.H. Depledge, P.C. de Ruiter, P. Doelman, T.L. Forbes, T.G. Hallam, S.P. Hopkin, V. Huhta, J.E. Kammenga, W. Kratz, C. Kula, R. Laskowski, H. Petersen, A.J. Reinecke, S. Rundgren, H. Siepel, C.A.M. van Gestel, J. van Wensem and J.A. Wiles.

Help from Désirée Hoonhout and Nan Kasanpawiro was indispensable when re-styling and organizing the final manuscript.

The editors also thank the Netherlands Integrated Soil Research Programme, who kindly provided a grant which not only funded the organization of the workshop, but also subsequently made the production of this book possible.

Nico van Straalen and Hans Løkke
Amsterdam and Silkeborg
February 1996

Introduction

1

Ecological approaches in soil ecotoxicology

NICO M. VAN STRAALEN AND HANS LØKKE

1.1 INTRODUCTION

In 1977 the Organization for Economic Co-operation and Development (OECD) established its chemicals testing programme. One of the aims was to prepare state-of-the-art reports on test methods and guidelines that could be used to predict the effects of chemicals on humans and the environment. Since then, a suite of test guidelines has been produced and implemented for international use, in particular within the OECD, the European Union, and the Environmental Protection Agency of the United States of America.

The physicochemical tests have been standardized to obtain information on environmental distribution, bioaccumulation, bioavailability, degradation pathways and fate, e.g. in soils or sediments. During the first decade of ecotoxicology, research focused mainly on the aquatic environment. Test procedures using algae, daphnia and fish have been highly developed and are used worldwide in routine testing of existing and new chemicals, and of effluents. More species have been added for advanced testing. Regulatory guidelines for classification of polluted soil and chemicals depend on the availability of ecotoxicological test methods and risk assessment procedures. At present, the assessment of ecotoxicological effects on soil organisms is only partly possible because the number of available soil ecotoxicity test guidelines is inadequate. Consequently, there is a need for the development of standardizable test procedures for assessing the ecological effects of chemicals in the terrestrial environment. In the EU project SECOFASE (Kula et al., 1995), test systems for assessing sublethal effects of chemicals on fauna in the soil ecosystem are developed, improved and standardized in a collaboration programme between ten European laboratories.

Risk assessment is a central theme in chemicals control, and methodologies are developed for international harmonization. Van Leeuwen and

Ecological Risk Assessment of Contaminants in Soil. Edited by Nico M. van Straalen and Hans Løkke. Published in 1997 by Chapman & Hall, London. ISBN 0 412 75900 4

Hermens (1995) have given an introduction to risk assessment, risk management and related issues. In current risk assessment procedures, the establishment of a so-called risk quotient, PEC/PNEC, is the central element, where PEC is the predicted environmental concentration, and PNEC is the predicted no effect concentration. Due to the complexity of natural ecosystems, the PNECs are usually derived from results obtained in the laboratory by toxicological experiments with selected single species. For most chemicals, data on terrestrial species are not available. The estimation of PEC is possible for soil ecosystems, although normally very simplistic models are used to derive average concentrations in soils and biota.

In the European Inventory of Existing Commercial Chemical Substances (EINECS) more than 100 000 chemicals are registered. It is unlikely that the required data will become available within the coming decades. However, a large proportion of these chemicals may not present any significant hazard potential. Even if the potentially dangerous compounds should be tested, a sufficiently large testing capacity is not available, and it would require enormous economic and scientific resources. Therefore, for a relatively large number of existing chemicals, predictions will provide an interesting alternative to actual testing (Nendza and Hermens, 1995).

For the generation of the missing data, quantitative structure–activity relationship (QSAR) procedures have been developed for parameters which control the fate and dispersal of a chemical in the terrestrial environment. For chemicals with a non-specific mode of toxic action, estimation models have been derived for aquatic toxicity (Nendza and Hermens, 1995). Similar relationships between the biological activity of a compound and its chemical structure or a physicochemical property may be developed for the terrestrial environment. Some procedures have also been suggested for the extrapolation of single species acute toxicity to chronic toxicity effects for other species. The application of QSAR estimates is still not valid for regulatory use, since even small changes in chemical structure may lead to different modes of action. Therefore, the selection of the correct model will be strongly dependent on the mode of action to be studied.

1.2 PRESENT METHODS FOR EXTRAPOLATION

Several extrapolation methods have been developed that are based on species sensitivity distributions as reviewed by Løkke (1994). The US EPA (1985) was the first to present an extrapolation method aiming at protecting 95% of aquatic genera. This procedure assumed that the distribution curve of sensitivities of the species has a log-triangular form. Kooijman (1987) developed a statistical approach to show how the difference in the sensitivities of species for a chemical can be used to predict the range of sensitivities for all other untested species. This method assumes that the sensitivities of the species are described by a log-logistic distribution. A concentration is determined at which the LC_{50} of the most sensitive of the species present in the community exceeds that concentration by a specified probability. The method of Kooijman was adapted by Van Straalen and Denneman (1989) for the calculation of a hazardous concentration (HC) for 5% of the species in a soil ecosystem, the so-called HC_5. Two modifications of this approach were suggested by Wagner and Løkke (1991) and Aldenberg and Slob (1993).

Recently, there has been much discussion about the theoretical basis of the distribution-based extrapolation methods (Hopkin, 1993; Forbes and Forbes, 1993a, 1994; Okkerman *et al.*, 1993; Smith and Cairns, 1993; Van Straalen, 1993; Løkke, 1994). Hopkin (1993) argued that the hazardous concentration for copper, derived from the species sensitivity distribution, was insufficient to provide the minimum dietary requirement of the animals in the soil. Løkke *et al.* (1995) analysed data for the herbicide glyphosate and applied correction factors for the influence of exposure concentrations. In most cases, the extrapolation is based on results which are obtained in the laboratory by spiking soil samples with metal salts containing the free active metal cations or other toxic metal ion complexes. The active metal species used in the laboratory are not directly comparable with aged deposits under natural conditions which mainly consist of metal oxides and other metal species of very low bioavailability and toxicity.

Forbes and Forbes (1993a, 1994) stated that statistical extrapolation presently does not offer an improvement over much simpler methods using arbitrary assessment factors. They raised questions about the power of the chosen distributions, and about the selection of species and their degree of representation. A study by Løkke *et al.* (1995) showed that in the case of glyphosate the single species toxicity test results fit well into log-normal distributions. However, species of largely different taxonomy belong to quite different log-normal distributions. Thus, microorganisms, plants and animals were distinguished as three groups although the two latter groups might have been combined in one log-normal distribution. The single extreme value of a nematode did not fit into any of the other log-normal distributions used in this study. It could be concluded that, in general, it may be useful to consider the question whether data covering large taxonomic distances should be subdivided into smaller groups of organisms before performing hazard or risk assessment by use of distribution-based extrapolation methods.

One of the underlying assumptions of the extrapolation methods is that species selection is random and that the selected species represent the organism group or ecosystem. Forbes and Forbes (1993a, 1994) claim that it may be difficult to design sampling schemes in which random rather than 'representative' samples are selected unless the species composition of the community, ecosystem or statistical population of interest has been determined. The extrapolation method simply tells us that with a given set of data there is a certain probability of finding a species which is affected by concentrations lower than the estimated HC_5 value. Therefore it is important that the underlying toxicity data cover the whole range of single species toxicities that may occur in an ecosystem or within an organism group. In this context the single species do not represent the ecosystem but rather a range of toxicities. Criteria for the selection of species were formulated by Eijsackers and Løkke (1992).

A proper selection of species still leaves us with the problem that the functional aspects of the ecosystem may not be protected by the HC_5 value. Furthermore, the interaction between species and with the abiotic environmental conditions may influence the impact of the chemical. The estimated 'safe concentrations' are based on conservative assumptions underlying the design of the single species tests and the estimates are probably over-protective. These aspects need further research, preferably by field validation of the assumptions. In Chapter 5 of this book such a procedure for a laboratory-to-field validation is described.

1.3 NEED FOR ECOLOGICAL PRINCIPLES IN ECOTOXICOLOGY

As outlined in the previous sections, the current risk assessment methodologies are based on simplistic assumptions such as the risk quotient, PEC/PNEC, and the assessment of the hazardous concentration to 5% of the species, HC_5, in an ecosystem. Forbes and Forbes (1994) concluded that the bulk of our knowledge of chemical pollutants is based on a minority of substances and species, and the call for more data is frequently heard among scientists and legislators alike. Without a better theoretical understanding of how effects at various levels of biotic and abiotic organization interact, we will stand little chance of making sense out of even the most standardized and harmonized of data. This holds true for aquatic as well as for terrestrial ecotoxicology.

Several authors have pointed out time and again that the development of the field of ecotoxicology is hampered by the fact that there is too little input from ecology (Koeman, 1982; Cairns, 1988; Maciorowski, 1988; Depledge, 1993; Forbes and Forbes, 1994). The argument usually is that there is too much emphasis on the development of test systems, on the routine use of these systems and on using test results for risk assessment. To ensure that ecotoxicology can meet the challenges from environmental policy – not only now but also in the future – there should be a continued effort to strengthen the scientific basis of the field.

When calling for more 'eco' in ecotoxicology one should, however, realize that ecology itself is a very heterogeneous field. One may find researchers involved with descriptive field work, theoretical modellers and experimentalists with physiological or even biochemical interests. In addition, experience teaches that the ecological questions discussed by ecotoxicologists do not receive much attention from fundamental ecologists. There is therefore a potential danger that ecotoxicology is using ecological theories and concepts that are considered as outdated among the ecologists themselves. It may therefore be useful to review some modern ecological developments and to ask which of these may be of relevance to ecotoxicology.

Before addressing this issue, we would like to look critically at ecotoxicology itself. Some questions asked by ecotoxicologists seem to return over and over without the prospect of an answer. It could be argued that these questions may not be the right ones and that we should ask other questions instead. To make the point clear we recommend having a critical look at the following issues.

1.3.1 SINGLE-SPECIES VERSUS MULTISPECIES TESTS

This is a question which has been raised many times and has led to many fruitless discussions. The argument is that laboratory experimental procedures are inadequate to understand the possible hazards of chemicals in the complex environment and that laboratory tests need to be replaced by multispecies tests or mesocosms (Cairns, 1986). We argue that these approaches are not necessarily mutually exclusive.

Due to their complexity, multispecies tests are regarded as unsuitable for regulatory purposes. The outcomes of multispecies tests may vary considerably, and the systems may be considered as black boxes because they do not provide insight into the interactions between the components of the system. It may be argued that a suite of properly designed single species tests, in combination with an advanced food-web model, may provide more information and a better understanding of the dynamics of ecosystems than multispecies tests. Multispecies tests, however, simulate much more ecological realism, especially if they allow an assessment of secondary effects. An extensive comparison of single and multiple species tests within terrestrial ecotoxicology still needs to be done. At the present stage, the conclusion may be that single and multispecies tests are not mutually exclusive, they rather may support each other and are useful tools to solve different questions.

1.3.2 EXTRAPOLATION

Ecotoxicology is often seen as a study of extrapolations: from test species to sensitive species, from species to ecosystem, from laboratory to field. The laboratory–field extrapolation problems arising from soil factors are addressed by C.A.M. van Gestel (Chapter 2) and D.J. Spurgeon (Chapter 12) in this volume. Some researchers have argued that the main issue for ecotoxicology is extrapolation from species to ecosystem (Suter *et al.*, 1985; De Kruijf, 1991). This type of extrapolation is hampered by the fact that the properties of higher hierarchical levels may not always be predictable from those of lower levels. The discussion often centres around the question of 'emergent' properties of the hierarchical levels.

Although extrapolation undoubtedly is an important item, we should avoid the tendency to reduce ecotoxicology to an extrapolation problem. There are many important questions which cannot be seen as extrapolation problems. The emphasis placed on the hierarchical structure of ecosystems (species, communities, biomes) in ecology textbooks may become inappropriate; some ecologists today do not even draw a distinct line between the classical species approach and the more holistic systems approach (Lawton, 1994).

1.3.3 FURTHER DATA

Ecotoxicologists usually agree on the fact that the database for ecotoxicology (and particularly for soil) is far too limited to draw any conclusions on the hazards of chemicals in the environment. This may be true in general, but the emphasis on this issue may lead away from a more important question: that is, the available data set is limited to too small a number of test species. Especially for soil, there is a potential risk that the necessary growth of the database will be conducted using the same few species (*Eisenia fetida, Folsomia candida*). In a similar fashion, most of the literature deals with a few chemical compounds, such as selected pesticides, heavy metals, polycyclic aromatic hydrocarbons (PAHs) and chlorinated organic compounds. Although these chemicals are of high priority, there is a need for investigating a larger proportion of the more than 100 000 chemicals registered for use in Europe.

Returning to the field of ecology, we see three main topics of discussion that are currently of importance to ecotoxicology; these are discussed below.

1.4 MOLECULAR TECHNIQUES FOR THE ANALYSIS OF POPULATION STRUCTURE

Several ecological laboratories have started to invest in know-how and equipment to implement DNA techniques borrowed from molecular genetics (Berry *et al.*, 1992). Several techniques have proven their usefulness to answer specific ecological problems. The use of the PCR (polymerase chain reaction) has made it possible to amplify very small amounts of DNA and to apply the techniques to invertebrates with a small body mass. One of the most popular approaches goes under the acronym RAPD (randomly amplified polymorphic DNA), in which an analysis is made of polymorphisms obtained after amplifying pieces of DNA with short primers of arbitrary sequence, using the PCR technique. Other techniques look at microsatellites with a variable number of tandem repeats (VNTR) or at length fragments of restriction enzyme digested DNA (restriction fragment length polymorphism; RFLP).

The use of molecular techniques may provide a powerful tool to detect changes within a population due to toxicants. These changes could be made visible as changes in the frequencies of some genotypes compared with others, so as a pollution-induced micro-evolutionary process. Others have used the term pollution-induced community tolerance (PICT) to denote a similar approach, although not on the population but on the community level (Blanck *et al.*, 1988).

Successful examples applying molecular techniques for genetic variation to pollution problems in the environment cannot yet be given. Until now, most of this work has been done using the genetic variability revealed by enzyme polymorphisms, visualized on electrophoresis gels (Verkleij *et al.*, 1985; Nevo *et al.*, 1986; Frati *et al.*, 1992; Tranvik *et al.*, 1993).

In a study on the springtail species, *Orchesella cincta*, Frati *et al.* (1992) compared nine different populations, sampled from metal-contaminated and reference sites in the Netherlands, Belgium, Germany and Italy. An analysis of 22 different loci showed that 18 of them were monomorphic at all sampling sites, while four showed significant divergence. An analysis of genetic distances using Nei's index demonstrated that the populations in north-western Europe were all very similar, and that there was no correlation between genetic and geographic distance (Figure 1.1). There was, however, a significant correlation between the frequencies of the alleles at the *Got* (glutamate–oxaloacetate transaminase) locus with the average resistance to cadmium in the populations. The study allowed the conclusion that metal-induced tolerance does not seem to affect the overall genetic variability, but may affect the frequency of some specific loci. The application of molecular techniques to this and other examples will be a great challenge for soil ecotoxicology.

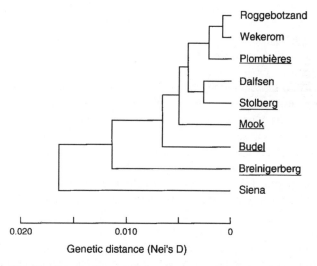

Figure 1.1 Genetic distance between nine European populations of the collembolan *Orchesella cincta*, as revealed by analysis of allozyme frequencies for 22 loci, of which four were polymorphic. Genetic distance is expressed using the index of Nei and the dendrogram was established using the UPGMA algorithm. The sites with names underlined have soils with metal contamination. (After Frati *et al.*, 1992.)

Molecular techniques in ecotoxicology are not only applied to reveal population structure, they are also used to construct biomarkers, bioprobes and bioreporter systems. A recent example is the use of a genetically modified cell line that will become luminescent when exposed to a dioxin-like molecule (Aarts *et al.*, 1993). Biomarkers may play an important role in ecotoxicological research to clarify the mechanistic links between effects at the molecular, biochemical and physiological levels (Van Gestel and Van Brummelen, 1996). Biomarkers may also be useful for exposure assessment in the field. Since the biomarker response is directly related to the bioavailability of chemical compounds, it may be used for screening purposes in polluted areas. Combined effects of mixtures of chemicals – even unknown chemicals – may be accounted for when using biomarkers. The practical use of biomarkers is, however, limited to well-defined areas because of the great inherent variability in the response. Biomarkers still need to be calibrated with regard to ecologically relevant end-points.

1.5 REACTION NORMS IN LIFE-HISTORY EVOLUTION

In life-history theory there is a growing attention for phenotypic plasticity of quantitative characters (Stearns, 1989). 'Reaction norm' is a technical term to denote the set of phenotypes that may develop from one genotype in different environments. Plasticity of a character (e.g. body size, wing pattern, sex) is assumed to arise from the action of regulatory genes that modify the expression of structural genes; the heritability of plasticity, as a trait in itself, is derived from the genotype – environment interaction (Scheiner and Lyman, 1989). Via (1993), however, questions the concept of plasticity as something which is subject to selection separate from the mean value of the character.

Sensitivity to toxicants may be considered as the standard reaction norm studied by ecotoxicologists. The concentration–response curve, when established for one genotype, is a way to express the phenotypic plasticity for that genotype with respect to toxicants in the environment. Following this way of thinking, toxicants are seen as just another stress factor, in addition to the classical environmental factors studied by ecologists, such as temperature, food, and nutrients. J.E. Kammenga, in Chapter 14 of this book, considers the reaction norm as the basis for derivation of threshold concentrations of toxicants.

The application of life-history theory to soil ecotoxicology is illustrated by several examples (Bengtsson *et al.*, 1985; Van Straalen *et al.*, 1989; Crommentuijn *et al.*, 1993). There are many parthenogenetic species in the soil invertebrate community, for which the reaction norm approach would be experimentally feasible. Analysis of clone-specific tolerances to toxicants would allow the distribution of sensitivities of natural populations to be established (Forbes and Forbes, 1993b).

Crommentuijn *et al.* (1995), in a comparative study of sensitivity of soil arthropods to toxicants, argued that not only the sensitivity itself is important for the response of a population, but also the difference between the exposure levels that lead to lethal and sublethal effects. The ratio between the LC_{50} and the NEC (no effect concentration) for reproduction was denoted by sublethal sensitivity index (SSI) (Figure 1.2). A high value for SSI implies that reproduction is inhibited at an exposure level far below the LC_{50}, while a small value for SSI implies that reproduction is maintained until the animal dies.

Among soil invertebrates, there seem to be different strategies for the priorities in energy allocation when the animal is put under stress. Studies using cadmium showed that the maintenance of reproduction, possibly at the cost of survival, is exemplified by the collembolan species *Folsomia candida* and *Orchesella cincta*, while the oribatid mite *Platynothrus peltifer* showed the reverse pattern (inhibition of reproduction, resistance in terms of survival). The most sensitive criterion on the individual level is not necessarily the most relevant on the population level, because the different life-history characteristics have different effects on the intrinsic rate of population increase. This is illustrated in Figure 1.3, which shows that for the collembolan *Folsomia candida* body growth reacted more sensitively to cadmium in soil, compared with population growth.

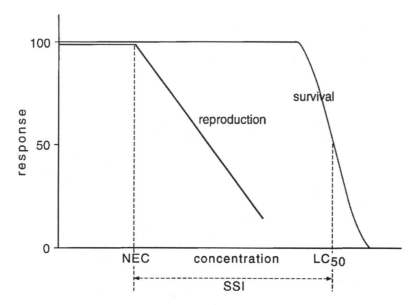

Figure 1.2 Theoretical concentration–response relationships showing the derivation of the sublethal sensitivity index (SSI), proposed by Crommentuijn *et al.* (1995), as the quotient of the threshold concentrations for survival (LC_{50}) and for reproduction (NEC).

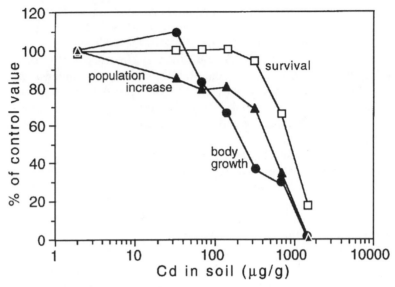

Figure 1.3 Concentration–response relationships for the effect of cadmium in soil on the collembolan *Folsomia candida*. (After Crommentuijn *et al.*, 1993.)

It may be concluded that the life-history strategy, which is reflected in the distance between reaction norms for reproduction and survival, should be taken into account when predicting effects on populations.

1.6 FOOD-WEB THEORY

A growing group of ecologists is analysing the properties of communities by means of food-webs, using various types of models (Pimm *et al.*, 1991; Moore *et al.*, 1993a; De Ruiter *et al.*, 1994). This approach starts by classifying an assemblage of species on the basis of their trophic status or feeding habits (Faber, 1991). The trophic relationships between the groups are then specified; this may be done in three different ways: trophic interactions only; trophic interactions including fluxes of materials; and trophic interactions including the consequences for the population size of each group. An interesting group of models are those developed for mineralization fluxes in below-ground food-webs (Hunt *et al.*, 1987; De Ruiter *et al.*, 1993; Moore *et al.*, 1993b). J.C. Moore in Chapter 7 of this book presents an analysis of the stability of soil community food-webs. Cohen *et al.* (1994) apply a food-web approach to analyse community changes following application of insecticides.

There are other food-web approaches in community ecology that may be very useful. Gutierrez *et al.* (1984) and Graf *et al.* (1990) formulated models including both physiological demands and trophic interactions; these models were used to simulate the dynamic interactions between climate, crop, pest and natural enemies in agricultural systems. This approach is also applicable to soil ecosystems (J.A. Axelsen, Chapter 11 of this book), and can be extended to incorporate several important species of the food-web in soil ecosystems. Another interesting community approach is exemplified in the study by Pahl-Wostl (1994), who made a sensitivity analysis of aquatic ecosystems using a summary parameter ('ascendency') to characterize the information in the food-web.

The food-web approach holds great potential for application in ecotoxicology. In principle, it would be possible to specify for each functional group in the web a distribution of sensitivities, like the distribution of sensitivities postulated for whole communities (Van Straalen and Denneman, 1989). With the food-web model, it may be possible to analyse the stability consequences of inhibiting certain sensitive groups of species. This would be an excellent way of fusing the species and systems approaches to ecotoxicology.

One of the problems with soil food-webs is that trophic relationships are not easily observed. Much has to be guessed from laboratory observations, gut content studies and tracer studies. In some cases however, special devices have been built – rhizotrons – that allow direct inspection of life below the surface. The food-web revealed by Gunn and Cherrett (1993) is an example of a such a rhizotron study (Figure 1.4). The web shown has many properties in common with webs established for terrestrial and aquatic ecosystems, but the soil web shows relatively few links between top species and intermediate species and among intermediate species. On the other hand, there are many connections of all species with the basal resources, detritus, carcasses and microbial biomass. Even groups which are often considered as top carnivores, such as centipedes, very frequently feed upon the basal resources. Another dynamic food-web was developed recently by Berg (1997), based on estimates for the densities of various functional groups in a pine forest floor. The model is able to predict the time course of nitrogen mineralization throughout the seasons. J. van Wensem (Chapter 10 of this book) discusses the risk assessment aspects of food-web models.

It remains unknown whether soil food-webs have properties that make them different from food-webs in other ecosystems. One important property seems to be the great degree of omnivory and the relative rarity of food specialization. Neutel *et al.* (1994) were able to show that the input of detritus from outside the system, and the recycling of material, add to the stability of food-chains, even without intraspecific competition. This supports the notion that a greater role of detritus in ecosystems is associated with greater stability.

1.7 THE COMPARATIVE APPROACH

There is a common theme in the three developments discussed above: the strength of the comparative approach. The importance of extending the basis of ecotoxicology by comparing populations, species and functional groups is

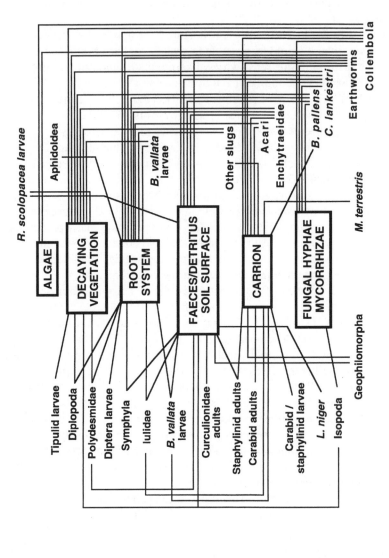

Figure 1.4 Trophic interactions between dead organic matter and various organisms as revealed by observations made in a subterranean rhizotron. (After Gunn and Cherrett, 1993.)

exemplified by Figure 1.5. Ecotoxicological effects may be viewed as substance–receptor interactions. Like other substance–receptor interactions in biology (e.g. the action of a hormone on its target organ, the recognition of an antigen by an antibody, etc.), the effect of the interaction depends on the properties of both the substance and the receptor. Environmental chemists have been successful in setting up quantitative structure–activity relationships (QSARs), which provide rules of thumb for the relationships between properties of a substance and its biological action. QSARs can even be used to derive maximum acceptable concentrations in the environment (Verhaar et al., 1994). Similarly, ecotoxicologists should study the properties of ecological receptors, not by comparing substances but by comparing species. If rules of thumb could be set up using this approach, these might be called quantitative species–sensitivity relationships (QSSRs) (Hoekstra et al., 1994).

The concept of an ecological receptor involves more than only toxicological sensitivity. Van Straalen (1994) pointed out that ecological receptors are characterized by one or more of the following three properties:

1. Ecological receptors are often exposed to high concentrations of a toxicant, due to their microhabitat choice, their feeding strategy or their activity at surfaces where toxicants accumulate.
2. Ecological receptors have an intrinsic sensitivity to a toxicant, due to the presence of a biochemical receptor, the ability to bioaccumulate or the inadequacy of the detoxication system.
3. Ecological receptors cannot easily recover their population size, for example due to a long life-cycle, a small population size or little genetic variability.

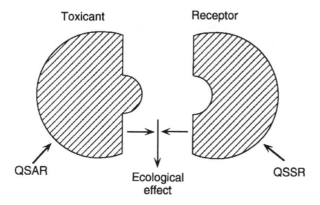

Figure 1.5 Ecological effects of toxicants may be seen as interactions between toxicants (characterized by quantitative structure–activity relationships, QSARs) and ecological receptors (characterized by quantitative species–sensitivity relationships, QSSRs).

Exposure, sensitivity and recovery together determine the vulnerability of a species to toxicants. In an ideal ecotoxicology course, the lecturer would teach his students the properties of ecological receptors, give them a checklist of QSSRs and send them away into the field, after which they would be able to predict where, and on which species, the introduction of a substance X into field A would be first recognizable. Although this may seem like a Utopian situation now, the argument demonstrates the importance of a better theory-based ecotoxicology including ecological generalizations.

1.8 CONTENTS OF THE BOOK

This book aims to introduce ecological principles in risk assessment of contaminants in soil and to give an overview of tools and new possibilities in the area. It also tries to identify some of the important gaps that remain in our understanding of terrestrial ecology which is of fundamental importance to ecotoxicology, and how current and future research may attack them.

Chapter 2 describes the difficulties involved in the extrapolation of results from laboratory toxicity tests to field conditions, with special emphasis on environmental conditions and bioavailability. A suite of laboratory and *in situ* bioassays with different soil organisms is reviewed.

Unlike zoologists, microbiologists are not yet in a position to select a universal test microorganism. As presented in Chapter 3, microbiologists describe soil microbial community structure and monitor its changes with the aid of modern molecular tools and tests of microbial function. Recommendations are made to select possible avenues of investigation to find potentially useful assessments of microbial activity which could be used as indicators of pollutant effects. A set of different techniques is reviewed.

Chapter 4 deals with the species concept in relation to 'biodiversity'. The determination of the 'hazardous concentration for 5% of the species', the HC_5 approach, is clearly dependent on a sound taxonomy on which to base the estimates of the total number of species. Long-term monitoring of changes in the species composition of sites is very difficult if the taxonomy changes through time. The author suggests the use of 'units of biodiversity' which might constitute a species group, race, genus or higher taxonomic category.

The theme of Chapter 5 is extrapolation from laboratory to field. Population and community parameters appropriate for laboratory–field comparisons are considered, e.g. population tolerance and pollution-induced community tolerance (PICT). A comparative procedure is described to evaluate both the adequacy of input data of risk assessment methods, and the field relevance of risk levels, such as the HC_5 concept.

In ecotoxicology there is a need for more interdisciplinary approaches. In Chapter 6, the links between structure and function in marine sedimentary and terrestrial soil ecosystems are dealt with, using contaminated marine sediments as examples. A **master process** is defined as a process that is of overriding importance in the coupling of system structure and function. It is concluded that purely structural measurements fail to tell the whole story, and focusing on guilds of species which control master processes is suggested.

Chapter 7 brings together the structural and functional aspects of ecosystems by linking models used to describe population dynamics and structural characteristics of ecosystems with models used to describe ecosystem processes. The conclusion is that energetics and dynamics are inextricably inter-related, and that the energetic organization of communities forms the basis of ecosystem stability. The challenge will be to adopt a whole-systems perspective, rather than studying single species or groups of species.

In ecology there are many possible outcomes from a given set of initial conditions. Predictions and risk assessment based on laboratory data do not readily fit with this view of the world. Chapter 8 shows that they can be placed in an ecological context, and thereby greatly enrich the value of the data obtained. Four pesticide case studies are reviewed that provide evidence for scale-dependency in ecotoxicological effects. The incorporation of more details on the spatiotemporal dynamics of pollutants, habitat and organisms to reduce uncertainties in risk assessment is suggested.

The question of spatial heterogeneity is further elucidated in Chapter 9 and seen in combination with dispersal mechanisms and local adaptations to contaminants. Different approaches to describe and predict the movement pattern of collembolans and bacteria in soils are compared. The inclusion of spatial and temporal heterogeneity, dispersal, and resistance in soil quality assessment is suggested.

Chapter 10 provides a general look at models that predict effects of chemicals on ecosystem structure and nutrient cycling, and at models for predicting the risk of secondary poisoning. The justification and limitations of models, and their application in risk assessment is discussed.

The prospects of ecosystem modelling in ecotoxicology are intriguing but modelling is very demanding in terms of biological information. In Chapter 11 a detailed mathematical simulation model combining population dynamics and energetics is presented as a tool for extension of results from laboratory data to field ecosystem effects. It is recommended that the physiological time-scale be applied in studies on the impact of chemicals on insects and other poikilotherm organisms.

In succeeding chapters different specific approaches to the assessment of ecologically relevant parameters are covered. Chapter 12 describes a field case study on the assessment of HC_5 of metals from earthworm toxicity data. By mapping toxicity data, it was found that laboratory data could not be related directly to the field, probably due to differences in bioavailability. Chapter 13 focuses on the importance of evaluating the effects on different life-history characteristics of springtails to understand the effects at the population level, and Chapter 14 presents a concept to evaluate toxicity data on the basis of the reaction norm for life-history data. The chapter exemplifies that the greatest toxic impact on fitness is not correlated with the critical effect level of the most sensitive traits. The theme of fitness is illustrated further in Chapter 15 with a focus on long-lived iteroparous invertebrates, i.e. snails and centipedes. It is described how the fitness costs may be estimated using data in the literature on life-histories and a matrix modelling approach to experimental data is suggested.

Finally, in Chapter 16 recommendations are given on how to focus future practice, and research and development activities to achieve better ecotoxicological risk assessment based on ecological principles. This chapter was drafted during a workshop, held in Papendal, Arnhem, The Netherlands, 2–5 October 1994, where all authors of this book presented their contributions for open discussion and revision.

1.9 SUMMARY

To improve the basis for ecological risk assessment of soil pollution, a strengthening of inputs from ecology into soil ecotoxicology is of utmost importance. Three major developments may be identified, which are taking place in the fundamental ecological sciences at present, and which seem to be very promising for ecotoxicology. These are:

1. the application of molecular techniques for the analysis of population structure,
2. the further development of life-history theory based on the concept of the reaction norm, and
3. the rise of community analysis using food-web models.

Some examples showing applications of these developments are given. A common theme in all three is the comparative approach. The comparison of populations, species, and functional groups may lead to quantitative species–sensitivity relationships, which may help to better characterize the ecological receptors of chemicals in the environment.

REFERENCES

Aarts, J.M.M.J.G., Denison, M.S., De Haan, L.H.J., Schalk, J.A.C., Cox, M.A. and Brouwer, A. (1993) Ah receptor-mediated luciferase expression: a tool for monitoring dioxin-like toxicity, in *Organohalogen Compounds*, Vol. 13 (eds H. Fiedler, H. Frank, O. Hutzinger, W. Parzefall, A. Riss and S. Safe), Riegelink, Vienna, pp. 361–4.

Aldenberg, T. and Slob, W. (1993) Confidence limits for hazardous concentrations based on logistically distributed NOEC data. *Ecotox. Environ. Safety*, **25**, 48–63.

Bengtsson, G., Gunnarsson, T. and Rundgren, S. (1985) Influence of metals on reproduction, mortality and population growth in *Onychiurus armatus* (Collembola). *J. Appl. Ecol.*, **22**, 967–78.

Berg, M.P. (1997) *Decomposition, nutrient flow and food web dynamics in a stratified pine forest soil.* PhD thesis, Vrije Universiteit, Amsterdam.

Berry, R.J., Crawford, T.J. and Hewitt, G.M. (eds) (1992) *Genes in Ecology*, Blackwell Scientific Publications, Oxford.

Blanck, H., Wängberg, S.Å. and Molander, S. (1988) Pollution-induced community tolerance – a new ecotoxicological tool, in *Functional Testing of Aquatic Biota for Estimating Hazards of Chemicals* (eds J.J. Cairns Jr and J.R. Pratt), American Society for Testing and Materials, Philadelphia, pp. 219–30.

Cairns, J.J. (1986) What is meant by validation of predictions based on laboratory toxicity tests? *Hydrobiologia*, **137**, 271–8.

Cairns, J.J. (1988) Putting the eco in ecotoxicology. *Reg. Toxicol. Pharmacol.*, **8**, 226–38.

Cohen, J.E., Schoenly, K., Heong, K.L., Justo, H., Arida, G., Barrion, A.T. and Litsinger, J.A. (1994) A food web approach to evaluating the effect of insecticide spraying on insect pest population dynamics in a Philippine irrigated rice ecosystem. *J. Appl. Ecol.*, **31**, 747–63.

Crommentuijn, T., Brils, J. and Van Straalen, N.M. (1993) Influence of cadmium on life-history characteristics of *Folsomia candida* (Willem). *Ecotox. Environ. Safety*, **26**, 216–27.

Crommentuijn, T., Doodeman, C.J.A.M., Doornekamp, A., Van der Pol, J.J.C., Rademaker, M.C.J. and Van Gestel, C.A.M. (1995) Sublethal Sensitivity Index as an ecotoxicity parameter measuring energy allocation under stress. Application to cadmium in soil arthropods. *Ecotox. Environ. Safety*, **31**, 192–200.

De Kruijf, H.A.M. (1991) Extrapolation through hierarchical levels. *Comp. Biochem. Physiol.*, **100C**, 291–9.

Depledge, M.H. (1993) Ecotoxicology: a science or a management tool? *Ambio*, **22**, 51–2.

De Ruiter, P.C., Moore, J.C., Zwart, K.B., Bouwman, L.A., Hassink, J., Bloem, J., De Vos, J.A., Marinissen, J.C.Y., Didden, W.A.M., Lebbink, G. and Brussaard, L. (1993) Simulation of nitrogen mineralization in the below-ground food webs of two winter wheat fields. *J. Appl. Ecol.*, **30**, 95–106.

De Ruiter, P.C., Neutel, A.M. and Moore, J.C. (1994) Modelling food webs and nutrient cycling in agro-ecosystems. *Trends Ecol. Evol.*, **9**, 378–83.

Eijsackers H. and Løkke, H. (eds) (1992) *SERAS – Soil Ecotoxicological Risk Assessment System. A European Scientific Programme to Promote the Protection of the Health of the Soil Environment.* Report from a Workshop held in Silkeborg, Denmark 13–16 January 1992. National Environmental Research Institute, 1992.

Faber, J.H. (1991) Functional classification of soil fauna: a new approach. *Oikos*, **62**, 110–17.

Forbes, T.L. and Forbes, V.E. (1993a) A critique of the use of distribution-based extrapolation models in ecotoxicology. *Funct. Ecol.*, **7**, 249–54.

Forbes, V.E. and Forbes, T.L. (1993b) Ecotoxicology and the power of clones. *Funct. Ecol.*, **7**, 511–12.

Forbes, V.E. and Forbes, T.L. (1994) *Ecotoxicology in Theory and Practice*, Chapman & Hall, London.

Frati, F., Fanciulli, P.P. and Posthuma, L. (1992) Allozyme variation in reference and metal-exposed natural populations of *Orchesella cincta* (Insecta: Collembola). *Biochem. Syst. Ecol.*, **20**, 297–310.

Graf, B., Baumgärtner, J. and Gutierrez, A.P. (1990) Modelling agroecosystem dynamics with the metabolic pool approach. *Mitt. Schweiz. Ent. Ges.*, **63**, 465–76.

Gunn, A. and Cherrett, J.M. (1993) The exploitation of food resources by soil meso- and macro invertebrates. *Pedobiologia*, **37**, 303–20.

Gutierrez, A.P., Baumgärtner, J.U. and Summers, C.G. (1984) Multitrophic models of predator–prey energetics. *Can. Ent.*, **116**, 923–63.

Hoekstra, J.A., Vaal, M.A., Notenboom, J. and Slooff, W. (1994) Variation in the sensitivity of aquatic species to toxicants. *Bull. Environ. Contam. Toxicol.*, **53**, 98–105.

Hopkin, S.P. (1993) Ecological implications of '95% protection levels' for metals in soils, *Oikos*, **66**, 137–41.

Hunt, H.W., Coleman, D.C., Ingham, E.R., Ingham, R.E., Elliot, E.T., Moore, J.C., Rose, S.L., Reid, C.P.P. and Morley, C.R. (1987) The detrital food web in a shortgrass prairie. *Biol. Fert. Soils*, **3**, 57–68.

Koeman, J.H. (1982) Ecotoxicological evaluation: the eco-side of the problem. *Ecotox. Environ. Safety*, **6**, 358–62.

Kooijman, S.A.L.M. (1987) A safety factor for LC_{50} values allowing for differences in sensitivity among species, *Water Res.*, **21**, 269–76.

Kula, H., Heimbach, U. and Løkke, H. (eds) (1995) *Progress Report 1994 of SECO-FASE, Third Technical Report. Development, Improvement and Standardization of Test Systems for Assessing Sublethal Effects of Chemicals on Fauna in the Soil Ecosystem.* National Environmental Research Institute, Denmark.

Lawton, J.H. (1994) What do species do in ecosystems? *Oikos*, **71**, 367–74.

Løkke, H. (1994) Ecotoxicological extrapolation: tool or toy? in *Ecotoxicology of Soil Organisms* (eds M.H. Donker, H. Eijsackers, and F. Heimbach), Lewis Publishers, Boca Raton, pp. 411–25.

Løkke, H., Christensen, B. and Møller, J. (1995) Extrapolation of the effects of glyphosate from the laboratory to the field. *Archiwum Ochr. Srodowiska*, **1**, 109–20.

Maciorowski, A.F. (1988) Populations and communities: linking toxicology and ecology in a new synthesis. *Environ. Toxicol. Chem.*, **7**, 677–8.

Moore, J.C., De Ruiter, P.C. and Hunt, H.W. (1993a) Influence of productivity on the stability of real and model ecosystems. *Science*, **261**, 906–8.

Moore, J.C., De Ruiter, P.C. and Hunt, H.W. (1993b) Soil invertebrate/micro-invertebrate interactions: disproportionate effects of species on food web structure and function. *Vet. Parasitol.*, **48**, 247–60.

Nendza, M. and Hermens, J. (1995) Properties of chemicals and estimation methodologies, in *Risk Assessment of Chemicals: An Introduction* (eds C.J. Van Leeuwen and J. Hermens), Kluwer Academic Publishers, Dordrecht, The Netherlands, pp. 239–92.

Neutel, A.M., Roerdink, J.B.T.M. and De Ruiter, P.C. (1994) Global stability of two-level detritus decomposer food chains. *J. Theor. Biol.*, **171**, 351–3.

Nevo, E., Noy, R., Lavie, A., Beiles, A. and Muchter, S. (1986) Genetic diversity and resistance to marine pollution. *Biol. J. Linn. Soc.*, **29**, 139–44.

Okkerman, P.C., Van de Plassche, E.J., Emans, H.J.B. and Canton, J.H. (1993) Validation of some extrapolation methods with toxicity data derived from multiple species experiments. *Ecotox. Environ. Safety*, **25**, 341–59.

Pahl-Wostl, C. (1994) Sensitivity analysis of ecosystem dynamics based on macroscopic community descriptors: a simulation study. *Ecol. Modelling*, **75/76**, 51–62.

Pimm, S.L., Lawton, J.H. and Cohen, J.E. (1991) Food web patterns and their consequences. *Nature*, **350**, 669–74.

Scheiner, S.M. and Lyman, R.F. (1989) The genetics of phenotypic plasticity. I. Heritability. *J. Evol. Biol.*, **2**, 95–107.

Smith, E.P. and Cairns, J. Jr (1993) Extrapolation methods for setting ecological standards for water quality: statistical and ecological concerns. *Ecotoxicology*, **2**, 203–19.

Stearns, S.C. (1989) The evolutionary significance of phenotypic plasticity. *Bioscience*, **39**, 436–45.

Suter, G.W. II, Barnthouse, L.W., Breck, J.E., Gardner, R.H. and O'Neill, R.V. (1985) Extrapolating from the laboratory to the field: how uncertain are you? in *Aquatic Toxicology and Hazard Assessment* (eds R.D. Carwell, R. Purdy and R.C. Bahner), American Society for Testing and Materials, Philadelphia, pp. 400–13.

Tranvik, L., Bengtsson, G. and Rundgren, S. (1993) Relative abundance and resistance traits of two Collembola species under metal stress. *J. Appl. Ecol.*, **30**, 43–52.

US EPA (US Environmental Protection Agency) (1985) *Water Quality Criteria; Availability of Documents*. Federal Register, vol. **50**, 30784–96.

Van Gestel, C.A.M. and Van Brummelen, T.C. (1996) Incorporation of the biomarker concept in ecotoxicology calls for a redefinition of terms. *Ecotoxicology*, **5**, 217–25.

Van Leeuwen, C.J. and Hermens, J.L.M. (1995) *Risk Assessment of Chemicals: An Introduction*, Kluwer Academic Publishers, Dordrecht, The Netherlands.

Van Straalen, N.M. (1993) An ecotoxicologist in politics. *Oikos*, **66**, 142–3.

Van Straalen, N.M. (1994) Biodiversity of ecotoxicological responses in animals. *Neth. J. Zool.*, **44**, 112–29.

Van Straalen, N.M. and Denneman, C.A.J. (1989) Ecotoxicological evaluation of soil quality criteria. *Ecotox. Environ. Safety*, **18**, 241–51.

Van Straalen, N.M., Schobben, J.H.M. and De Goede, R.G.M. (1989) Population consequences of cadmium toxicity in soil microarthropods. *Ecotox. Environ. Safety*, **17**, 190–204.

Verhaar, H.J.M., Van Leeuwen, C.J., Bol, J. and Hermens, J.L.M. (1994) Application of QSARs in risk management of existing chemicals. *SAR and QSAR in Environ. Res.*, **2**, 39–58.

Verkleij, J.A.C., Bast-Cramer, W.B. and Levering, H. (1985) Effects of heavy-metal stress on the genetic structure of populations of *Silene cucubalus*, in *Structure and Functioning of Plant Populations* (eds J. Haeck and J.W. Woldendorp), North-Holland Publishing Company, Amsterdam, pp. 355–65.

Via, S. (1993) Adaptive phenotypic plasticity: target or by-product of selection in a variable environment. *Am. Nat.*, **142**, 352–65.

Wagner, C. and Løkke, H. (1991) Estimation of ecotoxicological protection levels from NOEC toxicity data, *Water Res.*, **25**, 1237–42.

PART ONE

Extrapolation from Experiments

2   Scientific basis for extrapolating results from soil ecotoxicity tests to field conditions and the use of bioassays

CORNELIS A.M. VAN GESTEL

2.1 EXTRAPOLATING ECOTOXICITY TEST DATA

The risk assessment of chemicals for ecosystems is generally based on the results of laboratory toxicity tests. For an overview of tests available or under development for the soil ecosystem, refer to Van Straalen and Van Gestel (1993). To derive 'safe' levels of chemicals in soil ecosystems, a number of extrapolation methods have been developed. These extrapolation methods all assume a certain distribution (log-logistic, log-normal or triangular) of sensitivities of organisms in the field, and aim at deriving a so-called 95% protection level, i.e. the hazardous concentration for 5% of the species, the so-called HC_5 value. For a critical overview of extrapolation methods see Suter (1993).

One of the starting points of these extrapolation methods is that the sensitivity of the organisms tested in the laboratory is representative of the sensitivity of their equivalent in the field. Van Straalen and Denneman (1989) mention a number of arguments why field organisms could differ in sensitivity from laboratory organisms (Table 2.1). These arguments should be taken into account when the establishing laboratory-to-field extrapolation factors.

As shown in Table 2.1, some of these arguments may lead to a higher sensitivity for organisms in the field; others may have the opposite effect. Van Straalen and Denneman (1989) conclude that it cannot be decided whether there is a difference in sensitivity of organisms in laboratory and field conditions, and therefore set the laboratory-to-field extrapolation factor to 1.

This chapter will elaborate further mainly on the first two arguments mentioned in Table 2.1, with emphasis on the second. As it will appear from the overview presented, only a limited insight exists into the influence of

Ecological Risk Assessment of Contaminants in Soil. Edited by Nico M. van Straalen and Hans Løkke. Published in 1997 by Chapman & Hall, London. ISBN 0 412 75900 4

environmental conditions on the toxicity of chemicals to terrestrial organisms. There is also a lack of information on factors determining bioavailability, thus hampering the extrapolation of toxicity data between different soils, and from test soils used in laboratory studies to field soils. This and the scarcity of information on the toxicity of mixtures call for bioassay methods to evaluate contaminated soils. Special attention will therefore be devoted to laboratory and *in situ* bioassay methods.

2.2 INFLUENCE OF ENVIRONMENTAL CONDITIONS ON THE TOXICITY OF CHEMICALS

Very few studies have dealt with the influence of abiotic conditions on the toxicity of chemicals to terrestrial organisms. In some cases, the influence of soil pH, soil moisture content or temperature on the growth and/or reproduction of test organisms has been determined without relating results to toxicant stress. The influence of soil pH on metal toxicity is more often studied, but in these studies the effect of pH is indirect by affecting bioavailability. Such effects of soil properties on bioavailability will be addressed in section 2.3.1. This section will briefly discuss the effects of soil moisture content and temperature.

2.2.1 SOIL MOISTURE CONTENT

Studies on the influence of soil moisture content on the toxicity of pesticides for the field cricket *Gryllus pennsylvanicus* were first performed in the 1960s (Harris, 1964a,b, 1967). In all studies, a number of insecticides appeared to be (much) more toxic in wet soil compared with dry soil, and the toxicity generally increased with increasing soil moisture content. This may be related to the fact that uptake proceeds via the soil pore water (see section 2.3.1), and is lower in dry soil. For lipophilic chemicals, such as the pesticides studied by Harris, sorption to soil is fairly strong and pore water concentrations will therefore be almost independent of moisture content. It is therefore not completely clear why toxicity was increasing with increasing soil moisture con-

Table 2.1 Arguments used in establishing a laboratory-to-field extrapolation factor. (From Van Straalen and Denneman, 1989)

Argument	Difference[*]
1. In the laboratory, organisms are tested under optimal conditions	+
2. In the field, bioavailability of chemicals may be lower than in laboratory tests	−
3. In the field, organisms are exposed to mixtures of many chemicals	+
4. In the field, ecological compensation and regulation mechanisms are operating	−
5. In the field, adaptation to chemical stress may occur	−
6. Adaptation often entails costs in ecological performance	+

[*]+ Indicates a positive argument to maintain an extrapolation factor greater than 1.
− Indicates a negative argument.

tent; maybe soil moisture levels were still fairly low (below field capacity), and active water uptake by the test animals increased with increasing moisture availability. For some combinations of chemical and soil type, an opposite relationship between toxicity and soil moisture content was noted.

C.A.M. Van Gestel and A.M.F. Van Diepen (submitted) found only a slight effect of different soil moisture contents on the toxicity of cadmium for the collembolan *Folsomia candida* in an artificial soil. The Collembola showed delayed but higher reproduction at low soil moisture contents. At increased cadmium levels, reproduction was less affected by soil moisture content. Table 2.2 lists the EC_{50} values for growth and reproduction after 6 weeks of exposure.

C.A.M. Van Gestel and A.M.F. Van Diepen (submitted) also found that the internal effect concentrations (EC_{50}) of cadmium in *Folsomia candida* ranged between 90.3 and 242 $\mu g\ g^{-1}$ for the effect on growth, and between 11.7 and 52.4 $\mu g\ g^{-1}$ for the effect on reproduction at the different soil moisture contents. The highest EC_{50} value for the reproductive effect was found at the lowest moisture content, but the EC_{50} for the effect on growth was the lowest at this moisture level. From their results, two conclusions can be drawn: firstly, the difference in sensitivity may vary by a factor of 2 due to differences in soil moisture content, but it remains inconclusive whether a low or high moisture content reduces or increases the sensitivity of Collembola to cadmium; and secondly, internal effect concentrations are also different. The latter conclusion is rather unexpected, as internal concentrations were supposed to be similar, and independent of other factors.

Everts *et al.* (1991) report an increase in the toxicity of deltamethrin for the linyphiid spider *Oedothorax apicatus* at low and at high soil moisture contents. Perhaps the interaction of deltamethrin with the water regulation system of the spider (Everts, 1990; Jagers op Akkerhuis, 1993), resulting in an increased desiccation at low moisture contents and an increased water uptake at high moisture contents, might explain these findings.

Table 2.2 Effect of soil moisture content on the toxicity of cadmium for *Folsomia candida* in an artificial soil after 6 weeks of exposure. EC_{50} values are given with corresponding 95% confidence intervals (after Van Gestel and Van Diepen, 1997).

Soil moisture content		EC_{50} (in $\mu mg\ g^{-1}$ dry soil[*]) for the effect on	
% of dry weight	% of field capacity	Growth (fresh weight)	Reproduction
25	74	523 (408–670)	64 (41–99)
35	103	263 (178–434)	58 (29–115)
45	132	252 (176–365)	92 (80–106)
55	162	481 (351–658)	74 (-)

[*]Based on measured soil concentrations.

2.2.2 TEMPERATURE

For many biological (and chemical) processes temperature is a key factor and, within certain limits, these processes proceed faster with increasing temperature. However, only limited information is available on the effect of temperature on the toxicity of chemicals for terrestrial invertebrates. The following discussion therefore also includes information on the influence of temperature on the uptake and elimination of chemicals and the resulting internal body concentration.

Everts (1990) found increased toxicity of deltamethrin for *Oedothorax apicatus* at high temperatures and low air humidity. This could be due to an increase in the spider's sensitivity to drought caused by deltamethrin. An increased metabolism at higher temperatures was probably the explanation for the increased toxicity of azinphos-methyl and lindane in the isopod *Philoscia muscorum* at increasing temperatures, found by Demon and Eijsackers (1985).

Lower lead concentrations in earthworms (*Lumbricus terrestris*) in winter were reported by Bengtsson and Rundgren (1992), suggesting a significant effect of temperature on lead uptake in these organisms.

Janssen and Bergema (1991) studied the influence of temperature on the uptake and elimination of cadmium in the collembolan *Orchesella cincta* and the oribatid *Platynothrus peltifer*. Both species were exposed to cadmium via food. Uptake rate was increased in both species at 20°C compared with 10°C, which in *Orchesella cincta* was accompanied by an increased excretion rate; hence the equilibrium concentration was independent of temperature. In *Platynothrus peltifer*, elimination rate was not affected by temperature, so this organism showed a higher net cadmium uptake at the higher temperature.

Recently, the collembolan *Folsomia candida* was found to show a cadmium uptake pattern similar to that of *Orchesella cincta*; zinc uptake in this species, however, seemed to be affected by temperature (C.E. Smit, unpublished results). In these studies with Collembola, the final cadmium concentration in the organisms was not affected by temperature, but it took longer to reach an equilibrium level at lower temperatures due to the lower uptake and excretion rates.

2.3 BIOAVAILABILITY

Toxicity tests with soil organisms are performed in different (artificial) soil types, and different soil types are usually applied for each organism (Van Straalen and Van Gestel, 1993). This means that results of different tests cannot be compared directly. So, when applying the extrapolation methods mentioned earlier to soil toxicity tests, results of tests carried out in different soils must be normalized. For this purpose, equations derived for the relationship between background levels of chemicals in Dutch reference soils and clay and/or organic matter content are often used (Van Straalen and Denneman,

1989; Denneman and Van Gestel, 1990). It is realized by these authors that such equations are not ideal, as an important factor such as pH is not included; hence the Technical Committee on Soil Protection in the Netherlands (TCB, 1992) concluded that reasons for using these equations to correct for differences in bioavailability are lacking. Consequently, the question remains as to how results obtained in one test soil can be extrapolated to other types.

As stated earlier, bioavailability may differ between laboratory and field soils. Although it is often assumed that bioavailability is higher in laboratory studies in which the chemical is newly added to the soil, in the field situation the chemical may have been present for long periods, leading to a much stronger binding to the soil particles and consequently, reduced bioavailability. This second issue is addressed below.

2.3.1 EXTRAPOLATION BETWEEN SOILS

The fact that bioavailability and therefore toxicity may be affected by soil properties may on one hand lead to a plea for including as many soil types in tests as feasible (Jepson *et al.*, 1994), on the other hand it asks for unifying concepts in bioavailability.

Van Gestel and Ma (1988, 1990) determined the acute earthworm toxicity and sorption of a number of organic chemicals, five chlorophenols, 1,2,3-trichlorobenzene and 2,4-dichloro-aniline, in four different soils, using the earthworms *Eisenia andrei* and *Lumbricus rubellus* as test species. Some results of this study are shown in Table 2.3.

From the results presented in Table 2.3, it can be concluded that LC_{50} values determined in different soils may differ by a factor of 4.4–12.8. Upon recalculation to pore water concentrations, this difference is reduced to a factor of 1.2–2.4. From this and the results reported for the other chemicals studied, Van Gestel and Ma (1990) concluded that the acute earthworm toxicity of

Table 2.3 Toxicity of pentachlorophenol (PCP) and 1,2,3-trichlorobenzene (TCB) for *Eisenia andrei* and *Lumbricus rubellus* in four different soil types*, expressed as LC_{50} values based on total soil concentrations and pore-water concentrations calculated from sorption data. (From Van Gestel and Ma, 1990)

Species	Chemical	LC_{50} (mg kg⁻¹ dry soil)				LC_{50} (μmol l⁻¹ pore water)			
		KOBG	HOLT	OECD	WAPV	KOBG	HOLT	OECD	WAPV
E. andrei	PCP	84	142	86	503	2.5	4.4	5.7	2.6
	TCB	134	240	134	596	16	17	12	16
L. rubellus	PCP	1206	1013	362	4627	59	48	25	29
	TCB	115	207	195	563	14	14	17	15

*KOBG, HOLT, OECD and WAPV are soils with 3.7, 6.1, 8.1 and 15.6% organic matter and a pH (1 M KCl) of 4.8, 5.6, 5.9 and 3.6, respectively.

these chemicals is mainly determined by the concentrations in the soil pore water, which can be predicted from the sorption data. This is the first time the so-called 'pore-water hypothesis' (Figure 2.1) has been shown to be valid for the terrestrial soil. An important consequence of this finding is that knowledge of sorption behaviour of organic chemicals may allow for the extrapolation of earthworm toxicity data from one soil to another.

For many chemicals, data on the toxicity for soil organisms are lacking but toxicity data for aquatic organisms are available. The findings presented above provided a motivation for the use of data on aquatic organisms in the setting of standards for chemicals in terrestrial soils. So, applying the pore-water hypothesis, aquatic toxicity data were recalculated to soil data using sorption data (Denneman and Van Gestel, 1990, 1991; Van de Meent *et al.*, 1990). This seemed to be justified by the introduction of the equilibrium partitioning concept for the derivation of sediment quality criteria. In fact, the pore-water hypothesis and the equilibrium partitioning concept are the same, except for the fact that the latter also assumes a similar distribution of sensitivities of sediment- (or soil-) living and aquatic organisms. Both concepts further assume that soil organisms are mainly exposed through the aqueous phase of the soil (pore water) and that an equilibrium exists between concentrations in soil, in pore water and in organisms. A rationale for the application of the equilibrium partitioning concept in the derivation of sediment quality criteria is given by Di Toro *et al.* (1991). According to Van Straalen (1993b) this concept seems to offer a good starting point, also allowing for the extrapolation of toxicity data from one soil type to another. Van de Meent *et al.* (1990) agreed with this, but concluded that for many chemicals proper sorption data are lacking thus hampering the application of the equilibrium partitioning concept in deriving soil quality criteria.

Van Gestel (1992), in reviewing the literature on bioavailability of chemicals for earthworms, concluded that the earthworm toxicity of organic chem-

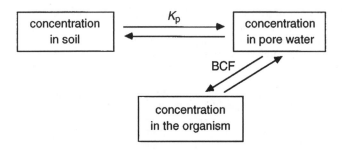

Figure 2.1 Pore-water hypothesis or equilibrium partitioning concept: the concentration of a chemical in soil organisms is considered to be related to the concentration in the pore water, which in turn is dependent on the sorption behaviour of the chemical. K_p, sorption coefficient; BCF, bioconcentration factor (= ratio of concentration in organism and concentration in pore water).

icals may be translated among soils using sorption data. For organic chemicals, structure–activity relationships describing the relationship between sorption to soil organic matter (K_{om}) or organic carbon (K_{oc}) and lipophilicity (K_{ow}) may be used to predict sorption constants. For ionizing organic chemicals, dissociation should be taken into consideration.

Belfroid (1994) further elaborated on the findings of Van Gestel and Ma (1988, 1990), by determining the relative importance of different routes of uptake for *E. andrei*. Based on studies in which earthworms were exposed to a number of chlorobenzenes in soil, in water and via contaminated food, she concluded that the relative importance of oral uptake compared with uptake from pore water increased with increasing lipophilicity of the chemical and increasing organic matter content of the soil (Table 2.4).

From Table 2.4 it can be concluded that the more strongly a chemical is adsorbed to the soil (that is at higher organic matter content and/or higher log K_{ow}), the more oral uptake contributes to the total body burden of earthworms. In the most adsorbing situation, oral uptake will, however, never contribute more than 50% to the total uptake. A similar finding was reported by Loonen (1994), who found that uptake by *Lumbriculus variegatus* of dioxins (TCDD, OCDD) in laboratory-treated sediment was higher in the presence of sediment compared with tests in pore water. The uptake was a factor of 1–5 higher than predicted on the basis of the equilibrium partitioning concept. This suggests that the equilibrium partitioning concept is no longer valid for highly lipophilic chemicals, especially in high organic soils or sediments.

Belfroid (1994), in summarizing her results, concluded that the difference is limited to a factor of 2, and that for a first estimate of bioavailability and the potential bioaccumulation of organic chemicals in earthworms, the pore-water hypothesis may still be a useful starting point. In contrast, Løkke (1994) stated that it is not justified that only the water soluble fraction of the chemical in question is bioavailable. This assumption is preliminary and may be true only for soft-bodied organisms. There is only limited information available on this assumption (see above); thus, further research is urgently needed.

The discussion above was entirely devoted to organic chemicals. Van Gestel (1992), in reviewing the literature for heavy metals, concluded that

Table 2.4 Relative contribution of two routes of uptake to the total body concentrations of three organic chemicals with different lipophilicity in the earthworm *Eisenia andrei* in a sandy soil with a low (3%) and a humic soil with a high (20%) organic matter content. (From Belfroid, 1994)

log K_{ow}	Accumulation in sandy soil		Accumulation in humic soil	
	Dietary	Pore water	Dietary	Pore water
2.0	0.0	100	0.25	99.7
5.0	1.2	98.8	7.6	92.4
7.0	11.0	89.0	45.2	54.8

many factors affect sorption to soil and uptake in organisms, and therefore extrapolation between soils seems not yet to be possible. Van Gestel *et al.* (1995), in a review on the influence of soil characteristics on the toxicity of metals for soil invertebrates, concluded that pH is the most important factor followed by soil organic matter content and cation exchange capacity. However, for each metal another combination of factors seems to determine bioavailability and further research – by a combined effort of soil scientists and ecotoxicologists – is needed before these factors can be quantified.

In a study of the influence of organic matter content and pH on the toxicity of cadmium to the collembolan *Folsomia candida* in an artificial soil, Crommentuijn (1994) related effect levels not only to total soil concentrations, but also to the water-soluble cadmium content of the test soils. The water solubility of cadmium appeared to be related to total soil concentrations by means of Freundlich sorption isotherms (Table 2.5).

From Table 2.5, it can be concluded that LC_{50} and EC_{50} increase with increasing soil pH and organic matter content. When expressed on the basis of water-soluble cadmium concentrations, LC_{50} and EC_{50}, however, do not show a consistent trend with either pH or organic matter. Considering the pore-water hypothesis, similar values for LC_{50} and EC_{50} should have been derived for the different soils, which is not the case. Cadmium concentrations in animals showed a good correlation with water-soluble cadmium levels in

Table 2.5 Effect of cadmium on survival, growth and reproduction of *Folsomia candida* in an artificial soil substrate at different pH and organic matter content (OM); LC_{50} and EC_{50} values are expressed on the basis of total and water soluble cadmium concentrations in the soils (in $\mu g \ g^{-1}$ dry soil). (From Crommentuijn, 1994)

pH	%OM	CEC*	LC_{50}		EC_{50} growth		EC_{50} reproduction	
		(mmol kg⁻¹)	Total	Water sol.	Total	Water sol.	Total	Water sol.
5.66	2.0	31	323	78	246	4.5	125	1.24
5.44	3.6	58	684	205	356	5.7	44	0.57
5.65	5.2	75	758	147	389	5.3	82	0.27
5.93	6.8	100	940	128	651	15	193	1.55
5.85	8.4	115	890	93	615	13	130	0.58
5.75	10	133	1261	184	824	22	193	1.19
3.12	10	124	743	88	302	13	102	0.96
3.73	10	132	1276	548	316	3.7	102	1.12
4.32	10	136	–⁺	–⁺	542	16	164	1.14
5.67	10	128	–⁺	–⁺	697	13	177	1.70
7.02	10	134	–⁺	–⁺	583	5.3	113	0.50
7.29	10	148	–⁺	–⁺	601	3.6	306	1.50

*Cation exchange capacity.
⁺LC_{50} exceeded the highest concentration tested (2500 μg Cd g^{-1}).

soil. Crommentuijn (1994) therefore concluded that internal LC_{50} and EC_{50} values were different for the different soils. Apparently, soil properties not only affect the sensitivity of Collembola by modifying bioavailability, but may also have a direct influence on the susceptibility of these organisms to cadmium. Based on these results, an alternative pore-water hypothesis for heavy metals may be formulated (Figure 2.2).

According to this alternative hypothesis, uptake and toxicity of a metal is assumed to be related to the free metal ion concentration in soil solution, which may be predicted from its speciation in the soil (solution). Both the speciation of the metal in soil and the uptake in soil organisms are influenced by the characteristics of the soil and the pore water. In addition, the suscepti-bility of the organism may also (indirectly) be affected by these characteris-tics. The validity of this hypothesis remains to be investigated.

2.3.2 BIOAVAILABILITY IN LABORATORY AND FIELD

According to Van Straalen (1993a) the question of bioavailability is often dis-regarded in toxicity tests. In such tests, chemicals are usually added freshly to the substrate, while in the field, equilibrium is established over a long period. This is confirmed by Boesten (1986), who found that the sorption of pesti-cides to soil can best be described assuming three classes of sorption sites. For the first class the establishment of an adsorption equilibrium requires some minutes, and for the second class some hours, but several months are needed for the third class. This may also explain why desorption may take a very long time, leading to a decreasing degradation rate of chemicals with time as shown, among others, in the case of polycyclic aromatic hydrocarbons (Förster *et al.*, 1993).

The theory of Boesten is supported by Loonen (1994), who found a reduced bioavailability of dioxins (TCDD, OCDD) for *Lumbriculus variegatus* with

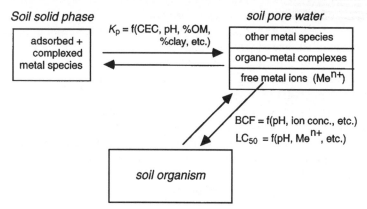

Figure 2.2 Pore-water hypothesis for the toxicity and bioaccumulation of heavy metals in soil invertebrates. (See legend to Figure 2.1 and text for further explanation.)

increasing incubation time of treated sediment. After 21 weeks of incubation, bioavailability was about two-fold lower than that found after 3 weeks of incubation. Similar results were reported by Sheppard and Evenden (1992), determining the bioavailability of uranium for *Lumbricus terrestris* at 1 and 2 years after treatment of soil. It appeared that bioaccumulation factor (BCF) values were much lower in the tests performed in the second year compared with the first year: in garden soil average BCF values were 0.24 and 0.089, respectively, and in limed sand 2.38 and 1.27, respectively. These values suggest a decrease in bioavailability with time. It should be noted, however, that the difference in BCF values in the first soil was mainly seen at concentrations between 20 and 300 $\mu g\ g^{-1}$, and not at lower concentrations (3–20 $\mu g\ g^{-1}$) or at 1000 $\mu g\ g^{-1}$, while the difference in the second soil was not very large (at the concentrations tested: 1–100 $\mu g\ g^{-1}$).

Verma and Pillai (1991) found that the bioavailability of 1,2,3,4,5,6-hexachlorocyclohexane (HCH) for earthworms decreased by a factor of 2 after 1 year of incubation in soil; for DDT, however, the bioavailability remained fairly constant over time.

Belfroid (1994) compared the uptake kinetics of chlorobenzenes in *Eisenia andrei* in artificial soil and field contaminated soil. Although there are differences in physicochemical properties between the two soils, it was apparent that uptake in the field soil was much lower than in the laboratory-treated soil, suggesting a decreased bioavailability in field contaminated soils.

Bioavailability in the field may not be constant. Forge *et al.* (1993), applying bioassays with *Colpoda steinii* on pore water obtained by centrifugation, observed higher levels of zinc, copper and nickel in pore water in summer than in winter. No such effect was found for cadmium and chromium, while for lead the reverse situation applied.

Jagers op Akkerhuis (1993) found that the uptake of deltamethrin in the linyphiid spider *Oedothorax apicatus* was dependent on walking activity, soil moisture content and substrate type. Particularly striking was his finding of an almost 100-fold higher uptake of deltamethrin sprayed on a soil covered with moss or fungi compared with bare soil.

Another problem related to the extrapolation from laboratory soils to the field pertains to the finding that, in toxicity tests with soil, renewal of test substrate is not usually possible, and the test chemical may degrade during the course of the experiment. Effective concentrations tend to be expressed on the basis of the amount of test chemical added, and do not take into account possible degradation; NOEC and EC_{50} values obtained in laboratory experiments may therefore be underestimated (Van de Meent *et al.*, 1990). Degradation may also occur in the field, but at different rates than in the laboratory, although the method of overcoming this difficulty remains currently unresolved.

Van Straalen *et al.* (1992) concluded that extrapolation of toxicity data from laboratory to field is valid only when the bioavailable concentration is constant. For chemicals emitted at intervals, such as pesticides, the possibilities for recovery should be included. Proposals are made to include the degradation rate (expressed as the half-life, DT_{50}) of the pesticide in the derivation of HC_5 values. This approach may be helpful, but it does not solve the problem of different degradation rates in the field and laboratory.

2.4 BIOASSAYS

For the evaluation of contaminated soil, it is often insufficient to rely on toxicity data determined for single chemicals. Mostly, contaminated soils contain a mixture of pollutants, the bioavailability of which may differ from that in laboratory soils. Therefore bioassays using soils from contaminated field sites are recommended for determining the actual toxicity and risk of contaminated soil (Denneman and Van Gestel, 1990; TCB, 1992).

In the previous sections, it was concluded that the bioavailability of organic chemicals and heavy metals may differ among soils and that there are indications of reduced bioavailability in field soils. Soil characteristics and time of incubation affect bioavailability. Bioassays may be used to test the theories on bioavailability presented in the previous section and to provide the insight needed to quantify the differences in bioavailability between laboratory and field soils.

An overview of bioassay methods for soil, sediment and water is given by Keddy *et al.* (1992), who stated that: 'bioassays provide a more direct measure of environmentally relevant toxicity of contaminated sites than do chemical analyses because the results are an integration of all environmental variables and contaminants' and 'there is ample evidence that the toxicity of contaminated sites, particularly when contaminant mixtures are present, can only be properly assessed using a battery of test species'. In Canada, bioassays are used to derive national criteria, to derive prioritized remediation objectives and to determine when remediation goals have been reached.

In an overview of test methods for soil, Sheppard *et al.* (1992) mentioned three types of bioassay:

1. Liquid extracts of soil are obtained for toxicity tests, often using methods developed for aquatic systems.
2. Toxicity tests in which a chemical is added to an artificial soil substrate.
3. Assessment of toxicity of whole field soils using plants or indigenous biota.

In the following, mainly the first and last bioassay types will be considered, and a brief overview will be given of bioassay methods available.

2.4.1 BIOASSAYS WITH MICROORGANISMS

A number of authors use the so-called Microtox test, using *Photobacterium phosphoreum* to assess the potential toxicity of contaminated soils (Ahlf *et al.*, 1993; Eiserman and Daei, 1993). Although most use soil eluates, Brouwer *et al.* (1990) also made the test applicable to solid materials, thus enabling determination of the toxicity of sediment and soil samples.

Vanhala and Ahtiainen (1994) concluded that the Microtox test applied to pore water is less sensitive than soil respiration or ATP content. However, they expect that the method developed by Brouwer *et al.* (1990) to apply the Microtox test to solid material will appear more sensitive. Both ATP content and soil respiration appeared to correlate better with total metal content in soil than with pore water concentrations.

Another bacterium used in the bioassay of effluents is *Escherichia coli* (Bitton *et al.*, 1992); this test may also be applied to eluate or pore water of contaminated soil.

A bioassay with *Bacillus cereus* applied to aqueous extracts and soil/water suspensions demonstrated that the effect of 2-naphthol on the dehydrogenase activity of this Gram-positive bacterium was modified by the presence of soil particles (Gunkel *et al.*, 1993). These authors also used *Pseudomonas putida* as a bioassay organism on aqueous extracts of polluted soil; in this test growth was the end-point. Ahlf *et al.* (1993) described the application of the bioassay with *Bacillus cereus* on aqueous extracts of contaminated soil. They also used a soil-contact test with the same bacterium, and concluded that this test seemed more realistic than one on aqueous extracts, mainly because the matrix effect of soil particles is included.

Dehydrogenase activity measured in field soils may show great variety, and Beyer *et al.* (1993) therefore concluded that a proper evaluation is not possible without analysis of other microbiological parameters, such as bio-mass, ATP and enzyme activity. This shows that microbiological parameters can only be suitable bioassay instruments when applied in unison with other measurements. It should further be noted that these parameters often rely on the indigenous microflora which may have adapted to the pollution, and thus may be less sensitive.

Recently, Frostegård *et al.* (1993) described a method to obtain further insight into possible shifts in the soil microbial community caused by expo-sure to heavy metals. This method uses the phospholipid fatty acid (PLFA) composition of a soil and employs the fact that different subsets of the soil microbial community have different PLFA patterns. The method appeared to be more sensitive than those monitoring ATP content, soil respiration or total PLFA content; hence it is a promising technique for the detection of changes in soil microbial communities resulting from exposure to heavy metals.

Within the framework of the Swedish MATS programme, a number of test guidelines, mainly for determining chemical effects on soil microflora, have been developed (Torstensson, 1993). A number of these may also be used as a 'bioassay':

- standard soil respiration
- ammonium oxidation (potential nitrification in soil amended with $(NH_4)_2SO_4$)

- potential denitrification
- heterotrophic nitrogen fixation in soil
- cyanobacterial nitrogen fixation in soil
- soil respiration curves (a method to test the abundance, activity and vitality of the microflora in forest soil)

The importance of natural variability for these tests is not yet clear. Furthermore, the aspect of genetic adaptation, which may affect the response of the indigenous microflora, is not mentioned.

A promising method to detect metal tolerance in soil bacterial communities has been described by Díaz-Raviña *et al.* (1994), using a thymidine incorporation technique. This appeared just as sensitive as the PLFA pattern method (Frostegård *et al.*, 1993) and was considered by the authors to provide a useful means of assessing the effect of heavy metals on soil bacterial communities over time.

2.4.2 BIOASSAYS WITH SOIL INVERTEBRATES

A number of bioassays with aqueous extracts of soil, using contaminated or decontaminated soil or dilutions of contaminated soil in non-polluted (artificial) soil have been described in the literature.

Samoiloff (1990) described a protocol for a bioassay of water and extracts (both organic and aqueous) of soil or sediment using the nematode *Panagrellus redivivus*. Growth of J2 juveniles is followed for 96 hours, and the number of animals not reaching the J4 stage is a measure of toxicity. The author considered this method especially suitable for toxicity ranking of samples.

A bioassay method using the ciliate *Colpoda steinii* to assess the toxicity of metal-contaminated soil was developed by Forge *et al.* (1993). This bioassay is performed using pore water samples obtained by centrifugation. They stated that this procedure has many advantages, but the main disadvantage is that it does not give an indication of the chemical identity of the toxin. Therefore, a combination of bioassays and chemical analysis should be recommended. They also found that growth of the *Colpoda steinii* was not more inhibited in water extracts of soil than in $CaCl_2$ extracts, although the latter contained higher metal concentrations; a protective effect of Ca^{2+} is suggested. When dried soil was re-wetted to obtain sufficient amounts of pore water, this seemed not to affect the toxicity of metals to *Colpoda steinii*, except for nickel.

Many test methods to determine the side effects of pesticides on beneficial arthropods have been described by Hassan (1985, 1992). Besides laboratory toxicity tests on artificial substrates, some bioassays are described in which arthropods (mainly beetles such as *Pterostichus cupreus*) are exposed to residues of pesticide spray in the field. In addition, field sampling (biomonitoring) after spraying with pesticides is suggested to assess possible side effects of the pesticide on spiders, carabids and Collembola.

Wiles and Jepson (1992) used two foliar and one soil bioassay to assess the effects of pyrethroid insecticides on aphid predators in cereals. In the soil bioassay, the animals are exposed to pesticide residues on soil contained in small tubs placed in the field. The animals are released in the tubs directly after spraying and mortality is assessed. Test species are *Bembidion lampros*, *Pterostichus melanarus*, *Coccinella septempunctatum* and *Leptyphantes tenuis*. A similar bioassay is described by Floate *et al.* (1989), using *Bembidion quadrimaculatum* and *Bembidion obscurellum*.

Wiles and Jepson (1992) concluded that *in situ* bioassays are cost effective and enable the design of new methodologies to answer specific questions concerning the effect of different compounds on test species and different substrates. They added that *in situ* bioassays enable the comparison of species susceptibility on a given substrate and pesticide toxicity on different substrates and may aid the interpretation of field trials. This does, however, require some insight into the route of exposure. They further noticed that 'bioassays may underestimate mortality as they often take into account one route of exposure. Conversely, they may overestimate mortality by exposing organisms to residues that they would not come into contact with in the field'. *In situ* bioassays are less reproducible than laboratory toxicity tests.

Paine *et al.* (1993) studied the bioaccumulation of PCBs in house crickets (*Achaeta domesticus* Linnaeus) incubated on polluted soils and compared this with PCB levels in field-caught animals. The animals were not brought into direct contact with the soil, but placed 1 cm above the soil, so the difference between the exposure conditions of the bioassay and native field animals remains unclear. In another bioassay using insects, the burrowing behaviour of the fruitfly *Ceratitis capitata* was used to assess the toxicity of pesticide residues in soil (Carante *et al.*, 1993).

Sheppard and Evenden (1994) described a bioassay method in which Petri dishes containing polluted soil are inserted into a box filled with clean soil. The colonization of the polluted soil with Collembola, Acari and other soil animals is assessed by Tullgren extraction of the Petri dishes and used as a measure of the severity of the pollution.

Corp and Morgan (1991) performed bioassays on metal-polluted soils using *Lumbricus rubellus* and compared the metal concentrations in these earthworms after 31 days' exposure with worms of the same species living in the contaminated soils. Metal concentrations in the native worms were generally higher than in the bioassay worms, but showed a comparable increase with increasing soil concentration. The authors therefore concluded that bioassays are only needed when it is not possible to find sufficient native worms on the contaminated soil. They further concluded that bioassays may be a useful instrument to obtain insight into the bioavailability of metals in series of contaminated soils. Finally, they made a plea for 'an earthworm standard of metal speciation' to replace classical extraction techniques using for example, acetic acid.

According to Morgan *et al.* (1992) earthworms may be good indicators of metal bioavailability, but they warn that genetic adaptation may give a false indication. Adapted animals may show a different bioaccumulation pattern and are therefore not suitable for biomonitoring.

A rather unusual test with earthworms was described by Mac *et al.* (1990) who used *Lumbricus terrestris* in a bioassay of sediments. Earthworms were exposed to sediment/water suspensions for 12 days; end-points were survival, weight change and fat content, while uptake of PCBs from the sediment was also assessed.

Callahan *et al.* (1991) described *in situ* bioassay methods using *Lumbricus terrestris* for the assessment of contaminated soil. After 7 days of exposure, survival, morbidity (coiling, stiffening, swelling, lesions) and body content of pesticides (DDT, chlordane, etc.) were determined. Morbidity parameters showed a good correlation with body burdens of DDT (including metabolites) and chlordane but not with soil concentrations of these pesticides. This bioassay was also described by Menzie *et al.* (1992) and compared with 14-day acute and 28-day subacute laboratory bioassays with *Eisenia andrei* and *Lumbricus terrestris*, respectively. In the first test, worms are exposed to a dilution series of polluted soil, whereas in the second test earthworms are only held in the polluted soil. In a review of the work described in these two papers, Callahan and Linder (1992) stated that 'in a toxicity-based strategy, bioassays and toxicity tests directly measure contaminant bioavailability, as well as associated adverse biological and toxic effect.' They concluded that there is no large difference between a 14-day bioassay with *Eisenia fetida* and a 28-day bioassay with *Lumbricus terrestris* and that results of the laboratory bioassays also show good comparison with those of 7-day field bioassays using the latter species. Finally, they stated that 'the 14 day OECD laboratory test guideline may be modified for soil contamination evaluation'.

Van Gestel *et al.* (1988, 1993) used bioassays with *Eisenia fetida* for the quality assessment of decontaminated soils. The bioavailability of heavy metals was the main end-point, and was determined by comparing BCF values found in different soils or by relating metal concentrations in the earthworms to background levels found in earthworms kept on non-polluted soils.

2.4.3 BIOASSAYS WITH HIGHER PLANTS

A number of bioassays for higher plants have been described in the literature, mostly relying on methods used for toxicity testing of chemicals, using soil or nutrient solutions as test substrate.

Wang and Keturi (1990) described the application of bioassays with plants to assess the quality of effluents. This method could also be useful in the evaluation of polluted soil using aqueous extracts or soil pore water. Seed germination was used as the test end-point, with rice, lettuce and tomato appearing to be the most suitable plants. A similar test using cress (*Lepidium sativum*) and oats (*Avena sativa*) was described by Ahlf *et al.* (1993), who applied these tests to aqueous extracts of contaminated soil.

On the basis of results of bioassays with different plant species on various soil types, Davies (1993) concluded that radish (*Raphanus sativus*) is the most suitable plant for biomonitoring derelict land, a linear relationship being found between zinc uptake in radish roots and soil concentrations.

Van Gestel *et al.* (1988) used bioassays with *Cyperus esculentus* to evaluate the influence of soil decontamination on the bioavailability of heavy metals. In a follow-up of this study, Van Gestel *et al.* (1993) used *Raphamus sativus* and *Lactuca sativa* as test plants instead of *Cyperus esculentus*. In the first study, BCF values were used to compare bioavailability, and it was concluded that decontamination may lead to an increased bioavailability of metal residues. In the second study, bioassays were also applied to a number of non-polluted soils to obtain a background level of metals in the test organisms. In this study, metal levels found in plants after soil decontamination often exceeded the background level. This was no longer the case when the decontaminated soil was enriched with compost to compensate for the loss of organic matter during the decontamination process and allowed to equilibrate for 6 months.

2.4.4 BIOASSAYS FOR SCREENING OF POLLUTED SOILS

Sheppard *et al.* (1992) concluded that the effects of soil properties on the toxicity of contaminants are poorly known, and that this information is essential for a meaningful interpretation of bioassays. Linder *et al.* (1992) stated that for a proper assessment of polluted soils, toxicity data, chemical analysis and ecological information should be combined. They insisted on making use of the knowledge of soil science when approaching contaminated soils. A list of potential tests is given, including possible restrictions.

Warren-Hicks *et al.* (1989) gave an overview of tests that could be used for soil and sediment. A dilution series of polluted soil should be subjected to:

- a 14-day acute *Eisenia fetida* test; and
- a seed germination test with *Lactuca sativa.*

For soil eluates another set of tests was recommended. A dilution series of soil eluate should be subjected to:

- a root elongation test with *Lactuca sativa*;
- a 48-hour acute test with *Daphnia magna/pulex*;
- a 96-hour growth test with *Selenastrum capricornutum*; and
- a 48-hour acute test with *Pimephales promelas.*

On the basis of an inventory of available methods, Keddy *et al.* (1992) concluded that, for screening purposes, the following combination of tests should be recommended:

- seedling emergence, using radish and lettuce, applied to soil samples;
- *Eisenia andrei*, survival, applied to soil samples; and
- *Selenastrum capricornutum*, growth inhibition, applied to pore water or leachate/eluate.

For final assessment, the same tests could be used, but an extension of the set with a number of tests is recommended; most of these tests still have to be developed. According to Keddy *et al.* (1992) the ideal set for the final assessment should contain:

- seedling emergence test
- earthworm reproduction test
- arthropod reproduction test
- a test with another soil invertebrate, e.g. a nematode
- a bacterial test
- an algal population growth test.

2.5 OUTLINE OF THE ELEMENTS OF EXTRAPOLATION

2.5.1 TEST CONDITIONS

In the laboratory, organisms are generally tested at optimal conditions to guarantee survival and optimal growth and/or reproduction in the controls. In the field, conditions may show considerable fluctuations and are never optimal. Temperature, soil moisture content and food availability show seasonal and even daily patterns. Deviations from the optimal conditions may affect the susceptibility of organisms. Furthermore, combined exposure of organisms to toxicants and sub-optimal conditions may render them more harmful for predation and diseases. Very little is known about these aspects.

From section 2.2, it can be concluded that only very little information is available on the influence of fluctuating temperatures and soil moisture contents on the toxicity of chemicals for terrestrial organisms. It is evident that for a refined risk assessment of chemicals in terrestrial ecosystems further research on these aspects is needed.

2.5.2 BIOAVAILABILITY OF ORGANIC CHEMICALS

For organic chemicals, the so-called pore-water hypothesis or equilibrium partitioning concept seems to offer good possibilities to extrapolate toxicity data from one soil to another. For dissociating chemicals, the effect of soil pH should be taken into account, whereas for non-dissociating chemicals sorption data may provide instruments for such a recalculation (Van Gestel, 1992). According to Di Toro *et al.* (1991), the calculation of sorption data should apply only to K_{ow} values determined by modern techniques, such as the slow-stirring method or the generator-column technique. For highly lipophilic chemicals, with log K_{ow} >5–6, the equilibrium partitioning concept may no longer be valid; this is especially the case in soils with high content of organic matter where sorption is strong (Belfroid, 1994).

The applicability of the pore-water hypothesis for soil was demonstrated by Van Gestel and Ma (1988, 1990) in acute toxicity tests with earthworms. It is, however, by no means certain that this hypothesis is also valid for other organisms. For Collembola, indications have been obtained that direct exposure to soil may be of much greater importance for metal uptake than exposure via the food (C.A.M. Van Gestel, unpublished results). Further research on the general applicability of the pore-water hypothesis for other organisms is needed. More research is also needed to investigate the validity of the pore-water hypothesis or equilibrium partitioning concept for sublethal effects, such as effects on growth and reproduction.

Recent literature data suggest that bioavailability in the field may be lower than in freshly contaminated laboratory test soils. The main reason for the difference in bioavailability is the longer contact time which has resulted in an increased sorption in the field soils. Further research into this aspect is urgently needed to bridge the gap between laboratory and field, and bioassays may appear to be useful tools for this.

There are also indications that bioavailability in the field may be modified by other factors, such as a moss or fungal cover of the soil surface (Jagers op Akkerhuis, 1993). This is a new problem, which may be especially relevant for surface-sprayed chemicals, such as pesticides. For these chemicals, toxicity is dependent on the uptake rate, which in turn also depends on the activity of the organism. Further research is needed to obtain a better insight into these aspects.

2.5.3 BIOAVAILABILITY OF HEAVY METALS

The bioavailability of heavy metals is affected by many factors. Soil pH seems to be the most important, followed by soil organic matter content and cation exchange capacity. For each metal, a different combination of factors seems to govern bioavailability (Van Gestel, 1992; Van Gestel *et al.*, 1995). It is therefore not possible to derive general rules for the extrapolation of toxicity data between soils. Neither is it possible to extrapolate toxicity data from (one) laboratory soil to field soils. Further research is necessary to obtain insight into all the possible influences that abiotic factors may have on the bioavailability of metals. A concerted action of environmental chemists and ecotoxicologists seems necessary, with some combination of speciation models describing the activity of metal ions in the soil solution and toxicity tests in different soils with controlled metal speciation seeming to be the best way to extrapolate metal toxicity among soils.

For metals, there are in addition indications that bioavailability in field soils is low compared with that in freshly contaminated test soils in the laboratory. Moreover, the bioavailability of metals in soil seems to fluctuate with time (Forge *et al.*, 1993). Finally, both pollutants and soil organisms may show a heterogeneous distribution in the soil, thus hampering a proper prediction of exposure. Such issues require further investigation, the best means being a cooperative approach by soil chemists and ecotoxicologists.

2.5.4 BIOASSAYS

In most polluted soils, as a mixture of chemicals is present, it is difficult to assess the potential risk of such soils on the basis of toxicity tests with individual chemicals. Bioassays seem to offer a good possibility for the risk assessment of polluted soils as their results offer an integration of bioavailability and mixture effects (Keddy *et al.*, 1992).

Many bioassay methods have been described in literature and, with most having been developed for aquatic systems, the majority use aqueous extracts or pore-water samples of the soil obtained by centrifugation. One aspect that might be overlooked by using these procedures is the possible modifying effect of soil particles on toxicity (Gunkel *et al.*, 1993).

In both Canada and the USA, studies have been performed on the most suitable combinations of bioassays for the assessment of contaminated sites. Because bioassay methods with terrestrial organisms are lacking in number, the sets of bioassays that are recommended consist of a combination of terrestrial and aquatic organisms. A weak point is that most are short-term tests, focusing on lethality as end-point. Keddy *et al.* (1992) noted this problem and recommended a further development of tests on sublethal end-points to be included. Sublethal end-points, however, may be strongly affected by abiotic conditions, such as the physicochemical properties and nutritional status of the soil. It will therefore be difficult to identify a valid control for comparison of results obtained in contaminated or decontaminated soils. Van Gestel *et al.* (1993) tried to avoid this problem by studying metal uptake rather than effects. By performing bioassays on non-polluted reference soils, an effective background level of metals in the bioassay organisms (earthworm, plants) was established, which was then compared with results of bioassays with these organisms on contaminated and decontaminated soils.

2.5.5 INTERNAL CONCENTRATIONS

Van Wensem *et al.* (1994) argued for the use of internal effect concentrations as a basis for extrapolation. This has the advantage that apparent differences in sensitivity which are in fact due to differences in bioavailability can be avoided. This approach is only applicable if internal effect concentrations are independent of abiotic conditions and are equal under laboratory and field conditions. Furthermore, this approach requires equilibrium to be reached between internal and external concentrations. Finally, when using BCF values for calculating internal concentrations it should be realized that these tend to decrease with increasing soil concentrations and, for essential elements such as zinc and copper, are by no means constant over a wide range of concentrations.

Hopkin (1993a) concluded that bulk analysis of acid digests of soil does not take bioavailability into account. Furthermore, it should be realized that fungi may exhibit considerable uptake of heavy metals through their hyphae and thus modify bioavailability. The distribution of metals in soil should also be taken into account in risk assessment. Hopkin concluded that, when considering these difficulties it would be better to base standards for chemicals (metals) in soil on internal body concentrations rather than on external soil concentrations.

Hopkin (1993b) also reviewed biomonitoring methods, and concluded that the internal concentration may be a valid measure of bioavailability. He reported a good relationship between internal metal concentrations of two isopod species (*Porcellio scaber* and *Oniscus asellus*) and the snail *Helix aspersa* sampled on the same locations; the internal concentrations did not correlate with total concentrations in soil.

From the results of Crommentuijn (1994), it may be concluded that internal effect concentrations are not always constant. For *Folsomia candida*, effect concentrations in the organisms appeared to be affected by soil organic matter content and soil pH. From these preliminary results, it is concluded that further research is needed to elucidate the influence of external factors on the internal effect concentrations of metals in soil invertebrates. These findings are supported by a recent study of the Dutch Health Council (Gezondheidsraad, 1994), which concluded that the internal dose concept requires further research to demonstrate its usefulness.

2.5.6 EXTRAPOLATION FROM LABORATORY TESTS TO FIELD CONDITIONS

In Table 2.1, six arguments were referred to which should be considered when establishing a laboratory-to-field extrapolation factor. Van Straalen and Denneman (1989) concluded that a difference between sensitivity of organisms in laboratory and field conditions could not be confirmed, and therefore set the laboratory-to-field extrapolation factor at 1.

This chapter has focused on two of the arguments listed in Table 2.1, namely test conditions (especially soil moisture content and temperature) and bioavailability. Data obtained from the literature on the influence of temperature and soil moisture content do not allow for a quantification of the effect of these factors on toxicity. With respect to the second argument, indications have been obtained for a reduced bioavailability of chemicals in the field. Available data, however, do not allow for quantification of the difference in bioavailability between laboratory and field. Such differences seem to depend on characteristics of the soil and the chemical involved, and on incubation time. This argues for a more realistic design of laboratory toxicity tests, allowing for a better equilibration of the chemical with the soil. On the other hand, more insight into the factors determining bioavailability is urgently needed. In conclusion, the evaluation made in this chapter provides no reasons to alter the laboratory-to-field extrapolation factor of 1 which was originally suggested by Van Straalen and Denneman (1989).

It has been stressed that further research is needed on the factors determining bioavailability of chemicals in soil. Such research may provide a basis for extrapolating toxicity data from one soil type to another. However, in order to extrapolate from laboratory results to field conditions, bioavailability data in the field are also needed. Bioassays may be a useful tool for assessing such bioavailability, but to obtain a clear insight into the aspects which govern bioavailability in the field, bioassay measurements should be combined with sorption (organic chemicals) and/or speciation studies (metals). In addition, bioassays should not only consider effects on the test organisms, but also include measurements of chemical concentrations in their body tissues. Only by a combination of bioavailability (speciation, sorption), bioaccumulation and toxicity (internal effect concentrations) studies, it will be possible to arrive at a quantification of laboratory-to-field extrapolation factors.

2.6 SUMMARY

In this chapter, the extrapolation of results from laboratory toxicity tests to field conditions is discussed, with special emphasis on environmental conditions and bioavailability. Environmental conditions, such as temperature and soil moisture content, may affect an organism's sensitivity to toxicants. From the few data available, however, it remains inconclusive as to how the impact of these conditions when extrapolating from laboratory to field conditions can be quantified. Differences in bioavailability may hamper the extrapolation of results obtained in one soil type to another. For organic chemicals, the so-called porewater hypothesis or equilibrium partitioning concept appears to be a good starting point for extrapolation between soils. However, its applicability – which has been demonstrated for earthworms – seems to be restricted to chemicals with log $K_{ow} \leq 5$. The validity of the pore-water hypothesis for other organisms than earthworms remains to be confirmed. This hypothesis cannot be used for heavy metals, as many factors determine metal speciation and their consequent bioavailability in soil. In addition, soil and pore-water characteristics may (indirectly) affect the organism's vulnerability. These aspects call for further research, preferably by a concerted action of soil chemists and ecotoxicologists. There are indications of a lower bioavailability of chemicals in the field than in the laboratory, especially due to differences in incubation time and chemical speciation. Data on this matter, however, are scarce and too fragmentary to allow for quantification of the difference in bioavailability between laboratory and field soils to be made. Bioassays provide a tool to bridge the gap between laboratory toxicity tests and the field situation, and may offer insight into bioavailability in both laboratory and field soils.

REFERENCES

Ahlf, W., Gunkel, J. and Rönnpagel, K. (1993) Toxikologische Bewertung von Sanierungen, in *Bodenreinigung. Biologische und chemisch-physikalische Verfahrensentwicklung unter Berücksichtigung der bodenkundlichen, analytischen und rentlichen Bewertung* (ed. R. Stegman), Economica Verlag, Bonn, pp. 275–86.

Belfroid, A.C. (1994) *Toxicokinetics of Hydrophobic Chemicals in Earthworms. Validation of the Equilibrium Partitioning Theory.* PhD Thesis, University of Utrecht.

Bengtsson, G. and Rundgren, R. (1992) Seasonal variation of lead uptake in the earthworm *Lumbricus terrestris* and the influence of soil liming and acidification. *Arch. Environ. Contam. Toxicol.*, **23**, 198–205.

Beyer, L., Wachendorf, C., Elsner, D.C. and Knabe, R. (1993) Suitability of dehydrogenase activity assay as an index of soil biological activity. *Biol. Fertil. Soils*, **16**, 52–6.

Bitton, G., Koopman, B. and Agami, O. (1992) Metpad™: a bioassay for rapid assessment of heavy metal toxicity in wastewater. *Water Environ. Res.*, **64**, 834–6.

Boesten, J.J.T.I. (1986) *Behaviour of Herbicides in Soil: Simulation and Experimental Assessment.* PhD Thesis, Agricultural University Wageningen.

Brouwer, H., Murphy, T. and McArdle, L. (1990) A sediment-contact bioassay with *Photobacterium phosphoreum. Environ. Toxicol. Chem.*, **9**, 1353–8.

Callahan, C.A. and Linder, G. (1992) Assessment of contaminated soils using earthworm test procedures, in *Ecotoxicology of Earthworms* (eds P.W. Greig-Smith, H. Becker, P. Edwards and F. Heimbach), Intercept Ltd, Andover, Hants, UK, pp. 187–96.

Callahan, C.A., Menzie, C.A., Burmaster, D.E., Wilborn, D.C. and Ernst, T. (1991) On-site methods for assessing chemical impact on the soil environment using earthworms: a case study at the Baird and McGuire superfund site, Holbrook, Massachusetts. *Environ. Toxicol. Chem.*, **10**, 817–26.

Carante, J.P., Battut, P., Lemaitre, C. and Dorier, A. (1993) La mouche mediterranéenne des fruits, un nouvel outil en ecotoxicologie. *Bull. Soc. Zool. Fr.*, **118**, 169–76.

Corp, N. and Morgan, A.J. (1991) Accumulation of heavy metals from polluted soils by the earthworm *Lumbricus rubellus*: can laboratory exposure of control worms reduce biomonitoring problems? *Environ. Pollut.*, **74**, 39–52.

Crommentuijn, G.H. (1994) *Sensitivity of Soil Arthropods to Toxicants.* PhD Thesis, Vrije Universiteit, Amsterdam.

Davies, B.E. (1993) Radish as an indicator plant for derelict land: uptake of zinc at toxic concentrations. *Commun. Soil Sci. Plant Anal.*, **24**, 1883–95.

Demon, A. and Eijsackers, H. (1985) The effects of lindane and azinphosmethyl on survival time of soil animals, under extreme or fluctuating temperature and moisture conditions. *Z. ang. Ent.*, **100**, 504–10.

Denneman, C.A.J. and Van Gestel, C.A.M. (1990) *Bodemverontreiniging en Bodemecosystemen: Voorstel voor C-(toetsings)waarden op Basis van Ecotoxicologische Risico's.* Report no. 725201001. National Institute of Public Health and Environmental Protection (RIVM), Bilthoven, The Netherlands (in Dutch).

Denneman, C.A.J. and Van Gestel, C.A.M. (1991) *Afleiding van C-waarden voor Bodem-ecosystemen op Basis van Aquatisch Ecotoxicologische Gegevens.* Report no. 7252001008. National Institute of Public Health and Environmental Protection (RIVM), Bilthoven, The Netherlands (in Dutch).

Díaz-Raviña, M., Bååth, E. and Frostegård, Å. (1994) Multiple heavy metal tolerance of soil bacterial communities and its measurement by a thymidine incorporation technique. *Appl. Environ. Microbiol.*, **60**, 2238–47.

Di Toro, D.M., Zarba, C.S., Hansen, D.J., Berry, W.J., Swartz, R.C., Cowan, C.E., Pavlou, S.P., Allen, H.E., Thomas, N.A. and Paquin, P.R. (1991) Technical basis for establishing sediment quality criteria for nonionic organic chemicals using equilibrium partitioning. *Environ. Toxicol. Chem.*, **10**, 1541–83.

Eiserman, R. and Daei, B. (1993) Evaluation of soil pollutions applying an ecotoxicological assay, in *Integrated Soil and Sediment Research: A Basis for Proper Protection* (eds H. J. P. Eijsackers and T. Hamers), Kluwer Academic Publishers, Dordrecht, pp. 313–14.

Everts, J.W. (1990) *Sensitive Indicators of Side-effects of Pesticides on the Epigeal Fauna of Arable Land.* PhD Thesis, Agricultural University Wageningen.

Everts, J.W., Aukema, B., Mullié, W.C., Van Gemerden, A., Rottier, A., Van Katz, R. and Van Gestel, C.A.M. (1991) Exposure of the ground dwelling spider *Oedothorax apicatus* (Blackwall) (Erigonidae) to spray and residues of deltamethrin. *Arch. Environ. Contam. Toxicol.*, **20**, 13–19.

Floate, K.D., Elliott, R.H., Doane, J.F. and Gillott, C. (1989) Field bioassay to evaluate contact and residual toxicities of insecticides to carabid beetles (Coleoptera: Carabidae). *J. Econ. Entomol.*, **82**, 1543–7.

Forge, T.A., Berrow, M.L., Darbyshire, J.F. and Warren, A. (1993) Protozoan bioassays of soil amended with sewage sludge and heavy metals, using the common soil ciliate *Colpoda steinii. Biol. Fertil. Soils*, **16**, 282–6.

Förster, H.-G., Tiberg, E., Knoblauch, H., Kaun, E., Genz, U., Fischer, W., Adam, R. and Brendel, M. (1993) Remaining PAH-contamination after microbial remediation: why does it exist and how to treat it, in *Integrated Soil and Sediment Research: A Basis for Proper Protection* (eds H. J. P. Eijsackers and T. Hamers), Kluwer Academic Publishers, Dordrecht, pp. 352–3.

Frostegård, Å., Tunlid, A. and Bååth, E. (1993) Phospholipid fatty acid composition, biomass, and activity of microbial communities from two soil types experimentally exposed to different heavy metals. *Appl. Environ. Microbiol.*, **59**, 3605–17.

Gezondheidsraad (1994) *Ecotoxicologie op Koers.* Netherlands Health Council, Report nr 1994/13 (in Dutch).

Gunkel, J., Rönnpagel, K. and Ahlf, W. (1993) Eignung mikrobieller Biotests für gebundene Schadstoffe. *Acta Hydrochim. Hydrobiol.*, **21**, 215–20.

Harris, C.R. (1964a) Influence of soil moisture on the toxicity of insecticides in a mineral soil to insects. *J. Econ. Entomol.*, **57**, 946–50.

Harris, C.R. (1964b) Influence of soil type and soil moisture on the toxicity of insecticides in soils to insects. *Nature*, **202**, 724.

Harris, C.R. (1967) Further studies on the influence of soil moisture on the toxicity of insecticides in soil. *J. Econ. Entomol.*, **60**, 41–4.

Hassan, S.A. (1985) Standard methods to test the side-effects of pesticides on natural enemies of insects and mites developed by IOBC/WPRS Working Group 'Pesticides and Beneficial Organisms'. *Bull. OEPP/EPPO Bull.*, **15**, 214–55.

Hassan, S.A. (1992) Guidelines for testing the effects of pesticides on beneficial organisms: description of test methods. *IOBC/WPRS Bull.* **XV/3**, 1–186.

Hopkin, S.P. (1993a) Ecological implications of '95% protection levels' for metals in soil. *Oikos*, **66**, 137–41.

Hopkin, S.P. (1993b) In situ biological monitoring of pollution in terrestrial and aquatic ecosystems, in *Handbook of Ecotoxicology* (ed. P. Calow), Blackwell Scientific Publishers, Oxford, pp. 397–427.

Jagers op Akkerhuis, G.A.J.M. (1993) *Physical Conditions Affecting Pyrethroid Toxicity in Arthropods*. PhD Thesis, Agricultural University Wageningen.

Janssen, M.P.M. and Bergema, W.F. (1991) The effect of temperature on cadmium kinetics and oxygen consumption in soil arthropods. *Environ. Toxicol. Chem.* **10**, 1493–501.

Jepson, P.C., Croft, B.A. and Pratt, G.E. (1994) Test systems to determine the ecological risks posed by toxin release from *Bacillus thuringiensis* genes in crop plants. *Mol. Ecol.*, **3**, 81–9.

Keddy, C., Greene, J.C. and Bonnell, M.A. (1992) *A Review of Whole Organism Bioassays for Assessing the Quality of Soil, Freshwater Sediment and Freshwater in Canada*. Report of the National Contaminated Sites Remediation Program, Environment Canada.

Linder, G., Ingham, E., Brandt, C.H. and Henderson, G. (1992) *Evaluation of Terrestrial Indicators for Use in Ecological Assessments at Hazardous Waste Sites*. United States Environmental Protection Agency, EPA/600/R-92/183.

Løkke, H. (1994) Ecotoxicological extrapolation: tool or toy? in *Ecotoxicology of Soil Organisms* (eds M.H. Donker, H. Eijsackers and F. Heimbach), Lewis Publishers, Boca Raton, Florida, pp. 411–25.

Loonen, H. (1994) *Bioavailability of Chlorinated Dioxins and Furans in the Aquatic Environment*. PhD Thesis, University of Amsterdam.

Mac, M.J., Noguchi, G.E., Hesselberg, R.J., Edsall, C.C., Shoesmith, J.A. and Bowker, J.D. (1990) A bioaccumulation bioassay for freshwater sediments. *Environ. Toxicol. Chem.*, **9**, 1405–14.

Menzie, C.A., Burmaster, D.E., Freshman, J.S. and Callahan, C.A. (1992) Assessment of methods for estimating ecological risk in the terrestrial component: a case study at the Baird and McGuire superfund site in Holbrook, Massachusetts. *Environ. Toxicol. Chem.*, **11**, 245–60.

Morgan, J.E., Morgan, A.J. and Corp, N. (1992) Assessing soil metal pollution with earthworms: indices derived from regression analysis, in *Ecotoxicology of Earthworms* (eds P.W. Greig-Smith, H. Becker, P. Edwards and F. Heimbach), Intercept Ltd., Andover, Hants, UK, pp. 233–7.

Paine, J.M., McKee, M.J. and Ryan, M.F. (1993) Toxicity and bioaccumulation of soil PCBs in crickets: comparison of laboratory and field studies. *Environ. Toxicol. Chem.*, **12**, 2097–103.

Samoiloff, M. (1990) The nematode toxicity assay using *Panagrellus redivivus*. *Toxicity Assessment*, **5**, 309–18.

Sheppard, S.C. and Evenden, W.G. (1992) Bioavailability indices for uranium: effect of concentration in eleven soils. *Arch. Environ. Contam. Toxicol.*, **23**, 117–24.

Sheppard, S.C. and Evenden, W.G. (1994) Simple whole-soil bioassay based on microarthropods. *Bull. Environ. Contam. Toxicol.*, **52**, 95–101.

Sheppard, S.C., Gaudet, C., Sheppard, M.I., Cureton, P.M. and Wong, M.P. (1992) The development of assessment and remediation guidelines for contaminated soils, a review of the science. *Can. J. Soil Sci.*, **72**, 359–94.

Suter, G.W. (1993) New concepts in the ecological aspects of stress: the problem of extrapolation. *Sci. Total Envir.*, Suppl., 63–76.

TCB (1992) *Advies Herziening Leidraad Bodembescherming I. C-toetsingswaarden en Urgentiebeoordeling*. Technical Committee on Soil Protection. The Netherlands (in Dutch).

Torstensson, L. (ed.) (1993) *Soil Biological Variables in Environmental Hazard Assessment. Guidelines*. Swedish Environmental Protection Agency, Solna.

Van de Meent, D., Aldenberg, T., Canton, J.H., Van Gestel, C.A.M. and Slooff, W. (1990) *Desire for Levels. Background Study for the Policy Document 'Setting Environmental Quality Standards for Water and Soil'*. Report no. 670101002, National Institute of Public Health and Environmental Protection (RIVM), Bilthoven, The Netherlands.

Van Gestel, C.A.M. (1992) The influence of soil characteristics on the toxicity of chemicals for earthworms: a review, in *Ecotoxicology of Earthworms* (eds P.W. Greig-Smith, H. Becker, P.J. Edwards, and F. Heimbach), Intercept Press, Andover, Hants, UK, pp. 44–54.

Van Gestel, C.A.M. and Ma, W. (1988) Toxicity and bioaccumulation of chlorophenols in earthworms, in relation to bioavailability in soil. *Ecotox. Environ. Safety*, **15**, 289–97.

Van Gestel, C.A.M. and Ma, W. (1990) An approach to quantitative structure–activity relationships (QSARs) in earthworm toxicity studies. *Chemosphere*, **21**, 1023–33.

Van Gestel, C.A.M., Adema, D.M.M., De Boer, J.L.M. and De Jong, P. (1988) The influence of soil clean-up on the bioavailability of metals, in *Contaminated Soil '88* (eds K. Wolf, W.J. Van den Brink, and F.J. Colon), Kluwer Academic Publishers, Dordrecht, pp. 63–5.

Van Gestel, C.A.M., Dirven-Van Breemen, E.M. and Kamerman, J.W. (1993) The influence of soil clean up on the bioavailability of heavy metals for earthworms and plants, in *Integrated Soil and Sediment Research: A Basis for Proper Protection* (eds H.J.P. Eijsackers and T. Hamers), Kluwer Academic Publishers, Dordrecht, pp. 345–8.

Van Gestel, C.A.M., Rademaker, M.C.J. and Van Straalen, N.M. (1995) Capacity controlling parameters and their impact on metal toxicity in soil invertebrates, in *Biogeodynamics of Pollutants in Soils and Sediments. Risk Assessment of Delayed and Non-linear Responses* (eds W. Salomons and W.M. Stigliani), Springer Verlag, Berlin, pp. 171–92.

Van Gestel, C.A.M. and Van Diepen, A.M.F (in press) The influence of soil moisture content on the bioavailability and toxicity of cadmium for *Folsomia candida* Willem (Collembola: Isotomidae). *Ecotox. Environ. Safety*.

Vanhala, P.T. and Ahtiainen, J.H. (1994) Soil respiration, ATP content, and *Photobacterium* toxicity test as indicators of metal pollution in soil. *Environ. Toxicol. Water Qual.*, **9**, 115–21.

Van Straalen, N.M. (1993a) An ecotoxicologist in politics. *Oikos*, **66**, 142–3.

Van Straalen, N.M. (1993b) Open problems in the derivation of soil quality criteria from ecotoxicity experiments, in *Contaminated Soil '93* (eds F. Arendt, G.J. Annokkee, R. Bosman and W.J. van den Brink), Kluwer Academic Publishers, Dordrecht, pp. 315–26.

Van Straalen, N.M. and Denneman, C.A.J. (1989) Ecotoxicological evaluation of soil quality criteria. *Ecotox. Environ. Safety*, **18**, 241–51.

Van Straalen, N.M. and Van Gestel, C.A.M. (1993) Soil invertebrates and micro-organisms, in *Handbook of Ecotoxicology* (ed. P. Calow), Blackwell Scientific Publishers, Oxford, pp. 251–77.

Van Straalen, N.M., Schobben, J.H.M. and Traas, T.P. (1992) The use of ecotoxicological risk assessment in deriving maximum acceptable half-lives of pesticides. *Pestic. Sci.*, **34**, 227–31.

Van Wensem, J., Vegter, J.J. and Van Straalen, N.M. (1994) Soil quality criteria derived from critical body concentrations of metals in soil invertebrates. *Appl. Soil Ecol.* **1**, 185–91.

Verma, A. and Pillai, M.K. (1991) Bioavailability of soil-bound residues of DDT and HCH to earthworms. *Curr. Sci.* **61**, 840–3.

Wang, W. and Keturi, P.H. (1990) Comparative seed germination test using ten plant species for toxicity assessment of a metal engraving effluent sample. *Water Air Soil Pollut.*, **52**, 369–76.

Warren-Hicks, W., Parkhurst, B.R. and Baker, S.S. (1989) *Ecological Assessment of Hazardous Waste Sites: A Field and Laboratory Reference.* United States Environmental Protection Agency. EPA/600/3-89/013.

Wiles, J.A. and Jepson, P.C. (1992) In situ bioassay techniques to evaluate the toxicity of pesticides to beneficial invertebrates in cereals. *Asp. of Appl. Biol.*, **31**, 61–8.

3   Is it possible to develop microbial test systems to evaluate pollution effects on soil nutrient cycling?

JOHN DIGHTON

3.1 THE NEED FOR SOIL MICROBIAL EVALUATION TECHNIQUES

Upper soil horizons are important natural resources for the growth of plants for food and industry (crops, animal produce, timber and biofuels). The maintenance of soil fertility, with minimal exogenous inputs, is vital for long-term sustainable plant productivity with viable economic returns.

The effects of pollution on soil organisms has the potential to disrupt nutrient cycling systems, reducing the availability of nutrients for plant growth. It is, therefore, desirable to have a test system which would reliably and quickly indicate when critical levels of pollutant loading have occurred and where key components of the soil microbial community are detrimentally affected to such a degree that nutrient cycling processes are disrupted. In order to do this successfully, we need to be able to identify the keystone target organisms, or relate their loss, or reduction in physiological competence, with reduction in nutrient cycling processes. The target aim is to have a simple tool to evaluate the state of activity or population of the keystone organism(s) *in situ* in the field situation.

The current state of knowledge of soil community structure, particularly of microorganisms, is such that we are not able to say what a desired community looks like, or how it should properly function. Thus, at present, the best we could hope to be able to do is to look for appropriate qualitative and quantitative tools to allow comparison between non-polluted and impacted soils to identify significant changes in soil biota. This concept has been used by Domsch *et al.* (1983) to evaluate safe levels of agrochemical use based on

Ecological Risk Assessment of Contaminants in Soil. Edited by Nico M. van Straalen and Hans Løkke. Published in 1997 by Chapman & Hall, London. ISBN 0 412 75900 4

sensitivity of the test system, percent deficit of system due to degree of impaction and reversibility of effect.

Ideal decision-making trees have been created for evaluating effects of plant protection products on soil organisms (OEPP/EPPO, 1994) but the value of these schemes lies with the selection of appropriate soil microbial evaluation techniques. In this chapter, it is not possible to produce a comprehensive review of the literature. The aim is to present some of the major methods available for the study of microbial populations and their function, with reference to some of the more recent or 'classic' papers and to assess the potential for these methods to be used in risk assessment related to pollution. In order to do this we need to identify key organisms in soil that may function as indicator species/groups or microbially mediated processes that can be easily measured. A provisional ranking of these, with respect to potential use, has been made by Domsch *et al.* (1983) and is outlined in Table 3.1.

3.2 WHO ARE THE KEY PLAYERS IN THE SOIL THEATRE?

In general we can consider the microbial components of soil as the bacteria and fungi. Actinomycetes are present, but studies on their diversity, abundance and function are few, so they will be excluded from this discussion. Bacteria and fungi are both involved in the decomposition process of breaking down dead plant and animal remains. Some components are used as energy and nutrient resources for microbial growth (immobilization) and others are released into the inorganic and organic nutrient pools in the soil solution (mineralization). Additionally, nutrients which are temporarily immobilized in microbial tissue can be subsequently mineralized on the

Table 3.1 Sensitivity of microbial test systems for risk assessment. (After Domsch *et al.*, 1983)

Sensitivity	Organisms/process
High	Population: nitrifiers
	Population: *Rhizobium*
	Population: actinomycetes
	Function: organic matter decomposition
	Function: nitrification
Medium	Population: bacteria
	Function: CO_2 production
	Function: O_2 uptake
	Population: fungi
	Function: denitrification
	Function: ammonification
Low	Function: N_2 fixation
	Population: *Azotobacter*
	Population: ammonifiers
	Population: protein degraders

death of the organism. The mineralized nutrients are then available for uptake by plants. In general, where resource quality of litter is low, in natural ecosystems (particularly forests and climatically limited systems with low pH), fungi appear to be more important organisms. Here, as a result of incomplete decomposition, the organic matter content of soil tends to be higher. Bacteria, however, appear to be more important in agriculture and grassland ecosystems, where resource quality is higher, litter turnover rates are greater and soil pH is higher.

Certain groups of fungi may also be important in assisting nutrient uptake by plants. Arbuscular and ectomycorrhizas are symbiotic associations between fungi and plant roots. The penetration of their fungal hyphae into soil, beyond the root nutrient depletion zone, enables mycorrhizas to increase the uptake of nutrient elements. The ability of some mycorrhizal species and groups to produce degrading enzymes also enables them to access organic sources of nutrients (Abuzinadah and Read, 1986; Abuzinadah *et al.*, 1986; Dighton, 1991), thereby short-circuiting the decomposition cycle of mineralization and uptake. Pollution effects on these fungal organisms can potentially reduce plant growth.

In addition to bacteria and fungi (microorganisms) there are a range of faunal groups represented in soil. These include Protozoa, Nematoda, Enchytraeidae, Lumbricidae and the arthropods Collembola, Acari and Insecta. Although these animals will not be discussed here in detail, it is important to recognize that they play a role in comminution of litter, assist in the decomposition process and interact significantly with the microorganisms influencing their growth (Newell, 1984a,b), nutrient mineralization rates (Anderson *et al.*, 1983) and nutrient uptake by mycorrhizal systems (Warnock *et al.*, 1982; McGonigle and Fitter, 1988).

3.3 THE FUNCTION OF MICROBIAL COMMUNITIES: A 'BLACK-BOX' APPROACH

If we look at the function of bacterial and fungal communities overall, we can produce a 'black-box' model of the decomposer system (Figure 3.1). From this figure, we can see that there are various parameters which could be measured – respiration (CO_2 evolution), litter (substrate) weight loss and nutrient mineralization – which could be used as a measure of activity of microbial communities and, hence, give an index of the effect of pollutants. These processes, however, also include the activities of soil fauna and so cannot represent the microbial communities alone. The application of biocides (Ingham *et al.*, 1989; Beare *et al.*, 1992) can be used selectively to remove faunal groups, to measure the function of microbial communities, but the release of grazing pressure and effect of biocides on the function of microbes may influence their activity. Thus, we have the possibility of using the following techniques to measure microbial activity indirectly.

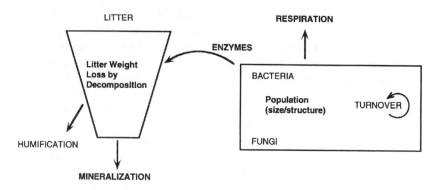

Figure 3.1 A 'black box' model of the decomposer system in soil, where the processes of respiration (metabolic activity), nutrient mineralization (result of metabolic activity) and substrate utilization (litter weight loss) may be considered without reference to the communities of organisms involved. The two end products of metabolism and mediators (enzymes) are measurable, and may provide an index of overall activity. Characterization of the population size/structure of bacterial and fungal communities may provide an index of disruption of the communities due to pollution and may identify indicator species.

3.3.1 SOIL RESPIRATION

As a result of microbial activity, CO_2 is released from the microbial population. Incubation of soil in enclosed microcosms or by the insertion of tubes with removable gas-tight lids into soil allows evolved CO_2 to be trapped, collected and quantified. The use of infra-red gas analysis (IRGA) to measure CO_2 evolved from soil allows rapid measurements to be made on large numbers of samples both in the laboratory and field (Ineson and Gray, 1980). Selective inhibitors such as streptomycin and cycloheximide may be used to separate the bacterial and fungal components of the respiration as for substrate induced respiration (Beare *et al.*, 1990).

3.3.2 DECOMPOSITION RATES

The decomposition of organic residues in soil is a main activity of the microbial pool. Measures of the rate of decomposition of organic material can be used as an index of the microbial activity. Specific substrates (e.g. leaf litters) may be placed in nylon mesh litter bags (McClaugherty *et al.*, 1985) of different mesh sizes, allowing access by microorganisms alone or microorganisms and soil animals, and placed in the natural environment. Measures of weight loss of the substrate with time give an index of decomposer activity.

Standardized substrates have also been used to compare decomposition rates between soils and soils under different treatments. Of these substrates cotton cloth has been the most widely used (Harrison *et al.*, 1988), an estimate of cellulose decomposition rate being measured as loss of tensile strength of the material. As the physical structure of cellulose in cotton is not similar to cellulosic material frequently encountered in natural substrates, and as pure cellulose rarely occurs alone in soil, there is considerable argument regarding the validity of use of this technique to measure decomposition rates (Howard, 1988). However, the method is easy to perform and can be used as a method of comparison between treatments within the same soil as an index of microbial cellulase activity. All measures of decomposition provide an index of microbial activity and relate to nutrient mineralization. Thus, they may be useful indices of pollution. The degree of impairment of decomposition, leading to a breakdown in system functioning, has, however, not been adequately quantified.

3.3.3 NUTRIENT MINERALIZATION

Changes in soil respiration may be related to changes in population size of the microbial pool or to its metabolic rate, but may not be related to its activity in terms of processes. The end result of organic matter decomposition by microbial activity is the release of nutrient elements in soluble labile pools as a result of degradation of organic components and turnover of the microbial pool. Zero tension lysimetry is often used to measure the release and loss of leaching of nutrient elements (mainly the macronutrients; ammonium and nitrate N, P, K, Ca, Mg, but may be used to measure any element as long as its concentration is high enough for detection, bearing in mind the limit of the lysimeter dimensions) from decomposing organic matter. Lysimeters may be placed in the field (Anderson *et al.*, 1990) or microcosm systems (Anderson and Ineson, 1984) may be used for laboratory incubations. In each case, the released inorganic nutrient elements are leached from the soil by water, collected and the chemical composition determined quantitatively. This method will provide an overall integrant of the effect of a soil pollutant on nutrient cycling. A limitation, however, is that lysimeters often lack living plants as nutrient sinks. Thus, the uptake of mineralized nutrients is not accounted for.

3.3.4 SPECIFIC PROCESSES

Denitrification can be measured in either microcosms in the laboratory or in soil cores isolated from bulk soil by insertion of a PVC drainpipe, using the acetylene inhibition technique of Ryden and Dawson (1982) and reviewed by Goulding *et al.* (1990). Acetylene is injected into the soil to inhibit reduction of N_2O to N_2. The N_2O diffusing into the headspace of the microcosm or soil core cover box is determined by gas chromatography. Non-symbiotic nitrogen fixation has been indicated as a method for an assay to comply with the US Toxic Substances Control Act for sensitivity of microbes involved in nitrogen cycling to environmental toxicants (Martinez-Toledo *et al.*, 1991). They describe methods for laboratory studies on unsterile soil where

Azotobacter chroococcum are quantified by enumerating bacterial cell numbers using dilution plating onto nitrogen-free medium as a selective medium. Nitrogenase activity is also measured using the acetylene reduction method of Hardy *et al.* (1968).

These process-oriented methods allow us to obtain an index of the effect of pollutants on processes mediated by soil organisms. Some of the techniques, such as lysimetry and respiration, are able to be carried out *in situ* in the field, whereas others (acetylene reduction) are usually carried out in the laboratory. In the latter case, extrapolation to the field is required. Since these measures are of function, the methods integrate over the community structure and species composition, but may reflect changes in the populations of the microorganisms. Microorganism population size is another determinant which could act as an index of pollution effects. The methods for determining microbial population size are numerous and each subject to practical or interpretational problems.

3.4 MEASURES OF MICROBIAL BIOMASS

The size of microbial components relative to that of other soil components can be seen in Table 3.2. In order to estimate biomass of either bacteria or fungi, small samples of soil (grams to tens of grams) are used. In seemingly uniform and homogenous soil systems (such as tilled agricultural soils) these small samples are hardly representative of the whole ecosystem. In natural ecosystems, particularly forests, the degree of heterogeneity in the soil is very much higher. In forested ecosystems patches may be of the order of millimetres in scale (microarthropod frass, microarthropod remains) to centimetres (leaf litter, root litter) to tens of metres (fallen tree trunks). Thus, the degree of replication of samples from natural ecosystems needs to be increased or stratified according to one patch type in order to obtain sensible and reproducible results.

Measures of biomass alone also do not distinguish between components of that biomass (species composition, or biodiversity), nor do they necessarily relate to physiological function as this is a function of biomass and species composition.

There are a number of ways in which both soil bacterial and fungal populations can be determined. These methods have recently been reviewed for applicability to soil and soil organic matter components (Frankland *et al.*, 1990; Gray, 1990).

3.4.1 DIRECT OBSERVATION

Fungal hyphae and bacteria may be quantified from soil suspensions which have been diluted by known amounts of hyphal lengths or bacterial cell numbers counted from a known aliquot of the diluted suspension. Two main

Table 3.2 Sizes of soil structures, soil microorganisms and soil fauna. (Data compiled from various sources)

Object		Diameter (or length)
Physical structures		
Coarse clay		1–2 μm
Silt		5–50 μm
Sand grains		0.1–1 mm
Soil microaggregates		0.001–1 mm
Soil crumbs		1–10 mm
Soil voids		0.01-several μm
Macropores		1-several mm
Invertebrate faecal pellets		100 μm
Microorganisms		
Bacteriophages		0.02–0.05 μm
Unicellular bacteria		0.1–3 μm
Yeast cells		5–10 μm
Fungal chlamydospores		200 μm
Fungal hyphae		2.5 μm
Soil invertebrates		
Protozoa		10–100 μm
Nematoda		40 μm
Collembola	- litter	3–4 mm length
	- humus	1–3 mm length
	- soil	1 mm length
Acari	- Mesostigmata	0.2–2 mm length
	- Cryptostigmata	0.2–1.5 mm length
	- Astigmata	0.2–1.2 mm length
	- Prostigmata	0.1–1.6 mm length
Enchytraeidae		c. 1 mm
Lumbricidae		2–10 mm

methods have been employed, soil smears (Trolldenier, 1973; Wynn-Williams, 1985) and membrane filtration (Hanssen *et al.*, 1974). The soil smear involves a known volume of the soil suspension being allowed to dry on a microscope slide; the membrane filtration technique involves the filtration of the soil suspension through a filter of pore size sufficiently small to allow retention of the organisms in question, the filter being then mounted on a microscope slide. In both cases the sample is stained to ease identification and visualization of the organism, and bacterial cell numbers or hyphal lengths are measured. These numbers can then be calculated to give biomass per unit soil mass or volume. Advances in the use of metabolically excited fluorescent stains have allowed discrimination between living and dead organisms, allowing estimates of live biomass to be made. Acridine orange and fluorescein diacetate (Fry, 1990) are common stains used, but Roser (1980) has used ethidium bromide and fluorescein isothiocyanate (FITC) to allow discrimination between bacteria and fungi as ethidium bromide stains only cell interiors (bacteria) and FITC, cell walls (fungi).

These methods provide good estimates of biomass, but require considerable staff resources and expertise in the use of microscopes, particularly for fluorescence staining. The membrane filtration methods are more rapid than soil smears and, once equipped, a laboratory can process large numbers of samples.

3.4.2 DILUTION PLATING

This method involves the suspension of soil in water or isotonic solution as above, appropriate dilution of the suspension and subsequent plating of an aliquot on an agar medium. After a period of incubation, colonies of organisms developing on the plate are counted. The number of organisms in the sample is calculated from the colony counts and the dilution of the suspension. The method is subject to a large number of assumptions based on the nature of the propagule giving rise to the colony, the selectivity of the medium to particular groups of organism, and the different incubation times required for different species. Due to questions regarding the nature of the propagule, it is probably a technique more applicable to bacteria than to fungi, where propagules may be either hyphae or spores, and a count of spores does not give an index of metabolically active fungal biomass. Selectivity of media may be a disadvantage of this technique as not all groups of bacteria or fungi are able to grow equally well on the same medium. Alternatively, this factor may be turned to an advantage as highly specific media will select for organisms with similar physiological functions, e.g. lignin medium will select only those organisms capable of lignin degradation.

3.4.3. SUBSTRATE-INDUCED RESPIRATION

In this technique, glucose is added to soil to saturate the enzyme system of the biomass. The initial rate of CO_2 production will be related to the microbial biomass present (Anderson and Domsch, 1978). The method relies on the determination of coefficients of conversion of CO_2 production per unit of microbial biomass (determined by direct measures) for each soil type used. However, the procedure can be run in large numbers, particularly if CO_2 determinations can be carried out using IRGA, and is amenable for use as a standard monitoring technique. In addition, by the use of selective inhibitors (streptomycin for bacteria and cycloheximide for fungi) the contribution of bacteria and fungal components of the total microbial biomass can be determined (Beare *et al.*, 1990).

3.4.4. COMPONENTS AND PRODUCTS

These techniques rely on the determination of components in the structure of the organism or products produced by them. Many of these chemicals are easily measured and, with conversion factors to link quantity of chemical per unit mass of microorganism, may be used as determinants of microbial population size.

(a) Enzymes

Of the products produced by microorganisms, enzymes are most commonly measured (Burns, 1978; Tabatabai, 1982). In measures of enzymes in bulk soil, there is no distinction between those produced by fungi and by bacteria. Enzymes decay in soil with time and activity has been shown to vary considerably with season (Harrison and Pearce, 1979; Rastin *et al.*, 1988), so the standardization of conditions needs to be taken into account. For these reasons, soil enzyme analysis was rejected as an appropriate technique for determining pesticide effects on soil microflora (Somerville *et al.*, 1985). However, general enzyme activity of populations of fungi and bacteria in soil and on decaying plant substrates was found to be simply measured using fluorescein diacetate hydrolysis (FDA) as an index (Schnurer and Rosswall, 1982). This index correlated well with respiration. In discussions of enzyme production by microflora in association with lignin decay, Sinsabaugh *et al.* (1993) and Sinsabaugh (1994) proposed that their ease of measurement makes them ideal candidates for measures of microbial activity, as many samples can be processed rapidly. At high levels of resolution they may be useful for identifying components of the microbial community. The use of total soil enzymes as an indicator of pollution by pesticides has been regarded as unsuitable by Gerber *et al.* (1991) due to the difficulty in relating activity to fractions of the microbial population and differing results at different physical conditions and substrate availability.

The detection of physiological capacity and, hence, community structure of bacterial communities can be gained by the use of multiple physiological and biological tests. One such test, using colorimetric determination of carbon substrate utilization (enzyme competence) has been streamlined in the BIOLOG microtitre plates. Individual carbon substrate utilization has been used as a measure of effects of pollutants, such as acetone mineralization by Van Beelen *et al.* (1991). The BIOLOG plate system consists of multiple-well plates with each well containing a different carbon source (one acts as a control with no carbohydrate). Growth of bacteria within a cell is detected by release of a violet end-product of respiration reaction with tetrazolium violet. Using a soil dilution as inoculum, a potential metabolic fingerprint of the bacterial community is revealed. Comparison between fingerprints can be analysed by cluster analysis and dendrograms of similarity and difference be constructed. Using this method, Winding (1994) showed that beech forest soil communities could be distinguished from agricultural and meadow soils and that distinct bacterial communities could be identified from different-sized soil aggregates. It is possible that the change in fingerprint of a single soil type in response to a pollution event could be rapidly detected by this method. However, the versatility of bacteria in the diversity of enzymes produced, the triggers for enzyme expression and the potential for inbuilt redundancy (Coleman *et al.*, 1994) in bacterial communities may mitigate against the usefulness of this technique to identify species changes in the bacterial community. In a microbial system, these methods could overcome the problem of debating the use of single versus mixed species tests. These physiological tests are somewhat species independent, giving an index of total bacterial activity, rather than function of single individual species.

(b) Chemical constituents of cells

For fungi, specific chemicals may be found in the biomass structure, which can be determined quantitatively and related to biomass; two of these are chitin and ergosterol. Chitin is present in fungal cell walls and can be degraded and extracted from soil as glucosamine and quantitatively determined by high performance liquid chromatography (HPLC). Chitin, however, is not unique to fungi and interference may come from that derived from soil invertebrates. Ergosterol, the most abundant sterol in fungi, is more specific than chitin/glucosamine and may be easily extracted from soil and quantitatively determined by HPLC (Grant and West, 1986; West *et al.*, 1987).

3.5 MEASURES OF SPECIES COMPOSITION

To evaluate the species composition of microbial communities, appropriate extraction techniques are required to adequately isolate all microorganisms from soil. Then it is necessary to be able to identify all species and compare species abundance between treatments. Suitable extraction methods such as the dilution plate method and direct observation have been discussed above and have been shown to be somewhat ineffective in isolating all species to the same degree of efficiency. However, using a dilution plate type of technique, Zhdanova *et al.* (1994) were successful in being able to describe the changes in community structure of soil microfungi as a result of chronic radiation dose, suggesting that melanization of fungal hyphae may impart some protective qualities. More recently, analysis of 16S rRNA has been used to evaluate diversity in natural bacterial communities (Devereux *et al.*, 1993; Hofle, 1993; Liesak *et al.*, 1993) and community hybridization technique used for studies of changes in microbial communities in soil (Ritz and Griffiths, 1994). Analysis of phospholipid fatty acids has also been used for bacteria (Korner and Laczko, 1992) and can be used to distinguish between fungal species also and has been used to evaluate pollution effects (Frostegård *et al.*, 1993). Many of these techniques provide information about species diversity and microbial community structure without being able to name individual species. Hence species diversity indices may be compared between treatments, but these cannot be related to function.

Studies of DNA profiles of soil (Torsvik *et al.*, 1993) show that bacterial populations are probably 200-fold higher than estimated by culturing techniques. Estimates by Hawksworth (1991) indicate that we may now be able to recognize some 5% of the potential world species list of fungi. If we consider the very small proportion of known fungal and bacterial species whose physiology have been investigated, can we confidently extrapolate the significance of loss or gain of fungal and bacterial species to a community, as a result of a pollution event, to a change in physiological function of that community? Similarly, with so many as yet unknown species of bacteria and fungi, present in such low numbers as to be unculturable with available techniques, what is the functional significance of the loss of a more readily observed (and thus assumed dominant) species? Will the function of this lost species be replaced by another (less dominant) species (Coleman *et al.*, 1994), implying inbuilt redundancy in the community?

3.6 CASE STUDIES OF MEASUREMENT OF POLLUTION EFFECTS ON SOIL MICROBIAL COMMUNITIES/FUNCTION

In investigations of the effects of heavy metals on soil microbial communities and their functioning, Reber (1992) developed a test to establish which bacterial species could use aromatic substances as carbon sources. He concluded that the diversity of bacteria was reduced in three out of five soils contaminated with copper, zinc, nickel and cadmium. Fritze *et al.* (1989) showed a significant reduction in soil respiration and fungal hyphal length in soil collected close to a copper–nickel smelter than further away, but little change in the rate of pine litter decomposition (Figure 3.2). Bacteria exhibiting the ability to degrade starch, pectin, xylan and cellulose decreased by a factor of 13–37 in sites close to the smelter. Bardgett *et al.* (1994) investigated the effects of Cu, Cr and As in pasture soils of New Zealand, showing that microbial biomass, using the substrate-induced respiration (SIR) technique, was significantly reduced in polluted soil, but this effect was mainly a reduction in the fungal rather than bacterial component of the microflora. Sulphatase activity was also reduced at concentrations of Cr >1000 mg kg^{-1} soil, although phosphatase, urease and invertase activities were not affected. It has been suggested that bacterial communities may not alter due to pollution, as species adapted to heavy metal, e.g. Cu-tolerant species, may be selected for in polluted areas (Huysman *et al.*, 1994). This would, however, not explain why Lettl (1985) found significant reductions in total numbers of heterotrophic bacteria in spruce forests subjected to high SO_2 inputs or that Wardle and Parkinson (1990a,b), measuring basal respiration, SIR and dilution plating, found greater effects of the herbicide glyphosate on bacteria than on fungi or actinomycetes.

Fungal basidiomycete fruit body occurrence has been used as an index of the impact of atmospheric pollution. Arnolds (1988) and Nauta and Vellinga (1993) have discussed the reduction of, particularly, ectomycorrhizal fungal fruiting structures as an index of SO_2 and NH_4 pollution. The association

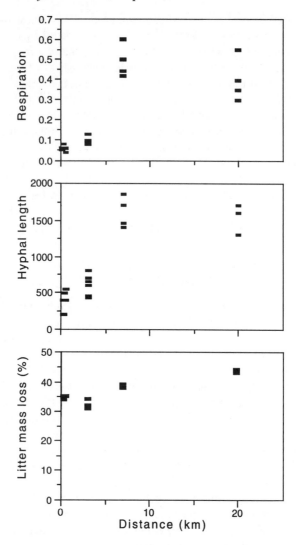

Figure 3.2 Measure of soil respiration, fungal mass and decomposition rate of Scots pine litter in relation to distance from a Cu–Ni smelter. Data from Fritze *et al.* (1989), showing that different measures may not correlate. (Respiration as % CO_2 evolved cm^{-3} soil; fungal hyphal length as m cm^{-3}; litter mass loss as %).

between occurrence of fruiting structures and the associated mycorrhizal fungus on the root has not, however, been fully verified. However, work from Termorshuisen (1990), Termorshuisen and Schaffers (1987) and Dighton and Skeffington (1987) have shown reduction in the mycorrhizal flora of trees in the field and a change in mycorrhizal species composition as a result of acid rain inputs. Much of the current understanding of this problem has been reviewed by Jansen and Dighton (1990). More recently work of Newsham *et al.* (1992a,b) has shown a significant reduction in saprotrophic fungal growth, change of species composition and reduction of litter decomposition following exposure to elevated SO_2. In relation to heavy metal pollution, some ecto-mycorrhizal fungal species may be both very tolerant of high levels of metal and accumulate Cu and Zn in the fungal mantle, thus preventing accumulation of toxic levels in the host plant (Denny and Wilkins, 1987a,b).

3.7 CONCLUSIONS

There are a number of techniques available to determine reliably and reproducibly the size of bacterial and fungal populations in soil, particularly the SIR method. This, however gives us no idea of the species composition of the population or community. Alternatively, isolation techniques and DNA extraction provide data on community structure, but not function. It is only by looking at the whole function of the community by employing methods such as litter decomposition rate or mineralization that the functional aspects of the microbial community can be assessed. Our current understanding of the inter-relationship between microbial biomass and function has recently been explored by Ritz *et al.* (1994).

The spatial distribution of microorganisms in relation to both heterogeneity within the soil system and potential area affected by the pollutant are important components to take into account. If the pollutant affects only a small area (centimetres to metres) the potential for rapid recolonization, after removal of the pollutant, is high. Large-scale pollution events, of the order of tens of metres to kilometres in extent, may result in low levels of recolonization. The movement of microorganisms and potential for soil fauna to act as vectors has not been well studied, although a recent review (Dighton *et al.*, 1993) gives some estimates of potential movement and, hence, reinvasion potential. Models of bacterial and viral dispersal in soil have been based on reconstituted soil columns (Yeates and Yeates, 1991; Yeates and Ouyang, 1992) and may not reflect the effect of soil structure and heterogeneity found in natural field soils – a problem of extrapolation from laboratory to field.

The inability to select one all-embracing technique as a microbial indicator of pollution is an indication of the state of the current knowledge and technical development in soil biology. We are not able to identify 'keystone' species as we do not know the extent of functional replacement. Many (most) soil microorganisms are unknown to us, so their function and abundance are also unknown; hence the consequences of loss of one or more species in either the short or, especially, the long term are difficult, if not impossible, to evaluate. This has led to the recommendation of a workshop on laboratory testing for pesticide effects on soil microflora (Gerber *et al.*, 1991) to address 12 areas for further intensive work to develop more robust tests (Table 3.3).

Table 3.3 Research needs for development of risk assessment measures for soil microbiology. (After Gerber *et al.*, 1991)

1	Rhizosphere studies
2	Microbiological aspects of nutrient cycling other than nitrogen
3	Non-symbiotic N fixation
4	Microfaunal/microfloral interactions
5	Soil-borne plant pathogen/microbial antagonist interactions
6	Soil structure influenced by microorganisms
7	Microbial activity in subsoils
8	Combined stress effects
9	Relationship between laboratory and field measures
10	Improved concepts for field evaluation of laboratory tests
11	Novel techniques (DNA techniques; antibody techniques; identification of 'stress' proteins)
12	Field validation under levels of naturally occurring stress conditions

As a general recommendation, based on currently available methods and knowledge, I would suggest:

1. Functional (black-box) process measures should be made (respiration, mineralization, litter decomposition) as an index of general microbial activity and nutrient cycling.
2. Key species or functional groups should be enumerated where it is known or suspected that they are targeted by the pollutant, e.g. ectomycorrhizas and acidifying pollutants and metabolic function of bacteria using the BIOLOG plate method.
3. Molecular methods should be used to describe the community structure. If this does not reflect changes due to the pollutant at least it will provide us with valuable new evidence of the species diversity in different soil types.
4. Studies should be set up to evaluate the comparative sensitivities of different measures of microbial community structure, population size and functional aspects to determine more reliable and rapid microbial indicators of pollution.

Unlike the zoologists, we are not yet in a position to select a universal test microorganism. This, in itself, may not be a bad thing. Universal test organisms must have a similar response to the pollutant as the other main organisms of the same trophic group present in the environment under study. How many of the current test organisms can reliably live up to this criterion? With modern molecular tools we are, however, in a better position to be able to describe soil microbial community structure and monitor its changes. If we add to that tests of microbial function, such as the BIOLOG plate, we may be able to interpret community structure changes with change in function of the community and get an impression of the magnitude of the effect of the pollution event on process rates. Further research linking these techniques is thus highly warranted.

3.8 SUMMARY

The development of a risk assessment system for soil based on the activities of microflora and nutrient cycling must include measures of both the nature of the microbial population and its activity. Deriving an accurate model is made difficult by the problems in quantifying soil microbial populations, assessing their community structure and identifying the key functions of each component of the community. The methods available for the quantification and qualification of microbial communities and measures of physiological processes are outlined. Recommendations are made to select possible avenues of investigation to find potentially useful assessments of microbial activity which could be used as indicators of pollutant effects. The main problem with our current understanding is that we are not able adequately to identify components of the community and cannot assess damage on the basis of degree of reduction in physiological processes measured. As such, a broad-based measure of overall soil activity and nutrient mineralization is suggested to be a potentially useful starting point. The use of zero tension lysimetry would allow gross nutritional effects to be measured in the laboratory and field.

REFERENCES

Abuzinadah, R.A. and Read, D.J. (1986) The role of proteins in the nitrogen nutrition of ectomycorrhizal plants. I. Utilization of peptides and proteins by ectomycorrhizal fungi. *New Phytol.*, **103**, 481–93.

Abuzinadah, R.A., Finlay, R.D and Read, D.J. (1986) The role of proteins in nitrogen nutrition of ectomycorrhizal plants. II. Utilization of protein by mycorrhizal plants of *Pinus contorta. New Phytol.*, **103**, 495–506.

Anderson, J.M. and Ineson, P. (1984) Interactions between microorganisms and soil invertebrates in nutrient flux pathways of ecosystems, in *Invertebrate–Microbial Interactions* (eds J.M. Anderson, A.D.M. Rayner and D.H. Walton), Cambridge University Press, Cambridge, pp. 59–88.

Anderson, J.M, Ineson, P. and Huish, S.A. (1983) Nitrogen and cation mobilization by soil fauna feeding on leaf litter and soil organic matter from deciduous woodlands. *Soil Biol. Biochem.*, **15**, 463–7.

Anderson, J.M., Leonard, M.A. and Ineson, P. (1990) Lysimeters with and without tree roots for investigating the role of macrofauna in forest soil, in *Nutrient Cycling in Terrestrial Ecosystems: Field Methods, Applications and Interpretation* (eds A.F. Harrison, P. Ineson and O.W. Heal), Elsevier, London, pp. 347–55.

Anderson, J.P.E. and Domsch, K.H. (1978) A physiological method for the quantitative measurement of microbial biomass in soil. *Soil Biol. Biochem.*, **10**, 215–21.

Arnolds, E. (1988) The changing macromycete flora in The Netherlands. *Trans. Br. Mycol. Soc.*, **90**, 391–406.

Bardgett, R.D., Speir, T.W., Ross, D.J., Yeates, G.W. and Kettles, H.A. (1994) Impact of pasture contamination by copper, chromium, and arsenic timber preservative on soil microbial properties and nematodes. *Biol. Fert. Soils*, **18**, 71–9.

Beare, M.H., Neeley, C.L., Coleman, D.C. and Hargrove, W.L. (1990) A substrate-induced respiration (SIR) method for measurement of fungal and bacterial biomass on plant residues. *Soil Biol. Biochem.*, **21**, 1027–36.

Beare, M.H., Parmelee, R.W., Hendrix, P.F. and Cheng, W. (1992) Microbial and faunal interactions and effects on litter nitrogen and decomposition in agroecosystems. *Ecol. Monogr.*, **62**, 569–91.

Burns, R.G. (1978) *Soil Enzymes*, Academic Press, London.

Coleman, D.C. (1994) Compositional analysis of microbial communities: is there room in the middle? in *Beyond the Biomass: Compositional and Functional Analysis of Soil Microbial Communities* (eds K. Ritz, J. Dighton and K. Giller), Wiley, Chichester, pp. 201–20.

Coleman, D.C., Dighton, J., Ritz, K. and Giller, K. (1994) Perspectives on the compositional and functional analysis of soil communities, in *Beyond the Biomass: Compositional and Functional Analysis of Soil Microbial Communities* (eds K. Ritz, J. Dighton and K. Giller), Wiley, Chichester, pp. 261–71.

Denny, H.J. and Wilkins, D.A. (1987a) Zinc tolerance in *Betula* spp. III. Variation in response to zinc among ectomycorrhizal associates. *New Phytol.*, **106**, 535–44.

Denny, H.J. and Wilkins, D.A. (1987b) Zinc tolerance in *Betula* spp. IV. The mechanisms of ectomycorrhizal amelioration of zinc toxicity. *New Phytol.*, **106**, 545–53.

Devereux, R., Kurtz, J. and Mundform, G. (1993) Molecular phylogenetic explorations of natural microbial community composition and diversity, in *Trends in Microbial Ecology* (eds R. Guerrero and C. Pedros-Alio), Spanish Society for Microbiology, Barcelona, pp. 387–90.

Dighton, J. (1991) Acquisition of nutrients from organic resources by mycorrhizal autotrophic plants. *Experientia*, **47**, 362–9.

Dighton, J. and Skeffington, R.A. (1987) Effects of artificial acid precipitation on the mycorrhizas of Scots pine seedlings. *New Phytol.*, **107**, 191–202.

Dighton, J., Jones, H.E., Robinson, C.H. and Beckett, J. (1993) *Dispersal of genetically modified microorganisms by abiotic factors and soil invertebrates.* Contract report to Department of the Environment (UK), Institute of Terrestrial Ecology, Grange-over-Sands, UK.

Domsch, K.H., Jagnow, G. and Anderson, T-H. (1983) An ecological concept for the assessment of side-effects of agrochemicals on soil microorganisms. *Res. Rev.*, **86**, 66–105.

Frankland, J.C., Dighton, J and Boddy, L. (1990) Methods for studying fungi in soil and forest litter, in *Methods in Microbiology: Techniques in Microbial Ecology* (eds R. Grigorova and J.R. Norris), Academic Press, London, pp. 343–404.

Fritze, H., Niini, S., Mikkola, K and Makinen, A. (1989) Soil microbial effects of a Cu–Ni smelter in south-western Finland. *Biol. Fert. Soil.*, **8**, 87–94.

Frostegård, A., Tunlid, A. and Bååth, E. (1993) Phospholipid fatty acid composition, biomass and activity of microbial communities from two soil types experimentally exposed to different heavy metals. *Appl. Environ. Microbiol.*, **59**, 3605–17.

Fry, J. (1990) Direct methods and biomass estimation, in *Methods in Microbiology: Techniques in Microbial Ecology* (eds R. Grigorova and J. R. Norris), Academic Press London, pp. 41–85.

Gerber, H.R., Anderson, J.P.E., Bugel-Mogensen, B., Castle, D., Domsch, K.H., Malkomes, H-P., Somerville, L., Arnold, D.J., Van de Werf, H., Verbeken, R. and Vonk, J.W. (1991) 1989 revision of recommended laboratory tests for assessing side-effects of pesticides on the soil microflora. *Toxicol. Environ. Chem.*, **30**, 249–61.

Goulding, K.W.T., Webster, C.P., Powlson, D.S., Poulton, P.R. and Bakar, R.A. (1990) Direct methods for estimating denitrification losses of nitrogen fertilizer, in *Nutrient Cycling in Terrestrial Ecosytems: Field Methods, Application and*

Interpretation (eds A.F. Harrison, P. Ineson and O.W. Heal), Elsevier, London, pp. 153–69.

Grant, W.D. and West, A.W. (1986) Measurement of ergosterol, diaminopimelic acid and glucosamine in soil; evaluation as indicators of microbial biomass. *J. Microbiol. Methods*, **6**, 47–53.

Gray, T.R.G. (1990) Methods for studying the microbial ecology of soil, in *Methods in Microbiology: Techniques in Microbial Ecology* (eds R. Grigorova and J.R. Norris), Academic Press, London, pp. 309–42.

Hanssen, J.F., Thingstad, T.F. and Goksoyr, J. (1974) Evaluation of hyphal lengths and fungal biomass in soil by a membrane filtration technique. *Oikos*, **25**, 102–7.

Hardy, R.W., Holstein, R.D., Jackson, E.K. and Burris, R.C. (1968) The acetylene–ethylene assay for N_2 fixation: laboratory and field evaluation. *Plant Physiol.*, **43**, 1185–207.

Harrison, A.F. and Pearce, T. (1979) Seasonal variation of phosphatase activity in woodland soils. *Soil Biol. Biochem.*, **11**, 405–10.

Harrison, A.F., Latter, P.M. and Walton, D.W.H. (1988) *Cotton Strip Assay: An Index of Decomposition in Soils*. Institute of Terrestrial Ecology, Grange-over-Sands, Cumbria, UK.

Hawksworth, D.L. (1991) The fungal dimension of biodiversity: magnitude, significance and conservation. *Mycol. Res.*, **95**, 641–55.

Hofle, M.G. (1993) Community structure and dynamics of bacterioplankton as revealed by low-molecular weight RNA analysis, in *Trends in Microbial Ecology* (eds R. Guerrero and C. Pedros-Alio), Spanish Society for Microbiology, Barcelona, pp. 391–5.

Howard, P.J. (1988) A critical evaluation of the cotton strip assay, in *Cotton Strip Assay: an Index of Decomposition in Soil* (eds A. F. Harrison, P.M. Latter and D.W.H. Walton), Institute of Terrestrial Ecology, Grange-over-Sands, Cumbria, UK, pp. 34–42.

Huysman, F., Verstraete, W. and Brookes, P.C. (1994) Effects of manuring practices and increased copper concentrations on soil microbial populations. *Soil Biol. Biochem.*, **26**, 103–10.

Ineson, P. and Gray, T.R.G. (1980) Monitoring the effect of 'acid rain' and sulphur dioxide upon soil micro-organisms, in *Microbial Growth and Survival in Extremes of Environment* (eds G.W. Gould and J.E.L. Corry), Academic Press, London, pp. 21–6.

Ingham, E.R., Coleman, D.C. and Moore, J.C. (1989) An analysis of food-web structure and function in a shortgrass prairie, a mountain meadow, and a lodgepole pine forest. *Biol. Fert. Soil*, **8**, 29–37.

Jansen, A.E. and Dighton, J. (1990) Effects of air pollutants on ectomycorrhizas. Commission of the European Community, *Air Pollution Research Report 30*, Brussels.

Korner, J. and Laczko, E. (1992) A new method for assessing soil microorganism diversity and evidence of vitamin deficiency in low diversity communities. *Biol. Fert. Soil*, **13**, 58–60.

Lettl, A. (1985) Sulphur dioxide pollution. 2. Influence of inorganic sulphur compounds on bacterial communities of forest soils. *Ekologia*, **4**, 121–33.

Liesak, W., Goebl, B.M. and Stackenbrandt, E. (1993) Molecular analysis of the bacterial diversity within an Australian Terrestrial soil sample, in *Trends in Microbial Ecology* (eds R. Guerrero and C. Pedros-Alio), Spanish Society for Microbiology, Barcelona, 69

Martinez-Toledo, M.V., Salmeron, V. and Gonzalez-Lopez, J. (1991) Effects of simazine on the biological activity of *Azotobacter chroococcum*. *Soil. Sci.*, **151**, 459–67.

McClaugherty, C.A., Pastor, J. and Aber, J.D. (1985) Forest litter decomposition in relation to nitrogen dynamics and litter quality. *Ecology*, **66**, 266–75.

McGonigle, T.P. and Fitter, A.H. (1988) Ecological consequences of arthropod grazing on VA mycorrhizal fungi. *Proc. Roy. Soc. Edinburgh*, **94B**, 25–32.

Nauta, M. and Vellinga, E.C. (1993) Distribution and decline of macrofungi in The Netherlands, in *Fungi of Europe: Investigation, Recording and Conservation* (eds D. Pegler, L. Boddy, B. Ing and P.M. Kirk), Royal Botanic Gardens, Kew, London, pp. 21–47.

Newell, K. (1984a) Interactions between two decomposer basidiomycetes and a collembolan under Sitka spruce: distribution, abundance and selective grazing. *Soil Biol. Biochem.*, **16**, 227–33.

Newell, K. (1984b) Interactions between two decomposer basidiomycetes and a collembolan under Sitka spruce: grazing and its potential effects on fungal distribution and litter decomposition. *Soil Biol. Biochem.*, **16**, 235–9.

Newsham, K., Frankland, J.C., Boddy, L. and Ineson, P. (1992a) Effects of dry deposited sulphur dioxide on fungal decomposition of angiosperm tree litter. I. Changes in communities of fungal saprotrophs. *New Phytol.*, **122**, 97–110.

Newsham, K., Boddy, L. Frankland, J.C. and Ineson, P. (1992b) Effects of dry deposited sulphur dioxide on fungal decomposition of angiosperm tree litter. III. Decomposition rates and fungal respiration. *New Phytol.*, **122**, 127–40.

OEPP/EPPO (1994) Decision-making scheme for the environmental risk assessment of plant protection products. *Bulletin OEPP*, **24**, 1–16.

Rastin, N., Rosenplanter, K. and Hutterman, A. (1988) Seasonal variation of enzyme activity and their dependence on certain soil factors in a beech forest. *Soil Biol. Biochem.*, **20**, 637–42.

Reber, H.H. (1992) Simultaneous estimates of the diversity and degradative capability of heavy-metal affected soil bacterial communities. *Biol. Fert. Soil.*, **13**, 181–6.

Ritz, K. and Griffiths, B.S. (1994) The potential application of a community hybridization technique for assessing changes in the population structure of soil microbial communities. *Soil Biol. Biochem.*, **26**, 963–71.

Ritz, K., Dighton, J. and Giller, K. (1994) *Beyond the Biomass: Compositional and Functional Analysis of Soil Microbial Communities*. Wiley, Chichester.

Roser, D. (1980) Ethidium bromide; a general purpose fluorescent stain for nucleic acid in bacteria and eucaryotes and its use in microbial ecology. *Soil Biol. Biochem.*, **12**, 329–36.

Ryden, J.C. and Dawson, K.P. (1982) Evaluation of the acetylene inhibition technique for the measurement of denitrification in grassland soils. *J. Sci. Food Agric.*, **33**, 1197–206.

Schnurer, J. and Rosswall, T. (1982) Fluorescein diacetate hydrolysis as a measure of total microbial activity in soil and litter. *Appl. Environ. Microbiol.*, **43**, 1256–61.

Sinsabaugh, R.L. (1994) Enzymic analysis of microbial pattern and process. *Biol. Fert. Soil*, **17**, 69–74.

Sinsabaugh, R.L., Antibus, R.K., Linkins, A.E., McClaugherty, C.A., Rayburn, L., Repert, D. and Weiland, T. (1993) Wood decomposition: nitrogen and phosphorus dynamics in relation to extracellular enzyme activity. *Ecology*, **74**, 1586–93.

Somerville, L., Greaves, M.P., Domsch, K.H., Verstraete, W., Poole, N.J., Van Dijk, H. and Anderson, J.P.E. (eds) (1985) *Recommended Laboratory Tests for Assessing the Side-Effects of Pesticides on the Soil Microflora*, Bristol, Long Ashton Research Station.

Tabatabai, M.A. (1982) Soil enzymes, in *Methods of Soil Analysis Part 2, Chemical and Microbial Properties*, (eds A.C. Pace, R.H. Miller and D.R. Keeney), American Society of Agronomy, Madison, pp. 903–42.

Termorshuizen, A.J. (1990) *Decline of Carpophores of Mycorrhizal Fungi in Stands of Pinus sylvestris*. PhD Thesis, University of Wageningen.

Termorshuizen, A.J. and Schaffers, A.P. (1987) Occurrence of carpophores of ecto-mycorrhizal fungi in selected stands of *Pinus sylvestris* in The Netherlands in relation to stand vitality and air pollution. *Plant Soil*, **104**, 209–17.

Torsvik, V, Goksoyr, J., Daaef, L., Sorheim, R., Michalsen, J. and Salte, K. (1993) Diversity of microbial communities determined by DNA reassociation technique, in *Trends in Microbial Ecology* (eds R. Guerrero and C. Pedros-Alio), Spanish Society of Microbiology, Barcelona, pp. 375–8.

Trolldenier, G. (1973) The use of fluorescence microscopy for counting soil micro-organisms, in *Modern Methods in the Study of Microbial Ecology* (ed. T. Roswall), *Bull. Ecol. Res. Comm. (Stockholm)*, **17**, 53–9.

Van Beelen, P., Fleuren-Kemila, A.K., Huys, M.P.A., Van Montford, A.C.P. and Van Vlaardingen, P.L.A. (1991) The toxic effects of pollutants on the mineralization of acetate in subsoil microcosms. *Environ. Toxicol. Chem.*, **10**, 775–89.

Wardle, D.A. and Parkinson, D. (1990a) Effects of three herbicides on soil microbial biomass and activity. *Plant Soil*, **122**, 21–8.

Wardle, D.A. and Parkinson, D. (1990b) Influence of the herbicide glyphosate on soil microbial community structure. *Plant Soil*, **122**, 29–37.

Warnock, A.J., Fitter, A.H. and Usher, M.B. (1982) The influence of a springtail *Folsomia candida* (Insecta, Collembola) on the mycorrhizal association of leek *Allium porrum* and the vesicular–arbuscular mycorrhizal fungus *Glomus fascic-ulatus*. *New Phytol.*, **90**, 285–92.

West, A.W., Grant, W.D. and Sparling, G.P. (1987) Use of ergosterol, diaminopimel-ic acid and glucosamine contents of soils to monitor changes in microbial populations. *Soil Biol. Biochem.*, **20**, 607–12.

Winding, A. (1994) Fingerprinting bacterial soil communities using Biolog microtitre plates, in *Beyond the Biomass: Compositional and Functional Analysis of Soil Microbial Communities* (eds K. Ritz, J. Dighton and K. Giller), John Wiley and Sons, Chichester, pp. 85–94.

Wynn-Williams, D.D. (1985) Photofading retardant for epifluorescence microscopy in soil micro-ecological studies. *Soil Biol. Biochem.*, **17**, 739–46.

Yeates, M.V. and Ouygang, Y. (1992) VIRTUS a model of virus transport in unsatu-rated soils. *Appl. Environ. Microbiol.*, **58**, 1609–16.

Yeates, M.V. and Yeates, S.R. (1991) Modelling microbial transport in the subsur-face: a mathematical discussion, in *Modelling the Environmental Fate of Microorganisms* (ed. C.J. Hurst), American Society for Microbiology, Washington, pp. 48–76.

Zhdanova, N.N., Vasilevskaya, A.I., Artyshkova, L.V., Sadovnikov, Yu. S., Lashko, T.N., Gavrilyuk, V.I. and Dighton, J. (1994) Changes in micromycete communi-ties in soil in response to pollution by long-lived radionuclides emitted in the Chernobyl accident. *Mycol. Res.*, **98**, 789–95.

Rà

PART TWO

Populations in Soil

4   Ecotoxicology, biodiversity and the species concept with special reference to springtails (Insecta: Collembola)

4.1 THE SPECIES PROBLEM IN ECOTOXICOLOGY

In systematic biology, no concept has been the subject of such heated debate as the species concept (Minelli, 1993)

Ecotoxicological research should provide information that will enable genetic diversity in ecosystems to be maintained or increased, especially in areas which in the past, present or future might be affected by pollution. Genetic diversity is often called 'biodiversity' and is usually measured by counting numbers of species. However, such an approach is scientifically unsound because for many groups of animals there is no reliable, internationally accepted definition of the threshold of morphological or genetic difference that is required between two organisms before they can be considered to be separate species.

In my opinion, this 'species problem' is one of the most important facing ecotoxicology and conservation. In this chapter, I will discuss various definitions of species, outline the problems of assessing biodiversity, and give an example of the taxonomic confusion that can arise in one group of soil animals, the springtails (Insecta: Collembola), in the absence of a rigorous definition of what constitutes a species.

Why is defining the species so important for ecotoxicology? I would like to highlight three examples.

1. The 'HC$_5$ approach' to setting critical concentrations for pollutants in soils relies on determining the 'hazardous concentration for 5% of the species' (Van Straalen, 1993a). This is clearly an impossible task without a sound taxonomy on which to base the estimates of total numbers of species.

Ecological Risk Assessment of Contaminants in Soil. Edited by Nico M. van Straalen and Hans Løkke. Published in 1997 by Chapman & Hall, London. ISBN 0 412 75900 4

2. The question 'How many species inhabit the Earth?' is difficult to answer. Estimates range from 3×10^6 to more than 30×10^6, of which fewer than 1.8×10^6 have been described (Margulis, 1992). Both estimates could be correct depending on where species boundaries are drawn.

3. Long-term monitoring of changes in the species composition of sites is very difficult if the taxonomy changes through time. Unfortunately, it is not required practice for voucher specimens to be deposited in museums for future workers to validate. Thus, work on the diversity of soil animals conducted in the past (e.g. Thompson, 1924; Morris, 1927; Ford, 1935) may be difficult to compare with modern surveys (e.g. Frampton *et al.*, 1992; Bardgett *et al.*, 1993; Ponge, 1993). The most abundant collembolan in the survey of Bardgett *et al.* (1993) was *Onychiurus procampatus* (up to 50 000 m^{-2} on Cumbrian uplands) a 'species' described by Gisin (a 'splitter' of taxa) in 1956. The species was identified by W.G. Hale who has published several papers supporting Gisin's narrow definition of species boundaries in *Onychiurus* (Hale, 1964, 1965, 1968). If the same survey were to have been conducted before the 1930s, the species would have been identified as *Onychiurus armatus* (Tullberg, 1869), a taxon in which many of Gisin's (and other authors') 'species' could be sunk (see Table 4.1). In a more recent study, Hågvar (1994) proposed that pollution impact could be indicated by studying changes in the log-normal distribution of dominance of species in communities of Collembola and oribatid mites. This approach can only be legitimate if the species under consideration are clearly defined.

4.2 DEFINING 'BIODIVERSITY'

4.2.1 THE SPECIES CONCEPT

Many books and papers have been written on the species concept. Excellent discussions of the issues involved can be found in Minelli (1993), O'Hara (1994), Rand and Wilson (1993), Whittemore (1993) and World Conservation Monitoring Centre (1992). The most extreme view is that 'species are individual organisms' but this approach has – not surprisingly – received little support. However, some species have been described from single (sometimes damaged) specimens, or juveniles which have later been shown to be examples of previously described species. One would hope that modern taxonomists would resist this approach.

Above the level of the individual organism, it is often the opinion of the taxonomist that determines the limits of a species, rather than the biology of the organisms. The most widespread definition, the **biological species concept**, was developed in the 1930s and 1940s. Under the biological species concept 'species are groups of actually or potentially interbreeding natural populations, which are reproductively isolated from other such groups' (Mayr, 1940). For practical reasons (least of all the level of understanding of

Table 4.1 Species of Genus *Protaphorura* Absolon, 1901 recorded from Britain in the check list for Collembola of Kloet and Hinks (1964) and the update of Gough (1978). Several of these species have been synonymized by Bödvarsson (1970), Pitkin (1980), Pomorski (1990) and others. There is no general agreement as to how many of the 38 'species' on this list should be considered as true biological species

1.	*Protaphorura alborufescens* (Vogler, 1895)
2.	*Protaphorura armatus* (Tullberg, 1869)
3.	*Protaphorura aurantiacus* (Ridley, 1880)
4.	*Protaphorura bagnalli* Salmon, 1959
5.	*Protaphorura bicampatus* Gisin, 1956
6.	*Protaphorura caledonicus* Bagnall, 1935
7.	*Protaphorura campatus* Gisin, 1952
8.	*Protaphorura debilis* (Moniez, 1890)
9.	*Protaphorura evansi* Bagnall, 1935
10.	*Protaphorura fimatus* Gisin, 1952
11.	*Protaphorura flavidulus* Bagnall, 1939
12.	*Protaphorura furciferus* (Börner, 1901)
13.	*Protaphorura halophilus* Bagnall, 1937
14.	*Protaphorura hortensis* Gisin, 1949
15.	*Protaphorura humatus* Gisin, 1952
16.	*Protaphorura imminutus* Bagnall, 1937
17.	*Protaphorura latus* Gisin, 1956
18.	*Protaphorura magnicornis* Bagnall, 1937
19.	*Protaphorura meridiatus* Gisin, 1952
20.	*Protaphorura nemoratus* Gisin, 1952
21.	*Protaphorura octopunctatus* (Tullberg, 1876)
22.	*Protaphorura procampatus* Gisin, 1956
23.	*Protaphorura prolatus* Gisin, 1956
24.	*Protaphorura pseudocellatus* Naglitsch, 1962
25.	*Protaphorura pulvinatus* Gisin, 1954
26.	*Protaphorura quadriocellatus* Gisin 1943
27.	*Protaphorura stachi* Bagnall, 1935
28.	*Protaphorura subaequalis* Bagnall, 1937
29.	*Protaphorura subarmatus* Gisin, 1957
30.	*Protaphorura sublatus* Gisin, 1957
31.	*Protaphorura subuliginatus* Gisin, 1956
32.	*Protaphorura s-vontoerneri* Gisin, 1957
33.	*Protaphorura thalassophilus* Bagnall, 1937
34.	*Protaphorura tricampatus* Gisin, 1956
35.	*Protaphorura trinotatus* Gisin 1961
36.	*Protaphorura tullbergi* Bagnall, 1935
37.	*Protaphorura uliginatus* Gisin, 1952
38.	*Protaphorura waterstoni* Bagnall, 1937

politicians – we are all 'ecotoxicologists in politics', van Straalen, 1993b), this would seem as good a definition as any, despite its drawbacks (Minelli, 1993; O'Hara, 1994).

However, outside large, well-studied groups such as birds and mammals, it is rare for the taxonomist to test for reproductive isolation in their species. In most cases this is impossible as most work is conducted on dead specimens. In relatively poorly studied groups such as Collembola, most species are 'morpho-species'. Within genera, the degree of difference between two individuals that constitutes evidence of reproductive isolation is heavily dependent on the opinion of the taxonomist (Lawrence, 1979). As we shall see later, this situation may lead to considerable difference in the levels of perceived biodiversity, due to human nature rather than nature! (Table 4.1).

4.2.2 MEASURING BIODIVERSITY

If we accept the biological species concept as a working hypothesis, there are still a number of problems associated with measuring biodiversity.

1. Many parts of the world are completely unexplored, particularly for soil organisms. Hence the order of magnitude range in estimates for the total number of species on the Earth (Margulis, 1992). Even in a country as well studied as Britain, there are several groups of animals for which there is little information on national distribution (e.g. mites, Collembola). This is compounded by the lack of user-friendly keys for identification for most groups in most countries (Behan-Pelletier, 1993).
2. National checklists of species can be inaccurate if they are based on uncritical repetition of literature references. Voucher specimens for such records may not be kept. Ecological papers often acknowledge the help of a named taxonomist in identifying their species but the outcome depends on whether the expert is a 'splitter'or a 'lumper', or even their scientific competence. Bellinger (1985) examined H.G. Scott's reference collection of Collembola in the US and found that only 24 of the specimens were certainly, and another 20 possibly, correctly identified. Of the remainder, 109 were placed in the wrong genus and seven others in the wrong family! Salmon (1959) came to similar conclusions regarding some of the 'species' of Collembola described by R.S. Bagnall.
3. Excessive splitting of taxa on the basis of minor differences in morphology or colour may result in the same biological species having more than one name. This is compounded if the same species is described independently by taxonomists in different countries. For example, the common and widespread collembolan *Entomobrya nivalis* (Linnaeus, 1758) was described under at least 81 species and subspecies names before 1960 (Salmon, 1964).

4.2.3 THE 'VALUE'OF DIFFERENT SPECIES

As scientists, we often shy away from value judgements and feel we must give cold advice to decision makers. However, we should recognize that in the eyes of the general public (and even some scientists!) a beautiful orchid or butterfly is perceived as being more important to conserve than an earthworm, slug or rat, even if the latter are rare. In the Philippines, the neanurid collembolan *Paralobella orousseti* is spectacularly coloured with yellow, red and white bands (Cassagnau and Deharveng, 1984). It is much easier to make a case for the conservation of this species than the more dourly coloured relatives in the same family. To a great extent we have to decide which species we wish to conserve as there is little evidence for 'biodiversity hotspots'. Sites which are 'good' for some rare species in a particular taxonomic group are not necessarily good for all groups (Prendergast *et al.*, 1993; Usher *et al.*, 1993).

An interesting approach is to rank sites on the basis of cladistic principles to conserve the maximum phylogenetic diversity (Vane-Wright *et al.*, 1991). The method does not rely on simple species counts, but on detailed knowledge of evolutionary relationships and worldwide distribution of taxa. For ecotoxicologists, such an approach may be necessary if we have to decide which site to allow landfill, pollution, etc., from a range of possible alternatives (called the 'agony of choice' by Vane-Wright *et al.*, 1991).

4.3 AN EXAMPLE OF THE SPECIES PROBLEM

4.3.1 INTRODUCTION

Collembola are among the most widespread and abundant soil animals (Hopkin, 1997). Population densities in excess of 20 000 m^{-2} are found in most types of habitat. Collembola are important components of tropical soils and may constitute one-third of all individual animals in rainforests (Stork and Eggleton, 1992; Stork and Blackburn, 1993). The distribution of Collembola is poorly known globally and it is possible that they could eventually be shown to be most diverse in temperate latitudes. The equatorial regions of our planet may not hold the greatest diversity of all groups of invertebrates (Platnick, 1991).

4.3.2 IDENTIFICATION

Within the Collembola, the family Onychiuridae are mostly small, poorly pigmented animals which live permanently in the soil and which have lost the ability to jump (Figure 4.1). For the purposes of this discussion, I shall focus on the genus *Protaphorura* which was formerly considered a subgenus of *Onychiurus*. Of the 334 'species' of Collembola on the official British list (Kloet and Hinks, 1964; Gough, 1978), 38 are in *Protaphorura* (Table 4.1). Within this genus, several characters have been used to separate species, the main ones being the presence, absence or positions of setae (chaetotaxy), and the numbers and arrangement of 'pseudocelli' on the different segments of the body (Figure 4.1). Pseudocelli are small circular weak points in the cuticle which rupture if the springtail is attacked, releasing a fluid that repels predators (Usher and Balogun, 1966; Rusek and Weyda, 1981).

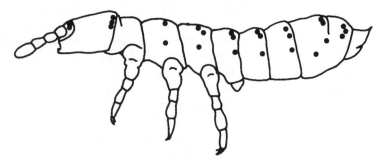

Figure 4.1 *Protaphorura armatus* (Tullberg, 1869) of approximately 1.5 mm in length (after Gisin, 1960) showing the positions of the pseudocelli (black dots).

Until the 1930s, relatively few species were recognized in *Onychiurus*. However R.S. Bagnall, and subsequently H. Gisin, described numerous species of *Onychiurus* from Britain and elsewhere in Europe, many on the basis of minute differences in chaetotaxy and numbers of pseudocelli (for examples see Bagnall, 1935; Gisin, 1960). A glance at Table 4.1 illustrates the impact that these two workers have had on the British list. Indeed, a list of world Collembola produced by K. Christiansen and P. Bellinger (personal communication) includes 143 species in *Protaphorura* alone. Gisin's approach has been criticised by several taxonomists including Bödvarsson (1970), Pitkin (1980) and Pomorski (1986).

The lack of agreement was highlighted by Pomorski (1990) who studied variations in pseudocelli and setae in laboratory cultures of *Protaphorura* and synonymized several species. Confining ourselves to species of *Protaphorura* on the British list, Pomorski (1990) considered the following to be 'good' species (with synonyms in brackets). *Protaphorura pannonicus* Haybach 1960 (= *P. trinotatus*), *P. subuliginatus* (= *P. subarmatus*), *P. armatus* (= *P. nemoratus, P. humatus, P. tricampatus*), *P. campatus* (=*P. procampatus*), *P. meridiatus* (=*P. pulvinatus*), *P. aurantiacus* (= *P. uliginatus, P. latus, P. prolatus, P. sublatus*). In other words, of the original 16 species, only six remain!

However, the 'splitting' of *Protaphorura* has been supported to some extent by Salmon (1959), Fjellberg (1980) and Hale (1964, 1965, 1968) the latter of which reported that several of Gisin's 'species' breed true in laboratory culture. A word of warning is in order, however. It is of course possible for two varieties to be given separate species status because they breed true in culture, even though they are not reproductively isolated.

Lawrence (1979) has pointed out that if all possible combinations of the characters used by Gisin were to exist, there are over 100 000 possible 'species' of *Onychiurus* that could be recognized. In the face of evidence demonstrating the variability of the numbers and arrangement of pseudocelli within single populations, and even on different sides of the same individual (Bödvarsson, 1970; Lawrence, 1979; Pitkin, 1980), it seems questionable to give many of these forms more than varietal status. For example, Table 4.1 is derived from the most recent checklists available for Britain. Kloet and Hinks (1964) and its update (Gough, 1978) are certainly used if a quick answer is needed to the question, 'How many species of Collembola are there in Britain?' 'Officially' there are 334 but it is clear from the work of Pomorski (1990) and others that there may be far fewer biological species.

Such problems are as bad, if not worse, in other genera of Collembola (e.g. *Isotoma*). These differences of opinion among taxonomists make international comparisons of biodiversity extremely difficult. For example, Fjellberg (personal communication) in his most recent checklist of Norwegian Collembola (an update of his 1980 key) recognizes 18 species in *Protaphorura*, many of them described by Gisin. I am not suggesting that the morphotypes of *Protaphorura* recognized by Fjellberg do not exist, merely that there are differences of opinion on whether they should be given specific status.

Could these problems be resolved by molecular methods? Frati *et al.* (1989, 1992) have studied genetic distance between certain species of Collembola. However, they conclude that while biochemical genetics can provide good resolution of qualitative and quantitative relationships between different species or populations, it cannot completely replace traditional systematic criteria which are usually cheaper, faster and simpler. The level of genetic distance which constitutes separate species status is still open to interpretation, although in a study on *Orchesella*, there was a good correlation between species boundaries based on coloration and those calculated from allozyme differences (Frati *et al.*, 1994).

4.4 PARTHENOGENESIS

A widely used ecotoxicological test measures the effects of chemicals on reproduction, growth and survival in parthenogenetic races of the springtail *Folsomia candida* Willem 1902. Crommentuijn (1994) studied the effects of cadmium, chlorpyrifos and fentin on four such races of Collembola (two from the Netherlands and one each from England and France). Although there were differences between races in susceptibility to some of the chemicals, these were small and it was concluded that *Folsomia candida* was a legitimate species to use for ecotoxicological testing.

Parthenogenesis has been found in several species of Collembola. In *Isotoma notabilis* Schäffer 1896 and *Mesaphorura krausbaueri* (Börner, 1901), the phenomenon has been proved experimentally in the laboratory (Petersen, 1971). *Willemia anophthalma* (Börner, 1901), *Isotomiella minor* (Schäffer, 1896), *Megalothorax minimus* (Willem, 1900) exist as female-only populations in some sites (Petersen, 1978) which is strong evidence for parthenogenesis. Several of these and other species are facultatively parthenogenetic and are able to exist as female-only or sexual populations (e.g. *Mesaphorura italica* Rusek 1971; Mateos, 1992).

The attraction of using parthenogenetic races of *Folsomia candida* is their ease of culture, and the ability to breed a strain from one individual. Such a strain should have less individual variation in response to chemicals than sexually reproducing populations. However, the problem still remains of relating the results of laboratory experiments to field populations of the same and other species. As the field evidence of Hågvar (1994) has shown, there are species-specific differences in the sensitivities of Collembola to pollutants. However, there is still a lack of experimental evidence of these differences and this is clearly an area in need of further research.

4.5 CONCLUSIONS

Within the Collembola, simple species counts may be too simplistic for giving a single value of biodiversity because of the difficulty of defining the limits of species. The new philosophy of degree of 'independent evolutionary history' or 'taxonomic distinctiveness' is useful (May, 1994). Thus, instead of numbers of species we could use 'units of biodiversity' which might constitute a species group, race, genus or higher taxonomic category. Ecotoxicologists need to recognize that there may be no simple answer to the question 'how many species are there?' at a particular site. Taxonomists need to recognize that the description of new species based on very minor differences in morphology can have far-reaching implications for the assessment of biodiversity (Angermeier, 1994).

I have presented the definition of the species concept as a problem. So let me end on a positive note with the following quote from O'Hara (1994):

> In writing about species in the Origin, Darwin often refers to them as 'permanent varieties' (1859, p. 475). I suggest that this is a particularly insightful phrase. Because evolutionary history is something we are still in the midst of, it will not always be possible for us to determine which varieties – which distinctive populations in nature – are temporary and which are permanent, and so our counts of species across space and through time will always have some measure of ambiguity in them that we cannot escape. If there is any consolation in this, it must be that the very existence of this ambiguity – the very fact that some organisms in nature cannot easily be grouped into species – is itself, as Darwin recognized, one of the most important pieces of evidence for the historical process we call evolution.

4.6 SUMMARY

Biodiversity of ecosystems is usually measured by counting numbers of species. The HC_5 approach to risk assessment of chemicals involves calculating the 'hazardous concentration for 5% of the species'. However, such approaches to conservation and ecotoxicology assume the existence of an internationally accepted definition of what constitutes a species. At present, this does not exist for some groups of organisms. In this chapter, various ways of defining species are described, the problems of assessing biodiversity are outlined, and an example is given of the problems that arise in trying to assess biodiversity of Collembola in the absence of a rigorous definition of what constitutes a species in one genus of this common and widespread group of soil invertebrates.

REFERENCES

Angermeier, P.L. (1994) Does biodiversity include artificial diversity? *Conservation Biology*, **8**, 600–2.

Bagnall, R.S. (1935) Contributions towards a knowledge of the Scottish Onychiuridae (Collembola). Part 1. *Scottish Naturalist*, July–August 1935, 111–17.

Bardgett, R.D., Frankland, J.C. and Whittaker, J.B. (1993) The effects of agricultural management on the soil biota of some upland grasslands. *Agr. Ecosyst. Environ.*, **45**, 25–45.

Behan-Pelletier, V.M. (1993) Diversity of soil arthropods in Canada: systematic and ecological problems. *Mem. Entomol. Soc. Canada*, **165**, 11–50.

Bellinger, P.F. (1985) The identity of H.G. Scott's Collembola in the Academy of Natural Sciences, Philadelphia, PA. *Entomol. News*, **96**, 78–82.

Bödvarsson, H. (1970) Studies of *Onychiurus armatus* (Tullberg) and *Folsomia quadrioculata* (Tullberg) (Collembola). *Opuscula Entomologica* Suppl. 36, 1–182.

Cassagnau, P. and Deharveng, L. (1984) Collemboles des Philippines. I. Les lobelliens multicolores des montagnes de Luzon. *Travaux du Laboratoire d'Ecobiologie des Arthropodes Edaphiques, Toulouse*, **5**, 1–11.

Crommentuijn, T. (1994) *Sensitivity of Soil Arthropods to Toxicants*. PhD Thesis, Vrije Universiteit, Amsterdam.

Darwin, C. (1859) *On the Origin of Species*. John Murray, London.

Fjellberg, A. (1980) *Identification Keys to Norwegian Collembola*. Norwegian Entomological Society.

Ford, J. (1935) The animal population of a meadow near Oxford. *J. Anim. Ecol.*, **4**, 195–207.

Frampton, G.K., Langton, S.D., Greig-Smith, P.W. and Hardy, A.R. (1992) Changes in the soil fauna at Boxworth, in *Pesticides, Cereal Farming and the Environment* (eds P. Greig-Smith, G. Frampton and T. Hardy), HMSO, London, pp. 132–43.

Frati, F., Fanciulli, P.P. and Dallai, R. (1989) Biochemical approach to the systematics of Collembola, in *3rd International Seminar on Apterygota* (ed. R. Dallai), University of Sienna, pp. 145–55.

Frati, F., Fanciulli, P.P. and Dallai, R. (1992) Genetic diversity and taxonomy in soil-dwelling insects: the genus *Orchesella*. *J. Heredity*, **83**, 275–81.

Frati, F., Fanciulli, P.P. and Dallai, R. (1994) Further acquisitions on systematic relationships within the genus *Orchesella* (Collembola, Entomobryidae) using allozymes. *Acta Zool. Fenn.*, **195**, 35–43.

Gisin, H. (1960) *Collembolenfauna Europas*. Museum d'Histoire Naturelle, Geneva.

Gough, H.J. (1978) Collembola. Supplement to Kloet, G.S and Hincks, W.D. (1964) A Check List of British Insects (Second Edition), Part 1: Small Orders and Hemiptera. *Antenna*, **2**, 51.

Hågvar, S. (1994) Log-normal distribution of dominance as an indicator of stressed soil microarthropod communities. *Acta Zool. Fen.*, **195**, 71–80.

Hale, W.G. (1964) Experimental studies on the taxonomic status of some members of the *Onychiurus armatus* species group. *Rev. Écol Biol Sol*, **1**, 501–10.

Hale, W.G. (1965) Post-embryonic development in some species of Collembola. *Pedobiologia*, **5**, 228–43.

Hale, W.G. (1968) A quantitative study of the morphological structures used as taxonomic criteria in the *Onychiurus armatus* group (Collembola, Onychiuridae). *Rev. Écol Biol Sol*, **3**, 493–514.

Hopkin, S.P. (1997) *Biology of the Springtails (Insecta: Collembola)*. Oxford University Press, Oxford.

Kloet, G.S and Hincks, W.D. (1964) A Check List of British Insects (Second Edition), Part 1 : Small Orders and Hemiptera. *Handbooks for the Identification of British Insects*. Royal Entomological Society of London, vol. **11**(1), pp. 1–119.

Lawrence, P.N. (1979) Observations on the taxonomy and ecology of *Onychiurus armatus* (Collembola: Onychiuridae) and their wider implications in agriculture and evolution. *Rev. Écol Biol Sol*, **16**, 259–77.

Margulis, L. (1992) Biodiversity – molecular biological domains, symbiosis and kingdom origins. *Biosystems*, **27**, 39–52.

Mateos, E. (1992) Las poblaciones de *Mesaphorura italica* Rusek, 1971 en suelos de Encinar Mediterraneo. *Actos do V Congresso Ibérico de Entomologia, Supplemento 3, Boletin da Sociedade Portuguesa de Entomologia*, 299–310.

May, R.M. (1994) Conceptual aspects of the quantification of the extent of biological diversity. *Phil. Trans. Roy. Soc. London*, **345B**, 13–20.

Mayr, E. (1940) Speciation phenomena in birds. *Am. Nat.*, **74**, 249–78.

Minelli, A (1993). *Biological Systematics – the State of the Art*, Chapman & Hall, London.

Morris, H.M. (1927) The insect and other invertebrate fauna of arable land at Rothamsted, Part II. *Ann. Appl. Biol.*, **14**, 442–64.

O'Hara, R.J. (1994). Evolutionary history and the species problem. *Am. Zool.*, **34**, 12–22.

Petersen, H. (1971) Parthenogenesis in two common species of Collembola: *Tullbergia krausbaueri* (Börner) and *Isotoma notabilis* Schäffer. *Rev. Écol Biol Sol*, **8**, 133–8.

Petersen, H. (1978) Sex ratios and the extent of parthenogenetic reproduction in some collembolan populations, in *First International Seminary on Apterygota*, Siena (ed. R. Dallai), Accademia delle Scienze di Siena detta de Fisiocritici, Siena, Italy, pp. 19–35.

Pitkin, B.R. (1980) Variation in some British material of the *Onychiurus armatus* group (Collembola). *Syst. Entomol.*, **5**, 405–26.

Platnick, N.I. (1991) Patterns of biodiversity: tropical vs. temperate. *J. Nat. History*, **25**, 1083–8.

Pomorski, R.J. (1986) Morphological–systematic studies on the variability of pseudocellae and some morphological characters in *Onychiurus* of the '*armatus*-group' (Collembola, Onychiuridae). Part I. *Onychiurus (Protaphorura) fimatus* GISIN 1952. *Polskie Pismo Entomologiczne*, **56**, 531–56.

Pomorski, R.J. (1990) Morphological–systematic studies on the variability of pseudo-celli and some morphological characters in *Onychiurus* of the '*armatus*-group' (Collembola, Onychiuridae). Part II. On synonyms within the '*armatus*-group' with special reference to diagnostic characters. *Annales Zoologici (Warszawa)*, **43**, 535–76.

Ponge, J.F. (1993) Biocenoses of Collembola in Atlantic temperate grass–woodland ecosystems. *Pedobiologia*, **37**, 223–44.

Prendergast, J.R., Quinn, R.M., Lawton, J.H., Eversham, B.C. and Gibbons, D.W. (1993) Rare species, the coincidence of diversity hotspots and conservation strategies. *Nature*, **365**, 335–7.

Rand, D.A. and Wilson, H.B. (1993) Evolutionary catastrophes, punctuated equilibria and gradualism in ecosystem evolution. *Proc. Roy. Soc. London*, **253B**, 137–41.

Rusek, J. and Weyda, F. (1981) Morphology, ultrastructure and function of pseudo-celli in *Onychiurus armatus* (Collembola, Onychiuridae). *Rev. Écol Biol Sol*, **18**, 127–33.

Salmon, J.T. (1959) Concerning the Collembola Onychiuridae. *Trans. Roy. Entomol. Soc. London*, **111**, 119–56.

Salmon, J.T. (1964) An index to the Collembola. *Bull. Roy. Soc. New Zealand*, **7**, 1–651.

Stork, N.E. and Blackburn, T.M. (1993) Abundance, body size and biomass of arthropods in tropical forest. *Oikos*, **67**, 483–9.

Stork, N.E. and Eggleton, P. (1992) Invertebrates as determinants and indicators of soil quality. *Am. J. Alternative Agric.*, **7**, 23–32.

Thompson, M. (1924) The soil population. An investigation of the biology of soil in certain districts of Aberystwyth. *Ann. Appl. Biol.*, **11**, 349–94.

Usher, M.B. and Balogun, R.A. (1966) A defence mechanism in *Onychiurus* (Collembola, Onychiuridae). *Entomol. Monthly Magazine*, **102**, 237–8.

Usher, M.B., Field, J.P. and Bedford, S.E. (1993) Biogeography and diversity of ground-dwelling arthropods in farm woodlands. *Biodivers. Lett.*, **1**, 54–62.

Van Straalen, N.M. (1993a) Soil and sediment quality criteria derived from invertebrate toxicity data, in *Ecotoxicology of Metals in Invertebrates* (eds R. Dallinger and P.S. Rainbow), Lewis Publishers, Chelsea, USA, pp. 427–41.

Van Straalen, N.M. (1993b) An ecotoxicologist in politics. *Oikos*, **66**, 142–3.

Vane-Wright, R.I., Humphries, C.J. and Williams, P.H. (1991) What to protect? – Systematics and the agony of choice. *Biol. Conserv.*, **55**, 235–54.

Whittemore, A.T. (1993) Species concepts: a reply to Ernst Mayr. *Taxon*, **42**, 573–83.

World Conservation Monitoring Centre (1992) *Global Biodiversity: Status of the Earth's living resources*, Chapman & Hall, London.

5 Effects of toxicants on population and community parameters in field conditions, and their potential use in the validation of risk assessment methods

LEO POSTHUMA

5.1 VALIDATION OF RISK ASSESSMENT

5.1.1 RISK ASSESSMENT AND TOXICANT EFFECTS IN THE FIELD

Ecotoxicological risk assessment methods need to be validated due to the high economic consequences of too strict and the high ecological consequences of too weak risk management (Slooff *et al.*, 1986; Hopkin, 1993; Van Straalen, 1993a,b; Van Straalen *et al.*, 1994). In this chapter the term 'validation' is used in the sense of evaluation of differences between laboratory-based predictions and field effects. A validated method in this sense predicts field effects within specified adequacy limits. Until now, the results of single-species (SS) laboratory tests have only been compared with toxic effects in enclosures or in multi-species (MS) experiments (Heimbach, 1992; Okkerman *et al.*, 1993). Comparisons of predicted effects with effects in natural systems have, however, not yet been made.

This chapter focuses on the effects of toxicants in natural systems. A scheme is proposed for the validation of risk assessment methods. The scheme consists of an approach in which comparisons are made between effects of toxicants in laboratory and field conditions. Each polluted site may be considered a large-scale field experiment, in which toxic effects can be studied. Differences between toxicant effects in laboratory and field conditions may be caused by the fact that only the field effects show an integrated

Ecological Risk Assessment of Contaminants in Soil. Edited by Nico M. van Straalen and Hans Løkke. Published in 1997 by Chapman & Hall, London. ISBN 0 412 75900 4

response of biota to the direct and indirect effects of toxicants, and the effects of other stress factors and species interactions.

Effects of toxicants in the field may be expressed with a variety of parameters, ranging from physiological biomarkers to community effects. In the following, toxicant effects in the field are quantified by population and community parameters. Population-level parameters are important, since they can be compared with toxicity parameters (such as NOEC, EC_{50}) calculated from single-species toxicity experiments. The latter data are often used as the 'input' data for risk assessment methods. The community parameters are important since risk assessment methods usually aim to protect communities against effects of toxicants such as loss of species diversity. Case studies on community-level effects of toxicants in the field may help to identify the ecological meaning of 'safe concentrations' being exceeded.

5.1.2 A PROPOSAL FOR A VALIDATION SCHEME

The characteristics of contemporary risk assessment methods are summarized in Table 5.1. These risk assessment methods allow for the calculation of concentrations that are (implicitly) regarded as 'safe' for the maintenance of community structure and function. One of them is the method proposed by Van Straalen and Denneman (1989), with its derivatives (Wagner and Løkke, 1991; Aldenberg and Slob, 1993). In this method, laboratory toxicity data (NOECs) are used as 'input' data for the calculation of a continuous 'concentration–risk' relationship. The calculation is based on the assumption that sensitivities among species follow a certain mathematical distribution (lognormal or log-logistic); so far there is no evidence to reject this assumption (Kooijman, 1987; Kammenga *et al.*, 1994). The continuous relationship

Table 5.1 Common features of contemporary risk assessment methods. (For an overview of the methods and choices, see Van Straalen *et al.*, 1994)

- Risks are calculated based on toxicity parameters established in single-species–single-toxicant laboratory toxicity tests for various species.
- The preferred toxicity parameter is the NOEC for an ecologically relevant characteristic, in particular reproduction.
- Correction factors for environmental parameters, such as '% soil organic matter', are used when necessary.
- Either a safety factor or a frequency distribution of sensitivities* is applied, and this depends on the amount of toxicity data.
- The methods do not suggest what will happen in an ecosystem if risk levels are exceeded

---

*The continuous frequency distribution describes the distribution of toxicant sensitivities of species in a community. This distribution can be used to calculate a 'concentration–risk' relationship, in which the risk at a certain environmental concentration is expressed as the proportion of species exposed beyond its NOEC for reproduction (Figure 5.1(b), left). The starting point for this chapter is that further details of risk assessment methods based on differences in species sensitivities are known by the reader (see Van Straalen and Denneman, 1989; Wagner and Løkke, 1991; Aldenberg and Slob, 1993).

between concentration (X) and predicted risk (HC_X, for hazardous concentration) was used to develop the validation scheme proposed here. To validate this 'concentration–risk' relationship, the most profitable comparisons with field observations can be made when toxic effects on community parameters are quantified along a pollution gradient in the field.

The proposed validation scheme consists of two steps, namely:

1. At the level of species, comparisons of toxic effects in laboratory and field conditions should be made to improve adequacy of input data (Figure 5.1(a)). This level of comparison has also been addressed by C.A.M. van Gestel (Chapter 2, this volume), who treated bioavailability differences between laboratory and field as a cause of difference; here, emphasis is placed on parameters that express toxic effects after multi-generation exposure.
2. At the level of communities, comparison of 'safe' concentrations with toxic effects in communities should be made to evaluate the adequacy of risk assessment methods (Figure 5.1(b)).

The steps can be used iteratively. Future achievements can be taken into account both for the accuracy of input data and for future improvements of the risk extrapolation methods.

The scheme is introduced with suggestions for appropriate effect parameters and with some numerical examples derived from case studies. In the examples, no attempt is made to explain the causes of (dis)similarity between laboratory and field effects. Due to the inventory character of the chapter, the list of parameters considered should not be interpreted as complete; various other possibilities may be considered in the future.

A refinement of the scheme can be made for philosophical reasons. Community-level comparisons should result in calibration of the method, to correct for consistent under- or overestimation of the risk-level in comparison with unacceptable effects in the field. Eventually, with another series of comparisons, the calibrated method should be validated. This refinement is not applied here due to the inventory character of the chapter.

5.1.3 AIMS AND OUTLINE OF THE CHAPTER

The aim of this chapter is to explore observations on effects of toxicants in field conditions to use them for the validation of risk assessment methods. The chapter aims to present a vision on how the validation problem can be tackled, not on a conclusion on the validity of the method itself. In the exploration, three subjects are introduced.

1. The design of field studies is considered, and some conclusions on the design of field studies are formulated (section 5.2).
2. An inventory is made on the choice of parameters to study toxic effects in the field, both at the level of the population (section 5.3) and at the level of communities (section 5.4).

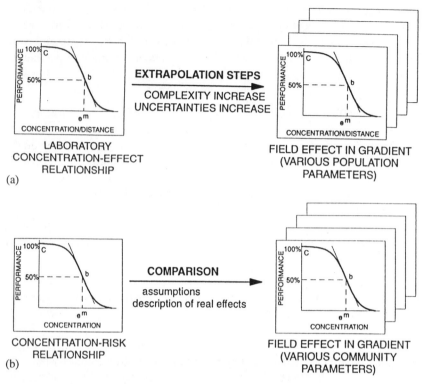

Figure 5.1 (a) The laboratory concentration–effect relationship for fitness, summarized through its NOEC for reproduction (left), should adequately predict fitness effects of toxicant exposure in the field (right). A gradient study on various characteristics associated to fitness in the field may result in a description of fitness effects on a continuous scale. The field response incorporates all uncertainty factors mentioned earlier (C.A.M. van Gestel, Chapter 2, this volume). Section 5.3 provides some results on population studies in the field. (b) Risk assessment methods derived from Van Straalen and Denneman (1989) result in the calculation of a sigmoid 'concentration–risk' relationship between ambient concentration and predicted adverse community effects (left). A gradient study in the field on various community characteristics associated to community stability and functioning can result in description of toxic effects on a continuous scale (right). Eventually, predicted effects (for example at the HC_5 level) can be compared to effect–concentrations in the field (EC_5). Section 5.4 provides some results on community studies in the field, and section 5.5 gives a numerical example for a lead-polluted site.

3. Benefits and pitfalls of the proposed method are explored in a case study on community effects in a gradient, where the comparison is made on the EC_5-level (section 5.5).

The reasoning is illustrated with examples that mostly refer to case studies on the field effects of metals. Metals can act as persistent chemicals, which implies that uncertainties that occur with non-persistent toxicants do not occur (see Van Straalen *et al.*, 1992). For easily degradable toxicants both the laboratory toxicity test results (and by that the risk assessment results) and the field effects partly depend on the half-life of the chemical. This makes the timing of observation a crucial, complicating factor, which is often insufficiently documented in practical studies (Hamers *et al.*, 1996).

5.2 APPROPRIATENESS OF STUDIES ON TOXICANT EFFECTS IN FIELD CONDITIONS

The appropriateness of case studies for the comparison of laboratory toxicity parameters with effects in field conditions is determined by: (i) the possibility to link field effects on population and community parameters to toxicants as causal agents; and (ii) the possibility to quantify effects at the desired concentration level (for example: EC_5 to HC_5). In this section, some main quality criteria are formulated to judge the quality of case studies. In addition to this, the number of field sites studied is important. With sufficient sites, mathematical models can be used to formulate a field concentration–effect curve to be compared with its laboratory equivalent. Two case studies are presented to illustrate an appropriate design of field studies.

5.2.1 MAIN QUALITY CRITERIA OF FIELD STUDIES

The most important criteria that determine the appropriateness of field studies are: (i) the effect must be attributable to the concentration of the toxicant; and (ii) the effect must be ecologically relevant. These criteria are treated here in general terms. The relevance of some possible parameters is discussed in sections 5.3 and 5.4.

To fulfil the first criterion, a causality analysis must be carried out. There should be: (i) a proper time-sequence of cause and effect; (ii) strength of association; (iii) specificity of association; (iv) consistency of replication; and (v) biological plausibility. These criteria are similar to Koch's classical postulates on the epidemiology of disease; the difference is that in ecotoxicological studies the epidemiology of toxic effects is studied. Suter (1992) and Posthuma and Van Straalen (1993) elaborated these criteria for field studies involving toxic effects at the population and ecosystem level that occur after long-term exposure. Most case studies do not mention explicitly the fulfilment of the criteria and causality is therefore often difficult to reconstruct. Many case studies neglect the possible influence of confounding factors that have attributed also to the observed response. A causal analysis would require that evidence be collected to prove that such factors can be neglected, for example by showing the absence of covariation between pollutant level and the additional factor.

The 'ecological relevance' of a parameter is difficult to establish, but some parameters are often considered to be associated to the persistence of populations or communities. Such parameters are to be chosen (Calow, 1994), and some have been worked out for population- and community-level analyses, in sections 5.3 and 5.4 respectively. Some relevant parameters cannot be used due to the strong influence of other factors, such as season. The parameter 'population size', for example, may not be useful, since it may vary strongly due to season and competitors.

Most of the useful population and community parameters require complex studies. Biomarkers expressing effects of toxicants at the physiological level could be a faster and less expensive alternative, but for many biomarkers the relationship with population and community persistence is still unclear (Huggett *et al.*, 1992; Peakall, 1992).

5.2.2 A MATHEMATICAL MODEL

The comparison of laboratory toxicity data or risk levels with field data is preferably made at certain specified concentration levels. At the species level, for example, toxicity parameters such as NOEC and EC_{50} are often used in laboratory studies, and one may wish to study field effects at these specific concentrations. At the community level, similarly, comparison of field effects with a risk level such as the HC_5 may be of specific interest. Field studies have, however, never been designed to describe the effects at sites with those specific concentrations, and thus inter- or extrapolation is always needed.

Inter- and extrapolation are possible if data for a series of locations, with different concentrations and effects, are given. The most profitable comparisons can be made when the field effects vary between 'control' performance and strong effects, similar to that found commonly in laboratory studies. Less profitable comparisons can be made if only few sites have been studied, since then the effects at the concentration of interest may be lacking.

For data in a range, sampled on a gradient of pollution, mathematical models can be applied for interpolation. A non-mechanistic deterministic model is sufficient for the present purpose; stochasticity could be incorporated in the future. The logistic model results in an empirical description of effects along a gradient. It does not allow, however, for an explanation of the effects.

(a) Population level

Concentration–response curves in laboratory conditions can be described by a generalized logistic model (see for example Haanstra *et al.*, 1985; Van Ewijk and Hoekstra, 1993). The logistic response curve is given by the formula

$$y = \frac{C}{1 + e^{-b(x-m)}}$$

where y = performance at toxicant concentration x (logarithmically transformed); C = value for y in control or reference condition; b = slope of the concentration–response-curve on a logarithmic scale; and $m = \ln(\text{EC}_{50})$, the concentration causing 50% reduction in y.

The performance of population parameters in the field may mimic the laboratory data, since a sigmoid response may also be expected: a normal field variability of performance at sites distantly in the gradient, rapidly increasing effects with decreasing distance to the pollution source, and some form of asymptote at the most polluted sites. The response pattern in field conditions may be influenced by other factors. These may cause random inter-site variability, or they may systematically co-vary with the pollutant concentration. To avoid these problems, effect parameters should be insensitive to local and temporal causes of variation, and to other factors than the pollutant. The logistic model has been applied in some examples.

(b) Community level

The above logistic model can also be applied for toxicant-induced changes of community parameters. It is reasonable to expect that community parameters change from natural variability of species diversity (or other community characteristics) at distant sites, via a gradual change towards an asymptote of 'low' performance, given homogeneity of other features along the gradient. Again, this will only hold for parameters that are largely unaffected by other factors than the toxicant. The response at each concentration can be compared with the predicted effects, for example at the level of the EC_5 versus HC_5.

Retrospective evaluation of existing gradient data may require some data handling before fitting the model to the data, due to large inter-site variability. Data handling can consist of fitting the model to concentration classes versus mean performance in the class, which is a procedure similar to the production of a stacked frequency distribution. This method was applied to some case studies below (e.g. the numerical example in section 5.5). Alternative mathematical methods to handle variability, such as the 'moving average method', or other smoothing techniques, may be applied. In the case study on the evolution of tolerance in microbial communities an elegant technique for the handling of inter-site variation not related to the pollutant concentration is incorporated in the procedure, the so-called PICT analysis (see section 5.4).

Case studies on an isopod, exposed in litter, and a worm, exposed in mineral soil, illustrate that a logistic model may be applied to obtain a continuous description of field responses along a gradient. In the examples, animals were exposed in laboratory conditions to field-collected substrate, so that differences in response can be attributed to differences in substrate quality and metal availability.

Case study 1: isopods in a heavy-metal gradient
Beyer *et al.* (1984) studied the mortality of woodlice (*Porcellio scaber*) in litter sampled at different distances from a zinc smelter (Figure 5.2). The curve

fitted to the observations showed that 10% mortality would occur at litter concentrations of Zn and Cd of c. 300 and 3 mg, respectively. The laboratory NOEC reproduction for zinc is c. 400 mg kg^{-1} (Van Capelleveen, 1987), closely similar to Beyer's observations. This similarity probably suggests that the availability of metals in artificially polluted substrate and in the field is similar. The toxicity data can also be compared with field effects expressed by the level of population tolerance. With a level of site pollution that causes 10% mortality within 8 weeks, with death occurring before maturity is reached (at c. 8 months; Donker *et al.*, 1993), one may expect selection for increased tolerance. Population tolerance was indeed observed in field populations living in such conditions (Donker and Bogert, 1991; see section 5.3.2 and Figure 5.7(b)).

Case study 2: worms in a heavy-metal gradient
Posthuma *et al.* (1994) studied reproduction effects in the worm *Eisenia andrei* in soil sampled in a gradient around a zinc smelter. Natural variability in metal and soil characteristics along the gradient prohibited a successful fit on the original data (no. of sites = 11). The logistic model was therefore fitted to class means for X and Y (no. of classes = 6, Figure 5.3). The cocoon production was reduced by 50% at an estimated distance from the emission

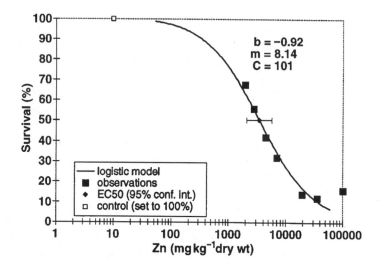

Figure 5.2 Concentration–effect relationship for survival of the isopod *Porcellio scaber* on litter from a zinc smelter gradient. The parameters of the sigmoid concentration–effect model are given in the figure (see section 5.2.2), the observed control survival was 70%, but for the fit it was assumed to be 100%. The survival data were read from a figure, the concentration data were recalculated from formula fitted to observations on distances and concentrations.

source of c. 1 km. At that site, the metal concentration for Zn alone was c. four-fold higher than the laboratory EC_{50}. This shows that toxic effects in this field soil are lower than in the artificially polluted soil at the same total Zn concentration. Spurgeon (see Chapter 12) showed that earthworm populations may indeed occur at field sites that should be uninhabitable by worms when judged by laboratory toxic effects. The difference between laboratory and field effects may probably be attributed to differences in metal availability between substrates. Additional observations for *E. andrei* indeed showed that availability, and possibly antagonism, among metals (Posthuma *et al.*, 1995), could be identified as the possible causes of the observed differences. It is also possible that earthworms at Spurgeon's study site have adapted to metal pollution, although there is no evidence for this.

Both case studies prove that the results of bioassays with gradient substrate, in comparison with the laboratory tests on the same fitness characteristic, may reveal the critical differences between laboratory and field effects.

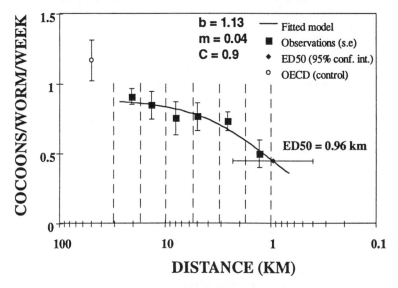

Figure 5.3 Toxic effects of metals in a semi-field bioassay with the worm *Eisenia andrei*, in soil sampled at 11 sites in a zinc smelter gradient. The studied parameter was cocoon production. Due to variability of soil quality, it was necessary to pool data within classes (indicated with broken lines) for successful estimation of the parameters of the sigmoid model. Model parameters are shown in the figure. Since there are various metals present, the model was fitted to the data as a distance–effect relationship, from which the ED_{50} (effect distance) was calculated as the distance at which, in a homogeneous soil type, the metals would have caused 50% reduction of cocoon production. The ED_{50} was after that recalculated to the mixture of metals present at this site, from distance–concentration relationships. The cocoon production that was simultaneously determined in a standard laboratory test in OECD artificial soil is also shown (control).

Knowledge of such factors, such as differences for the availability of the toxicant, may help to design toxicity tests that accurately predict effects for the same fitness characteristic in the field.

Theoretical and practical considerations show that the characteristics of an ideally complete gradient study are as follows.

Concerning the sampling strategy

- Samples should be taken along the whole gradient, at least with a number of samples that exceed the number usually applied in laboratory studies, since substrate variability is an added factor.
- The control situation should be well defined, considering natural variation through replicated sampling.
- Sample density should be high at sites where the change of toxic effects is highest, if necessary, preliminary information should be obtained from 'range-finding' sampling; and
- For all sampling sites concentration data for the toxicant and biotic responses should be measured, and measurements should be made on other factors that may vary along the gradient.

Concerning the choice of species under investigation

- Only species that disperse on a scale (much) smaller than the scale of the gradient should be incorporated (see for example P.C. Jepson, Chapter 8, this volume, and G. Bengtsson, Chapter 9, this volume).

Concerning the population or community parameters under investigation

- The parameters should be insensitive to environmental variables other than the pollutant.

5.3 MEASUREMENT OF TOXICANT EFFECTS IN POPULATIONS IN FIELD CONDITIONS

Improvement of the validity of risk assessment methods needs to start with improving the accuracy of the input data, i.e. the results of single species toxicity experiments should be relevant for the field. Relevance for the field is eventually determined by the net effect on fitness, since toxicants operate on fitness through various of its components simultaneously (e.g. J.E. Kammenga, Chapter 14, this volume; see also Van Straalen *et al.*, 1989; Crommentuijn *et al.*, 1993; Posthuma *et al.*, 1993b). Although comparisons can be made for separate components of fitness (see examples in section 5.2), an alternative type of parameter is needed to express the net response in the field. Population tolerance should be considered as an appropriate parameter to measure net fitness effects in the field. Its measurement is elaborated in section 5.3.1, and this is followed by some practical comparisons of laboratory toxicity data and the occurrence of population tolerance in the field (section 5.3.2).

5.3.1 FIELD EFFECTS EXPRESSED AS POPULATION TOLERANCE

Luoma (1977) proposed using pollution-induced population tolerance as an indicator of toxicant effects at the population level. Population tolerance implies that a population has been exposed to such a degree that natural selection has occurred or is still occurring. The use of population tolerance for the validation of laboratory toxicity data is, however, not so easy as Luoma suggested, since it involves many (possibly associated) population characteristics.

(a) Requirements for field studies

Application of the population tolerance approach in a gradient study is grossly similar to the community-tolerance approach proposed by Blanck *et al.* (1988) (see section 5.4, Figure 5.8). Specimens are collected at various sites with different toxicant concentrations, including a reference site. The concentration of the toxicant, preferably its available concentration, is measured. In addition, (possibly) confounding factors are also measured, since local site characteristics not associated to the toxicant concentration may affect population tolerance. Tolerance is quantified in an exposure experiment with the suspected toxicant as treatment, and observations are made on toxic effects on fitness characteristics, such as reproduction. Correlation between the toxicant concentration and tolerance shows that the differentiation can be attributed to the toxicant (Endler, 1986), in particular when correlations with the confounding factors are absent. The application of a correlation approach requires that at least three populations should be compared (see also Eberhardt and Thomas, 1991).

Population tolerance can be established with field-captured specimens, or with laboratory-reared offspring. The latter are preferred, however, since the observed pattern can most probably be attributed to genetic adaptation (Posthuma and Van Straalen, 1993), which is less sensitive to random effects of field variability than physiological acclimation. The evidence collected in case studies, however, suggests that population tolerance is in practice most often attributed to a combination of acclimation and adaptation (Klerks and Weis, 1987; Posthuma and Van Straalen, 1993).

(b) Complex adaptation

This is illustrated by a case study in springtails. Many population characteristics may be simultaneously changed in tolerant populations, ranging from allozyme frequencies to life-history characteristics. The simultaneous response of various characteristics can be inferred from life-history theory (Michod, 1979; Charlesworth, 1980), field experiments (Reznick *et al.*, 1990), and the observation that life-history components have different relationships to fitness in different species (J.E. Kammenga, Chapter 14, this volume). Evolutionary effects of toxicants on various characteristics will induce 'complex adaptation' (Ernst, 1983; Posthuma *et al.*, 1993b). A conceptual summary of complex adaptation was constructed from a case study on the springtail *Orchesella cincta* (Figure 5.4). Populations from metal-polluted sites have, in comparison with reference populations, a high metal excretion efficiency (Posthuma *et al.*, 1992; Figure 5.5), a high growth rate, a short juvenile period, a similar weight at first reproduction, and a high mortality under clean conditions (Posthuma, 1990; Posthuma *et al.*, 1993a,b). These changes were observed in laboratory-reared offspring and were attributed to genetic adaptation rather than to acclimation. Genetic adaptation depends on the initial presence of genetic variation. In a reference population, genetic

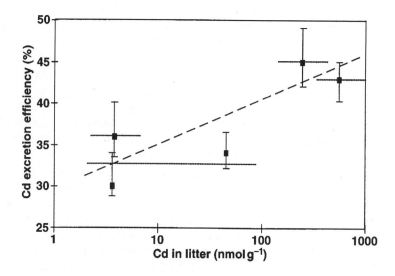

Figure 5.4 Population differentiation for cadmium excretion efficiency in *Orchesella cincta*. Site pollution is given on the x-axis, through the cadmium concentrations of the local substrate. From left to right, the populations originated from two reference sites, a zinc smelter site (c. 100 years of emissions), an abandoned lead mine site, and a lead smelter site (c. 150 years emission, and possible influences of surface ore spots). Horizontal lines join the populations with a similar metal excretion efficiency ($P > 0.05$); populations not joined show population differentiation related to site pollution ($P < 0.05$), as judged from one-way ANOVA followed by a multiple-comparisons test. Other site characteristics also influence population differences, as expressed by the difference between the two left-most reference populations. (Data from Posthuma *et al.*, 1992.)

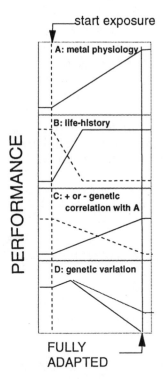

Figure 5.5 Conceptual and hypothetical relationships between population characteristics that may show a simultaneous response to selection for tolerance. x-axis: time (generations), y-axis: population performance measures. The figure is a summary of theoretical considerations, and literature data. For further explanation, see text.

variation was shown for various tolerance and life-history characteristics (Posthuma *et al.*, 1993a; Posthuma and Janssen, 1995). However, the presence of genetic variation for ecologically relevant characteristics may not be common for all species (Falconer, 1981), and many species may show genostasis due to a lack of appropriate genetic variation (Bradshaw, 1991).

The conceptual summary (Figure 5.5) shows the possible associations between the characteristics. Selection by toxicants may induce responses of the population mean (Y) over various generations (time, X) for various characteristics simultaneously (A–E). Selection may operate on metal regulation mechanisms that are directly involved in tolerance (graph A). Selection may also directly operate on fitness characteristics (graph B), since exceeding the NOEC for such a characteristic implies selection favouring a life-history that is optimal in the polluted conditions. In *O. cincta* both an increase (for example, growth rate) and a decrease of population mean (for example, juvenile period) was observed (Posthuma *et al.*, 1993b). The difference in response rate between graphs A and B says that each characteristic may show its own response rate (rate A < rate B), which depends both on the intensity of selection and on the amount of genetic variation.

Selection may also operate on a characteristic in an indirect way through genetic correlations (graphs A and C). Evolutionary change of such characteristics ceases when the selection on the characteristic directly selected for (A) ceases. The rate of change is determined by the strength of the genetic correlation and the presence of selection for the characteristic directly selected. The broken line shows a reduction of a fitness characteristic caused by a negative genetic correlation with tolerance, and this effect is often called a **cost of tolerance** (Wilson, 1988, Posthuma and Van Straalen, 1993). In *O. cincta*, a cost of tolerance may consist of an increased mortality rate of tolerant specimens in clean conditions. Since these populations are also characterized by a high body growth rate, this may be a consequence of increased ageing of fast-growing animals (Ernsting *et al.*, 1993). An alternative explanation may be that the tolerant animals suffer from a lack of Zn or other micronutrients in clean conditions, so that early death may occur due to micronutrient shortages (Postma *et al.*, 1995).

Genetic variation may change for all characteristics affected by the selection forces, directly or indirectly. In the initial stages of exposure, an increase of additive genetic variation can occur due to the expression and frequency increase of 'new' genes (Holloway *et al.*, 1990a,b). In later stages a loss of variation may occur, either due to selection alone, or due to the loss of variability common in populations after a bottleneck (Maruyama and Fuerst, 1985). In *O. cincta*, no genetic variation for excretion efficiency was found in a tolerant population, whereas it was present in a reference population (Posthuma *et al.*, 1993a).

(c) Expression of population tolerance in other species

Posthuma and Van Straalen (1993) made an inventory of the occurrence of adaptation in terrestrial invertebrate species. (Only a gross summary on the ecologically relevant responses is given here; for detailed information see the original article.) Population tolerance could be attributed to genetic adaptation in only a few species, but an association between the exposure concentration at the capture sites and tolerance was often found. This was judged by all the characteristics mentioned in the conceptual summary, often with only one or two of the possible characteristics in a case study. Most of the studied characteristics showed an improved functioning of the population in exposed conditions (response type: Figure 5.4, graphs A and B), although some examples also suggested costs of tolerance (Figure 5.4, graph C). Donker *et al.* (1996) showed an increased tendency for high mortality in metal-tolerant populations of the isopod *Porcellio scaber*. In summary, both profitable and adverse ecologically relevant effects of multi-generation exposure were found in field populations, for both tolerance and life-history characteristics. The necessary investigations are, however, complex, and may only be feasible for a small proportion of the species.

Responses of less ecological relevance are often easier to detect, which is important if one wants to compare laboratory and field effects on a regular basis. Genetic studies, which often do not require laboratory exposure experiments, have been used as alternative possibility to detect population responses in the field. Three techniques have been applied, viz. allozyme studies, mitochondrial DNA (mtDNA) analyses and heritability studies. Attempts to identify general effects of exposure on genetic variation have not been successful (Verkleij *et al.*, 1985; Frati *et al.*, 1992; Tranvik *et al.*, 1994). In these studies the allele frequencies at loci for common soluble enzymes were studied. Apparently, such loci are not affected by exposure directly or indirectly. Attempts focusing on single loci have been more successful. Lower (1975), for example, demonstrated changed allozyme frequencies at particular loci for the fruitfly in a gradient around a lead smelter. In the springtail *O. cincta*, tolerance was correlated with the change of allele frequencies at the glutamate–oxaloacetate transaminase (*Got*) locus (Frati *et al.*, 1992). In the congeneric species *O. bifasciata*, however, Tranvik *et al.* (1994) found no correlations of allelic variants with site pollution in a metal gradient. H. Siepel (personal communication) studied the allelic variation of the highly variable esterase loci in the parthenogenetic mite *Tectocepheus velatus*, and showed a loss of variability, and thus (probably) selection promoting certain clones. The allozymes that are apparently correlated to tolerance may be physically associated to tolerance genes on the genome, but they may also have a role in tolerance. In that case, they may serve as a biomarker of tolerance that can be used in comparisons with laboratory toxicity data, but the latter is difficult to establish (Koehn, 1978). Only in aquatic species have some indications on the role of alleles in tolerance been obtained, either through correlation analyses between allele frequencies and tolerance, or by tolerance studies with the alleles *in vitro* (Lavie and Nevo, 1982, 1986; Chagnon and Guttman, 1989; Gillespie and Guttman, 1989; Hawkins *et al.*, 1989; Benton and Guttman, 1990).

Recently, an alternative parameter for genetic effects was suggested. Murdoch and Hebert (1994) used mtDNA to detect changes of genetic variation in various pollutant-stressed populations of aquatic animals. Low genetic variation in mtDNA was found in populations from polluted sites compared with nearby unpolluted sites (Figure 5.6). The observed pattern may have been caused by selection operating on tolerance, or by population bottlenecks that affected only the exposed populations.

Finally, studies on the genetic variation of continuously variable characteristics, such as fitness characteristics, have suggested that selectively important genetic variation may be exhausted due to long-term selection in the field. Apart from the results with *O. cincta*, Shaw (1988) showed a loss of genetic variation for 10 out of 14 fitness characteristics in the moss *Funaria*

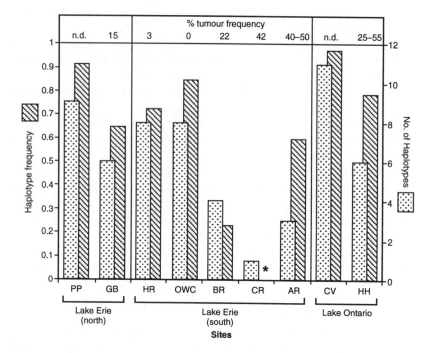

Figure 5.6 The loss of genetic variation in mitochondrial DNA (mtDNA) caused by a combination of a historical bottleneck and by selection, probably both induced by pollution. Genetic variability was measured in a fish species, the brown bullhead, inhabiting various sites in the Canadian Great Lakes, near industrial areas and in reference areas. The species under investigation has a low mobility, causing gene-flow and recovery to be low. Asterisks indicate significant loss of genetic variation compared with the nearby reference site ($P < 0.05$). Two measures of genetic variation are shown (haplotype frequency and number of haplotypes). The reference sites are: PP, HR and OWC, and CV; these sites have low tumour frequencies (or not determined); the polluted sites have high tumour frequencies.

hygrometrica. The techniques required to detect such effects, however, limit the use of heritability as an indicator of field effects.

In summary, population tolerance can be detected with an array of techniques, which may be complicated for ecologically relevant characteristics, or easier for characteristics of which the ecological relevance is still unclear. Despite this, laboratory toxicity data should be compared with field effects such as population tolerance, since there is a need to know the net effect of toxicants on fitness. The net effect is the only relevant criterion determining population persistence. For technical reasons, obviously the comparisons can now only be made for a restricted set of appropriate species.

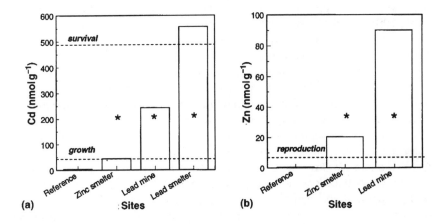

Figure 5.7 Site characteristics (expressed through local concentrations of a single metal only) and the occurrence of population tolerance (indicated with asterisks) exhibited as genetic adaptation, in relation to laboratory NOECs for various fitness characteristics, for (a) the springtail *Orchesella cincta* and (b) the isopod *Porcellio scaber*. Adaptation indicates the presence of (past) effects of the toxicant on fitness. NOEC data were obtained from (a) Van Straalen *et al.* (1989) and (b) Van Capelleveen (1987). Adaptation and site pollution data were obtained from (a) Posthuma (1990) and (b) Donker and Bogert (1991).

5.3.2 LABORATORY–FIELD COMPARISONS CONCERNING POPULATION TOLERANCE

The 'ideally complete' gradient approach to the study of toxicant adaptation has not been applied in any known population differentiation study of soil organisms (Posthuma and Van Straalen 1993). The comparison of laboratory toxicity parameters with the occurrence of population tolerance can be made for only few species, and on a site-by-site basis (Figure 5.7). For the isopod *Porcellio scaber* exceeding the NOEC$_{reproduction}$ of zinc apparently resulted in mortality (Figure 5.2) and genetic adaptation (section 5.2.2). For the springtail *O. cincta*, genetically adapted populations were found at sites where local soil concentrations exceeded the NOEC$_{body\ growth}$ for cadmium. The NOEC$_{survival}$ was exceeded only at a lead smelter site.

In both examples, laboratory–field comparisons are made based on total metal concentration in the substrate, without correction for the availability of the toxicant in the different situations (see C.A.M. van Gestel, Chapter 2, this volume) or for the effects of other metals or stress factors that are present at the field sites. Metals other than Zn or Cd, respectively, may have contributed to the occurrence of the adaptation response, since field sites are often polluted by mixtures of metals. Neither the separate contribution of each metal, nor the effect of mutual interferences, can be established since almost nothing is known on joint toxicity of toxicants in the soil compartment (Posthuma *et al.*, 1995). Nonetheless, the findings suggest that population tolerance may evolve at ambient concentrations close to, or exceeding the NOEC of a single toxicant for a fitness characteristic in laboratory conditions.

It is tempting to conclude that exceeding the NOEC for a fitness characteristic – and this need not be reproduction alone – may lead to population tolerance through genetic adaptation in populations in which genetic variation is present. This conclusion was confirmed with the midge *Chironomus* spp. Laboratory-reared populations exposed at the level of the NOEC for a fitness characteristic showed altered population characteristics after several generations of exposure (Postma and Davids, 1995; Postma *et al.*, 1995). No correction for availability is needed here, since the NOEC and the effects of the multi-generation exposure were determined in similar conditions. This conclusion is attractive from a theoretical point of view, since selection operating on continuously variable characteristics such as tolerance and body growth is expected to induce a continuous shift of the population mean of the affected characteristics.

5.3.3 CONCLUSION ON FIELD PARAMETERS AT THE POPULATION LEVEL

The comparisons made between population tolerance and laboratory NOEC values showed that tolerance measures can be applied in practice, although population tolerance may be complicated to measure. More examples are needed to detect whether adaptation always occurs at sites where the (laboratory) NOEC for a fitness characteristic is exceeded. It is obvious that, in future comparisons, exposure concentrations should be expressed in available fractions both for the laboratory and the field (see section 5.2.2, and C.A.M. van Gestel, Chapter 2, this volume).

Apart from tolerance evolution, other types of population characteristics may appear useful for comparisons of laboratory toxicity data and field effects. Population size may be a possibility, but a numerical example (see section 5.5 and Figure 5.11a–c) shows that this parameter is sensitive for environmental factors not directly associated to exposure concentration. Further possibilities for population parameters are not treated here.

5.4 MEASUREMENT OF TOXICANT EFFECTS IN COMMUNITIES IN FIELD CONDITIONS

Next to the study of the field relevance of input data, the second step in validation of risk assessment methods consists of an evaluation of the similarity of risk levels with effects in the field. The field response should be expressed using community parameters, which are able to express damage to community structure and function. Damage to the chosen parameters should obviously be unacceptable from a regulatory point of view.

In this section, various parameters that can act as indicators of toxic effects at the community level are treated. Indicators may refer to effects on community function (e.g. pollution-induced community tolerance; PICT) or to community structure (species diversity). A numerical example of structural changes along a pollution gradient is given in section 5.5.

5.4.1 FIELD EFFECTS EXPRESSED AS INCREASED COMMUNITY TOLERANCE

Blanck *et al.* (1988) proposed the use of community tolerance as an indicator of toxic effects at the community level. Toxicant stress will cause elimination of sensitive species from a community. Community tolerance will increase due to individual acclimation, genetic adaptation and the loss of the most sensitive species. The idea was called PICT, and a gradient approach was explicitly preferred for field studies. The idea was originally developed for algal communities. In these communities, effects of various concentrations of a toxicant on photosynthesis rate were used to quantify tolerance.

(a) Measurement of PICT

PICT can be measured using the following procedure (Figure 5.8). First, communities are sampled in a pollution gradient. The concentrations of toxicants are measured for each site. One sampling site should have a background toxicant level. Second, the samples are artificially exposed to the toxicant for which the community is suspected to have evolved increased tolerance. This is done with a concentration–effect approach, to establish the sensitivity of the community of each site. The response can be measured for any ecologically relevant parameter. The performance of the control treatment is set to 100% for each site, since the community's sensitivity rather than its local size is measured. This is an essential step in reducing inter-site variability not related to pollution. Finally, the community sensitivities obtained for each site are compared. The comparison may reveal that the sensitivity of the community decreases with increasing exposure (Figure 5.8). The PICT procedure is an exact copy of an optimal approach in the study of population tolerance (section 5.3); in the latter, however, the methodology has not yet developed into a fast and practically applicable method such as PICT.

Blanck *et al.* (1988) consider PICT a useful approach for the study of community effects of toxicants, since:

● It is a more sensitive measure than population tolerance, since community tolerance depends on the loss of sensitive species, besides adaptation and acclimation;

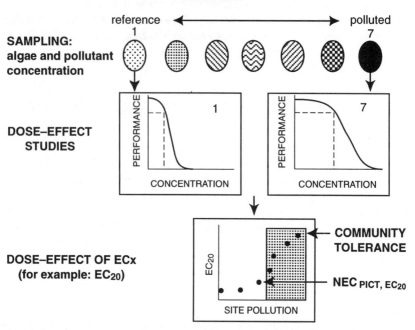

Figure 5.8 The concept of pollution-induced community tolerance (PICT), as defined by Blanck *et al.* (1988). For further explanation, see text.

- Photosynthesis rate is biologically meaningful and fast and easy to measure compared to population tolerance (section 5.3);
- Tolerance can be attributed to the toxicant that was also used in the concentration–effect tests of each community, whereas other sources of community divergence are either not reasonable, or they are correlated with the toxicant under investigation;
- PICT was associated with the occurrence of ecologically relevant community changes measured with conventional techniques, such as biomass production and species composition.

The PICT method was tested in artificially exposed communities and in exposed field communities (Blanck and Wängberg, 1988; Molander *et al.*, 1990; Wängberg *et al.*, 1991). PICT was more sensitive than the conventional techniques, except for one experiment (Molander and Blanck, 1992). A PICT test on communities affected by a mixture of toxicants showed that the role of both toxicants in community tolerance could be identified (Molander *et al.*, 1992). Other experiments suggested that the starting position of a community influenced the outcome of the PICT test (Wängberg and Blanck, 1990), that communities with a similar starting point showed sometimes different PICT development (Molander *et al.*, 1990), and that indirect effects of the toxicant on algal grazers may influence the development of PICT (Molander *et al.*, 1990, 1992).

(b) A PICT method for the terrestrial environment?

In soil, communities of algae cannot be used to quantify community tolerance. Another group of species should be chosen, with an appropriate parameter to quantify tolerance changes. The possibility to study PICT in communities of soil invertebrates, using for example isopod, springtail and worm species, is severely limited by the complex studies needed to quantify tolerance in each species separately (see section 5.3). The possibility to study PICT in microbial communities, however, is more attractive (Díaz-Raviña *et al.*, 1994). Microbial communities have been shown to respond to toxicants in field conditions with parameters regarding biomass or activity (e.g. Swift *et al.*, 1979; Domsch *et al.*, 1983). Some methods have been applied in field gradient studies (e.g. Tyler, 1975, 1984), but large inter-site variation of the parameter may prohibit fitting concentration–response curves. The PICT method may overcome this heterogeneity problem by quantifying tolerance for each site in a concentration–effect approach.

Díaz-Raviña *et al.* (1994) studied increased community tolerance in microorganism communities that were pre-exposed for 9 months to various metals added to soil. Three metal concentrations and a control were applied. The medium concentration was chosen at a concentration causing 50% reduction of the soil ATP level. Tolerance was quantified by the rate of thymidine incorporation (a measure of growth) in isolated bacteria. Extraction of bacteria from soil was necessary since thymidine binds to soil. Plate counts at different metal concentrations in the culture medium were made as an additional measure of community tolerance. Increased community tolerance (PICT) was established for Cu, Cd, Zn and Ni but not for Pb. Various cases of multiple tolerance were found, in which amendment of a metal caused tolerance for various other metals.

Problematic in these observations is that only a small proportion of the bacterial community is extracted, and that an even lower proportion of species ($<5\%$) usually grows on a specific medium. Despite this, there was a positive correlation between PICT established with thymidine incorporation and with colony counts. This reinforces the conclusion that PICT has occurred. Since only three concentrations and a control were used, there was insufficient information to establish a continuous relationship between soil concentration and the degree of PICT. Instead, the authors fitted a linear relationship to the three tested concentrations, for each metal. This resulted in estimates for threshold concentrations above which PICT starts to evolve. These threshold concentrations were higher than the HC_5 values of the studied metals in soil calculated by Van Straalen (1993b) from invertebrate toxicity data. The calculation procedure for the threshold, however, overestimates the concentration at which community tolerance really starts to develop. There is no evidence that microbial PICT evolves below the risk level HC_5, although measurements at soil concentrations near the HC_5 are required to strengthen this conclusion.

In an earlier study on field effects of lead, Doelman and Haanstra (1979) sampled soil at various reference, smelter and mine sites. Concentration–effect relationships for all sampling sites were determined for lead concentration in the culture medium versus the number of colony-forming units. Two culture media were used. Results recalculated from tabulated data show that the concentration–response curves were sigmoid, and that PICT occurred (Figure 5.9). In the polluted soils the HC_5 of lead (Van Straalen, 1994) was exceeded. In a similar approach, Doelman *et al.* (1994) compared the sensitivity–resistance index among soils from clean and polluted sites. The index method is similar to PICT, but fewer concentrations are used in the artificial exposure experiments for each site. The sensitivity–resistance index was reduced in polluted soils, which implies a relative increase of tolerant strains, as expected when PICT evolves. Doelman *et al.* (1994) also indicated the (socio)ecological relevance of PICT. They showed that tolerance evolves at the expense of the degradation capability of the microorganism community for the breakdown of organic pollutants (such as is needed in the remediation of soil loaded with

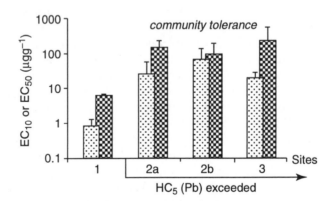

Figure 5.9 PICT for soil microorganisms. Soil samples were taken in various reference areas (1), at a smelter site in top soil and deeper soil (2a and 2b) and in metal-mining areas (3). The number of colony forming units was counted on media with different metal concentrations, in two culture media. Concentration–effect curves were fitted to the data for each site, and mean and standard deviations of EC_{10} and EC_{50} were calculated for replicated sites. Both the EC_{10} (light bars) and the EC_{50} (dark bars) were highest at the polluted sites, indicating PICT. The HC_5, estimated from SS toxicity data for soil invertebrates, was exceeded at sites 2 and 3.

organic toxicants). Among other causes, this may be a consequence of a reduced number of microbial species.

Both examples prove that PICT can apparently be established in terrestrial ecosystems with the use of microorganisms. Gradient sampling at polluted field locations is feasible, and may lead shortly to the identification of the field-NOEC of PICT that can be compared with risk levels.

The methods to quantify tolerance can be extended. Van Beelen *et al.* (1990, 1991), Van Beelen and Fleuren-Kemilä (1993), Van Beelen and Van Vlaardingen (1994) and Vonk and Matla (1993) reported on fast methods to quantify concentration–effect relationships for the breakdown of added, fast-degradable substrate in slurries made from fresh soil samples, at various concentrations of added toxicants. These methods can readily be applied in terrestrial PICT studies. These methods do not require growth of the colony, such as in thymidine incorporation and colony growth, but they focus on the physiological activity of the existing community.

A critical note can, however, be made. In all methods, the parameters are strongly influenced by the behaviour of the 'fast' representatives of the soil microorganism community. The use of a single method to quantify community tolerance is, therefore, not recommended, to avoid the outcome representing only part of the soil microorganism community. Two or three independent methods that reveal the same tolerance pattern are a strong indication that the observations really show PICT. Both Díaz-Raviña *et al.* (1994) (plate count and thymidine incorporation) and Doelman and Haanstra (1979) (two culture media) have applied such duplication of methods.

5.4.2 FIELD EFFECTS EXPRESSED WITH PARAMETERS FOR COMMUNITY STRUCTURE

For microorganisms both structural and functional parameters have been investigated along gradients, mostly without a PICT-like approach. For practical reasons case studies on invertebrates usually focused on structural parameters, in particular on the loss of species. Various field studies on changed structural parameters have been published, and some possibilities are evaluated below.

(a) Log-normal distribution of dominance

Several authors have suggested that communities of aquatic species from reference sites usually have a log-normal distribution of population sizes, whereas stressed communities showed an increase of the high abundance classes (Gray and Mirza, 1979; Gray, 1981; Pearson *et al.*, 1983). These authors considered a skewed distribution of population sizes to be an indicator of stress in a community. Eventually, a stressed community may return to a log-normal distribution of population sizes at a new level of stability, but with less species than before the pollution event. A similar reasoning has been made at the population level, where stress may result in a skewed distribution of tolerance until the population is fully adapted to the new environmental conditions (Ernst, 1983; Van Straalen *et al.*, 1986).

Hågvar (1994) tested the theory for soil invertebrate communities. The log-normal distribution was observed in various non-stressed situations. In stressed situations he observed considerable changes in population size of various species. These changes were, however, not expressed in the frequency distribution of population sizes. To solve this sensitivity problem, dominance structure instead of population size was used. Dominance values for all species at a site were calculated as the contribution of the species, in individual numbers, to the total community. This transformation resulted in skewed frequency distributions of dominance for the invertebrate communities in two gradients of metal pollution, namely the Kastad Pb gradient (see section 5.5), and a gradient around a brass mill (data from Bengtsson and Rundgren, 1988).

The principle may be applicable to other toxicant-exposed terrestrial invertebrate communities. In the examples studied by Hågvar, the dominance structure was changed on sites where the local metal concentration exceeded the HC_5. There were no observations made at sites with a soil concentration similar to the HC_5.

(b) The proportion of thelytokous species

Siepel (1994) proposed to use the proportion of thelytokous species in a community as an indicator of stress. Thelytokous organisms reproduce asexually and produce only female offspring. In an analysis of life-history tactics among soil invertebrates, Siepel argued that thelytoky will be favoured over other life-history tactics in a constant environment with a strong selection pressure. He argued that genetic variation at such 'constant' sites is unimportant for population persistence, and that tolerant females may produce tolerant clones. Such clones will rapidly out-compete less tolerant clones and species in the polluted area. Siepel analysed the data on the invertebrate community from the brass mill site studied by Bengtsson and Rundgren (1988), and found a sigmoid pattern of increase of the proportion of thelytokous species from c. 70% to almost 100%. The change of dominance structure identified by Hågvar (1994) has apparently been accomplished by the increase of thelytokous species.

(c) Species diversity measures

Hågvar (1994) also quantified diversity indices in the above mentioned studies. The Shannon–Wiener index, for example, reduced gradually from 3.16 to 2.41 in the brass mill gradient. In the literature, various diversity indices have been proposed to quantify species diversity, varying from counting the number of species, to more complex measures in which both species number and the distribution of individuals among species are incorporated. In principle, all these methods can be used to quantify changed community structure, but here only one method has been used to analyse community changes at a polluted site. This numerical example is treated in section 5.5.

5.4.3 CONCLUSION ON FIELD PARAMETERS AT THE COMMUNITY LEVEL

The examples show that effects measured in field communities can be used to validate risk levels. The chosen parameters thereby show the ecological meaning of exceeding risk levels: evolution of PICT, for example, leads to a loss of functional variability in the microorganism community. The examples do not yet show, however, that risk levels are exact estimates of threshold concentrations for adverse effects, since in most case studies no observations were made at or near the HC_5 concentration. In addition, it is obvious that alternate parameters can be applied to study community effects of pollutants. Further studies are needed to solve these omissions.

In future studies, it should be borne in mind that the relationships between risk levels and field effects are not fixed for several reasons. Firstly, progress in ecotoxicology may result in a new way to express toxicity, for example by using available concentrations to correct for exposure differences between laboratory and field conditions. Internal effect concentrations have also been suggested as a better alternative for expressing effects than external total toxicant concentrations (Van Wensem *et al.*, 1994). Newly developed ways to express toxicity should be used in both the expressions of risk levels and of field effects before comparisons of risk levels with field effects are made.

Secondly, effects in the field can be expressed by various community parameters, which may have different sensitivities for the toxicant (Figure 5.10). For the most sensitive community parameter, obviously, the fixed risk levels will be exceeded first. This focuses again on the question which parameters should be used for the validation of risk assessment methods. The answer to this question requires insight in the ecological meaning of exceeding the risk level for each parameter. Presently, it is implicitly assumed that protecting most of the species against adverse effects will maintain community structure and function, and both aspects are covered with the parameters 'PICT' and 'species diversity'. With the presently used methods, the examples on these parameters showed that effects were detected at sites where the HC_5 for the pertinent metals was (greatly) exceeded. An exception was the absence of PICT for Pb in soil amended with Pb for 9 months. From the literature data, however, it cannot yet be determined whether an ambient concentration that is slightly higher than the HC_5 would induce the adverse responses in all types of soil.

5.5 A NUMERICAL EXAMPLE OF A DETAILED GRADIENT STUDY, WITH HC_5 TO EC_5 COMPARISON

Many field gradients are characterized by the presence of various toxicants. This may complicate the comparison of HC_5 to EC_5 needed for the validation of risk assessment methods, since methods to cope with joint toxicity have not often been applied for terrestrial species. One example of community effects induced by a single toxicant is available. Hågvar and Abrahamsen (1990) studied the soil invertebrate community in a gradient polluted by Pb, near Kastad,

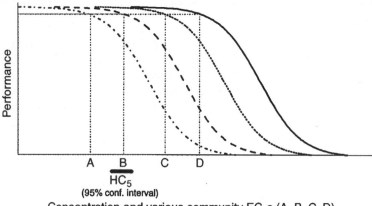

Concentration and various community EC_5s (A, B, C, D)

Figure 5.10 The relationship between a predicted hazardous concentration (e.g. the HC_5) and concentration–effect relationships of community parameters along field gradients. Some community parameters may show no response at the HC_5 level (c, d), whereas others may be more sensitive (a) or similar (b). The HC_5 should be calibrated against ecologically relevant community parameters. The HC_5 is depicted with confidence limits (Aldenberg and Slob, 1993), but the value may vary due to future developments of risk assessment methods.

Norway. The Kastad site is small (c. 50 m²), and has been polluted with Pb (and traces of some other toxicants) since the ice ages (Låg *et al.*, 1970).

Soil cores (*n* = 98) were taken along six lines crossing the area, into the undisturbed forest (for a distance of c. 1 m). Animals were extracted from the cores, identified and counted. Metal concentrations were determined in samples from the cores.

The data (originally kindly provided by S. Hågvar) were analysed with the sigmoid concentration–effect model, both through direct analysis of the original data per soil core, and after calculation of means per pollution class (*n* = 10). The population size of some species was studied, and an analysis was made on species diversity expressed by species number found per soil core. Only total lead contents were used as pollution data in the present numerical example, since the risk is now also calculated from total lead contents.

5.5.1 POPULATION RESPONSES ALONG THE GRADIENT

In section 5.3, on the analysis of field effects at the population level, it was suggested that population size may respond too rapidly to local environmental factors (predators, season) to be useful for comparisons with laboratory toxicity data. The data from the Kastad site allow for an analysis of effects of lead on population size, to evaluate this suggestion with field data.

The population size of some species (or higher taxa) along the concentration gradient was studied in relation to lead concentration in the soil (logarithmic scale). The observations show that there is a large inter-site variability (Figure 5.11, graphs A–C), so that smoothing methods must be applied for recognition of patterns. In the analyses, class means were calculated to reduce inter-site variability. All data were analysed with the sigmoid concentration–response model. For some species, the number of individuals per soil core indeed showed a sigmoid pattern of reduction along the gradient. A species that shows such a response is *Isotoma notabilis* (a springtail, Figure 5.11, graph A): the model fitted equally well to the observations and the class means. The genera *Suctobelba* (mites; Figure 5.11, graph C) and the Brachychtoniidae (mites, not shown) are higher taxa that exhibited a sigmoid response pattern, but occasionally the fit was successful on class means only. Other species, or higher taxa, showed a non-sigmoid response, such as *Folsomia quadrioculata* (springtails, not shown), *Schwiebea* cf. *talpa* (mites, not shown), *Oppia bicarnata* (mites, not shown), and *Oppiella nova* (mites; Figure 5.11, graph B). In these cases, a sigmoid response pattern is obviously absent. The increase in numbers at low Pb concentrations is apparently associated with the reduction in numbers of species competing for similar resources, or with a reduction of predators. Such species increase in numbers, until toxic effects of Pb become the dominant stress factor, and their sigmoid response begins.

These observations clearly show that population size is influenced by factors other than site pollution alone, and that such responses can be large in the field; even increases of population size with increasing lead concentration were observed. This shows that laboratory– field comparison with the parameter 'population size' can easily be disturbed by other factors than the toxicant concentration alone.

5.5.3 COMMUNITY RESPONSES ALONG THE GRADIENT

Species diversity was expressed with a simple diversity parameter, namely the number of species present in a soil core. Population size was not incorporated in the diversity parameter, since it is highly variable. The species diversity was calculated separately for oribatid mites [O], mesostigmatid mites [M] and springtails (Collembola [C]) and for the sum of O, M and C. These species diversity patterns have a sigmoid shape for the separate taxa (not shown) and for the total number of species (sum of O, M, C; Figure 5.11, graph D).

The observation that the pattern is grossly similar to a sigmoid pattern is the first indication from field observations that the assumption of bell-shaped distribution of species sensitivities (on log-scale) may hold true in field conditions. This assumption was only supported by laboratory data (Kooijman, 1987; Kammenga *et al.*, 1994). The evidence for the sigmoid shape is, however, not convincing when judged by statistical criteria; a straight line would fit equally well to the original observations, and probably even to the class

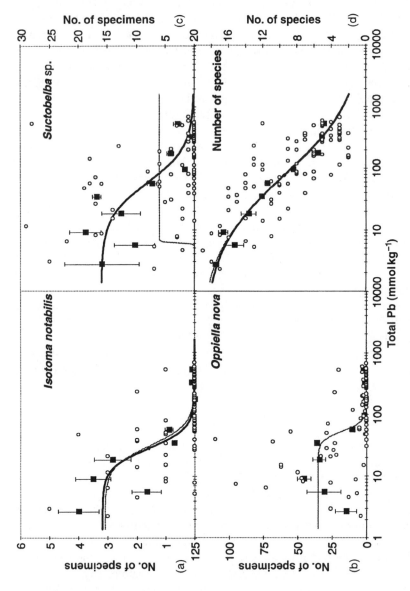

Figure 5.11 Concentration–effect relationships of numbers of individuals per species (or higher taxon) per soil core (a, b, c) and of number of species per soil core (d), in a lead-polluted gradient. Data were fitted to original observations (open cirlces) and class means (squares; bars denote standard errors). (Data were kindly provided by S. Hågvar.)

means. Nonetheless, a sigmoid response model was chosen for further analyses of the data, since intuitively any curve should have an equilibrium species diversity in the clean borders of the gradient. The use of the sigmoid model can be justified by the slightly sigmoid pattern in the observations of the class means. An alternative model should have been applied if there was information in the data to show, for example, an initial increase in species diversity at low lead concentrations.

The EC_5 (95% confidence intervals) of the community structure (sum O, M, C) was calculated and compared with HC_5 values for lead calculated from toxicity studies with soil invertebrates, with the NOEC as toxic end-point (Aldenberg and Slob, 1993; Van Straalen, 1994). The concentration where 5% of the species is lost through lead intoxication, directly or through altered nutrition or competition patterns, is c. 1.2 mmol Pb kg^{-1} soil. This value is higher than the estimated HC_5 values (Figure 5.12), but the confidence limits overlapped. Therefore, the HC_5 in this case study seems an accurate estimate of the concentration above which an effect at the community level will occur. Exceeding the HC_5 (expressed as total metal concentration) appears to coincide with a reduction of species diversity. The difference between the HC_5 and the EC_5 is small in view of the array of uncertainties involved in the calculation of the former value. Bioavailability differences, species interactions, nutrient situation and other stress factors may play a role (C.A.M. van Gestel, Chapter 2, this volume). Furthermore, one should realize that the EC_5 estimated here is based on extrapolation. First, the sigmoid curve had to be extrapolated into the clean area, where observations are insufficient to quantify clean area species number. Second, this extrapolation is influenced by the symmetry imposed by the model, and thus by the presence of a few species in the most polluted soil cores. The latter species may be not exposed, or they are tolerant or very mobile. Since the EC_5 depends strongly on the control performance, it is crucial to be able to estimate or measure the reference condition accurately.

In conclusion, the calculations clearly suggest the value of the approach of studying community changes along a field gradient. Whether the small difference between HC_5 and EC_5 found here is indicative for the adequacy of risk levels to predict field effects remains to be established.

5.6 VISIONS OF FUTURE DEVELOPMENTS

5.6.1 THE NEED FOR FURTHER FIELD STUDIES

The vision on future developments presented in this chapter considered a method to evaluate both the adequacy of input data of risk assessment methods, and the field relevance of risk levels. The reasoning focused on the method originally proposed by Van Straalen and Denneman (1989). The method can, however, also be applied to evaluate other methods for setting critical concentrations (e.g. Bengtsson and Tranvik, 1989; Hopkin, 1990).

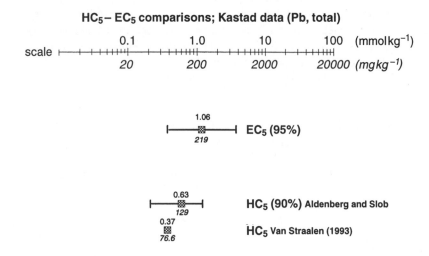

Figure 5.12 Comparison of the risk level HC_5 with the EC_5 of loss of species diversity in an invertebrate community inhabiting a longstanding lead-polluted gradient.

This proposal for validation of risk assessment methods was tested against some case studies on toxicant effects in the field. Only a restricted set of case studies was explored, so that only a gross summary of field effects can be made. First, at the level of the input data, the case studies suggested that ambient concentrations that exceed the NOEC (derived in laboratory conditions) seem to lead to population responses in the field. Secondly, ambient concentrations that exceed the HC_5 seem to induce community tolerance and species loss. These data are not robust yet, and need further analyses of case studies. Nonetheless, the proposed method for the validation of risk assessment methods is applicable, and may help to improve risk assessment methods.

A proposal was made for field parameters that can be used to judge population and community effects, and that should be studied in more detail. At the population level, attention was focused on adaptation phenomena (changes of tolerance- and life-history characteristics and of genetic variation) and on the methods to study these parameters accurately. At the community level, attention was focused on population tolerance (PICT, functional) and species diversity (structural). This choice of parameters may be incomplete, and ecological knowledge may help to choose additional parameters that are appropriate to quantify field effects of toxicants.

The comparisons of laboratory data and field effects may result in rejection or refinement of risk assessment methods: they may show similarity or dissimilarity. The degree to which dissimilarity can be accepted is a regulatory problem, and an answer is not formulated here. Governmental agencies should indicate the acceptable magnitude of dissimilarity, although ecotoxicologists have the responsibility to suggest a reasonable value that considers the variability among field sites. Similarity of laboratory and field effects leads to a validated method. Similarity is expected when all uncertainties of the risk assessment methods neutralize each other, as was originally assumed by Van Straalen and Denneman (1989). A validated method, which is apparently insensitive for any kind of additional influence, can be used for setting soil quality criteria for general policy use. If it is found, for example, that the range of dissimilarity between predicted and field effects lies within the range of a factor of 5 in the upward and downward direction, then this may be regarded as (regulatorily) acceptable in the light of the series of uncertainties involved in risk assessment. Such information may be enough for a policy decision to promote, reject, or refine the risk assessment methods. (Of course it is of scientific interest to know how the uncertainties result in 'similarity'.)

Dissimilarity may lead to diversification of risk assessment methods, through consideration of the most influential factors identified from case studies. In turn, this may lead to site-specific approach in risk assessment. Influential factors can be, for example, bioavailability (addressed by C.A.M. van Gestel, Chapter 2, this volume), joint toxicity of various toxicants, influences of other stress factors, etc.

The proposed validation scheme strongly depends on the availability of ecologically sound field studies. Retrospective evaluation of literature data is often hampered by incompleteness of the field data, either through too few replicates, or through inadequate observations. This is to be expected, since the studies were not specifically designed to allow for the proposed comparisons. A gradient study specifically designed for the validation purpose should focus on ecosystem responses of a series of variables, ranging from microbial activity to the species diversity structure of the soil animal community. In such multi-parameter studies it will appear that all parameters will exhibit their own field sensitivity. This will result in different locations of the concentration–response curves on the concentration axis (Figure 5.10). With such information, predicted effects or risk levels can be calibrated to field effects for ecologically relevant parameters. It is recommended that such multi-variable field studies be done. As a start, the proposed validation scheme can (and will) be further elaborated by retrospective evaluation of literature data.

5.6.2 PROBLEMS TO BE SOLVED

Various problems remain to be solved in the proposed scheme. A philosophical problem is whether the performance in reference conditions can accurately be estimated, even with large sample sizes and observations at large distances from the pollution source. Human activities of various kinds may already have altered performance measures, e.g. of species diversity, to below values that would occur without any human-induced disturbance. This problem requires a common approach for all cases, since EC_5 and the HC_5 are strongly influenced by the measured and estimated reference performance (see Figure 5.11, graph D, where reference species diversity was not accurately estimated).

Another problem that needs investigation is the joint effects of toxicants. Pragmatically, the magnitude of dissimilarity between predicted and field effects can be judged under the assumption of concentration additivity of the toxicants in the mixture (Van Straalen, 1994). Similar to the calculation of SS-toxic units one may calculate community toxic units. (An SS-toxic unit [based on simple similar action] expresses the concentration of a toxicant through a dimensionless value obtained by standardizing the concentration on toxicity end-points.) Concentrations expressed through SS-toxic units are usually calculated as fractions of the EC_{50}; in communities, standardization may be obtained by expressing concentrations in fractions of the HC_5.

Apart from the simultaneous presence of various toxicants, a serious problem may arise in the interpretation of field data due to the presence of other factors that co-vary with the level of exposure. It has been observed, for example, that the nematode community in a gradient around a zinc smelter showed no reduction in size near the factory since the adverse metal effect was compensated by a stimulating effect of the correlated pH increase along the gradient (T. Schouten, unpublished results). Covariation of pollutant concentration with other factors can be attacked in two ways. The first is a method that can be applied in future studies, and it consists of measurements of all possible confounding factors that reasonably could influence the biota. If such factors are shown to be irrelevant, then the effects can reasonably be attributed to the toxicant. With retrospective evaluation of literature data, in which no information is given on confounding factors, only inferences can be made on the relationship between laboratory and field effects. In the latter case, it is only the recognition of a trend in a series of observations that can be used to validate risk levels.

Finally, the case of non-persistent pollutants needs further consideration, since the predicted effects also depend on the breakdown rate of the toxicant. Application of the proposed validation method is expected to show that the field effects may, on the one hand, be less serious due to the disappearance of the chemical followed by recovery processes. On the other hand, many non-persistent pollutants are applied as pesticides with specific target species, and this may cause severe effects in specific parts of the community.

5.6.3 CONCLUSION

Despite the above-mentioned shortcomings, retrospective evaluation of literature data and analysis of future gradient studies are likely to reveal differences between field effects and risk levels, depending on the parameter and the adequacy of the methods. These differences should, however, be regarded as a guideline to the identification of the site-specific factors that caused the difference, for example those regarding availability of the toxicant. With knowledge of such factors, risk assessment methods will gain adequacy, but they may probably also develop in the direction of a more site-oriented approach. The proposed gradient studies may be used both in the validation of existing risk assessment methods, and in the development into a site-specific methodology.

5.7 SUMMARY

The knowledge of toxic effects of chemicals in populations and communities in field conditions can be used to evaluate the adequacy of risk assessment methods. Adequacy depends on accurate input data and on the ecological relevance of estimated risk levels. A comparative method is proposed to evaluate accuracy and relevance of field data. The method can be used in an iterative way to improve adequacy of risk assessment. Population and community parameters appropriate for laboratory–field comparisons are suggested. Gradients around pollution sources are identified as the optimal source for field data. Some case studies are numerically explored. At the population level, toxic effects appear to occur when the field concentration is grossly similar to or exceeds the laboratory NOEC for a fitness characteristic; genetic adaptation was used as the yardstick to identify adverse effects. At the community level, toxic effects appeared to occur when the field concentration exceeds a risk level known as the HC_5. A numerical example of community effects in a gradient showed close similarity between HC_5 and the EC_5 of loss of species diversity in a soil invertebrate community. Literature data and new observations on toxic effects in the field should be collected to determine whether this similarity is robust. Such studies can provide the data to validate risk assessment methods in the future, and may help to show the ecological meaning of exceedance of risk levels.

ACKNOWLEDGEMENTS

Sigmund Hågvar is acknowledged for his willingness to offer me his original data of the Kastad study. Colleagues of the Laboratory for Ecotoxicology of the RIVM, of the laboratory for Ecology and Ecotoxicology of the Vrije Universiteit Amsterdam, and participants of the SERAS meeting are acknowledged for their stimulating discussions and their suggestions. The two reviewers and the editors are acknowledged for their comments, which improved the quality of the chapter.

REFERENCES

Aldenberg, T. and Slob, W. (1993) Confidence limits for hazardous concentrations based on logistically distributed NOEC toxicity data. *Ecotox. Environ. Safety*, **25**, 48–63.

Bengtsson, G. and Rundgren, S. (1988) The Gusum case: a brass mill and the distribution of soil Collembola. *Can. J. Zool.*, **66**, 1518–26.

Bengtsson, G. and Tranvik, L. (1989) Critical metal concentrations for forest invertebrates. *Water Air Soil Pollut.*, **47**, 381–417.

Benton, M.J. and Guttman, S.I. (1990) Relationship of allozyme genotype to survivorship of mayflies (*Stenonema femoratum*) exposed to copper. *J. N. Am. Benthol. Soc.*, **9**, 271–6.

Beyer, W.N., Miller, G.W. and Cromartie, E.J. (1984) Contamination of the O_2 soil horizon by zinc smelting and its effect on woodlouse survival. *J. Environ. Qual.*, **13**, 247–51.

Blanck, H. and Wängberg, S.-Å (1988) Induced community tolerance in marine periphyton established under arsenate stress. *Can. J. Fish. Aquat. Sci.*, **45**, 1816–19.

Blanck, H., Wängberg, S.-Å and Molander, S. (1988) Pollution-induced community tolerance – a new ecotoxicological tool, in *Special Technical Publication 988*, American Society for Testing and Materials, Philadelphia, PA, pp. 219–30.

Bradshaw, A.D. (1991) Genostasis and the limits to evolution. *Phil. Trans. R. Soc. Lond. B.*, **333**, 289–305.

Calow, P. (1994) Ecotoxicology: what are we trying to protect? *Environ. Toxicol. Chem.*, **13**, 1549.

Chagnon, N.L. and Guttman, S.I. (1989) Differential survivorship of allozyme genotypes in mosquitofish populations exposed to copper or cadmium. *Environ. Toxicol. Chem.*, **8**, 319–26.

Charlesworth, B. (1980). *Evolution in Age-structured Populations*, Cambridge University Press, Cambridge.

Crommentuijn, T., Brils, J. and Van Straalen, N.M. (1993) Influence of cadmium on life-history characteristics of *Folsomia candida* (Willem) in an artificial soil substrate. *Ecotox. Environ. Safety*, **26**, 216–27.

Díaz-Raviña, M., Bååth, E. and Frostegård, Å. (1994) Multiple heavy metal tolerance of soil bacterial communities and its measurement by a thymidine incorporation technique. *Appl. Environ. Microbiol.*, **60**, 2238–47.

Doelman, P. and Haanstra, L. (1979) Effects of lead on the soil bacterial microflora. *Soil Biol. Biochem.*, **11**, 487–91.

Doelman, P., Jansen, E., Michels, M. and Van Til, M. (1994) Effects of heavy metals on microbial diversity and activity as shown by the sensitivity–resistance index, an ecologically relevant parameter. *Biol. Fertil. Soils*, **17**, 177–84.

Domsch, K.H., Jagnow G. and Anderson T.-H. (1983) An ecological concept for the assessment of side-effects of agrochemicals on soil microorganisms. *Res. Rev.*, **86**, 66–105.

Donker, M.H. and Bogert, C. (1991) Adaptation to cadmium in three populations of the isopod *Porcellio scaber*. *Comp. Biochem. Physiol.*, **100C**, 143–6.

Donker, M.H., Zonneveld, C. and Van Straalen, N.M. (1993) Early reproduction and increased reproductive allocation in metal adapted populations of the terrestrial isopod *Porcellio scaber*. *Oecologia (Berlin)*, **96**, 316–23.

Donker, M.H., Raedecker, M.H. and Van Straalen, N.M. (1996) The role of zinc regulation in the zinc tolerance mechanism of the terrestrial isopod *Porcellio scaber*. *J. Appl. Ecol.* **33**, 955–64.

Eberhardt, L.L. and Thomas, J.M. (1991) Designing environmental field studies. *Ecol. Monogr.*, **6**, 53–73.

Endler, J.A. (1986) *Natural Selection in the Wild*, Princeton University Press, Princeton, New Jersey, USA.

Ernst, W.H.O. (1983) Ökologische Anpassungsstrategien an Bodenfaktoren. *Ber. Deutsch. Bot. Ges.*, **96**, 49–71.

Ernsting, G., Zonneveld, C., Isaaks, J.A. and Kroon A. (1993) Relationships between growth, reproduction and survival in an insect with indeterminate growth. *Oikos*, **66**, 17–26.

Falconer, D.S. (1981) *Introduction to Quantitative Genetics*, 2nd edn, Longman, New York.

Frati, F., Fanciulli, P.P. and Posthuma, L. (1992) Allozyme variation in reference and metal-exposed natural populations of *Orchesella cincta* (L.) (Insecta, Collembola). *Biochem. Syst. Ecol.*, **20**, 297–310.

Gillespie, R.B. and Guttman, S.I. (1989) Effects of contaminants on the frequencies of allozymes in populations of the central stoneroller. *Environ. Toxicol. Chem.*, **8**, 309–17.

Gray, J.S. (1981) Detecting pollution induced changes in communities using the log-normal distribution of individuals among species. *Mar. Pollut. Bull.*, **12**, 173–6.

Gray, J.S. and Mirza, F.B. (1979) A possible method for the detection of pollution-induced disturbance on marine benthic communities. *Mar. Pollut. Bull.*, **10**, 142–6.

Haanstra, L., Doelman, P. and Oude Voshaar, J.H. (1985) The use of sigmoidal dose response curves in soil ecotoxicological research. *Plant Soil*, **84**, 293–7.

Hågvar, S. (1994) Lognormal distribution of dominance as an indicator of stressed soil microarthropod communities. *Acta Zool. Fenn.*, **195**, 71–80.

Hågvar, S. and Abrahamsen G. (1990) Microarthopoda and Enchytraeidae (Oligochatea) in naturally lead-contaminated soil: a gradient study. *Environ. Entomol.*, **19**, 1263–77.

Hamers, T., Notenboom, J. and Eijsackers, H.J.P. (1996) *Validation of Laboratory Toxicity Data on Pesticides for the Field Situation.* National Institute of Public Health and the Environment, Bilthoven, The Netherlands, Report no. 719102046.

Hawkins, A.J.S., Rusin, J., Bayne, B.L. and Day, A.J. (1989) The metabolic/physiologic basis of genotype-dependent mortality during copper exposure in *Mytilus edulis. Mar. Environ. Res.*, **28**, 253–7.

Heimbach, F. (1992) Correlation between data from laboratory and field tests for investigating the toxicity of pesticides to earthworms. *Soil Biol. Biochem.*, **24**, 1749–53.

Holloway, G.J., Povey S.R. and Sibly, R.M. (1990a) The effects of new environment on adapted genetic architecture. *Heredity*, **64**, 323–30.

Holloway, G.J., Sibly, R.M. and Povey S.R. (1990b) Evolution in toxin stressed environments. *Funct. Ecol.*, **4**, 289–94.

Hopkin, S.P. (1990) Critical concentrations, pathways of detoxification and cellular ecotoxicology of metals in terrestrial arthropods. *Funct. Ecol.*, **4**, 321–7.

Hopkin, S.P. (1993) Ecological implications of '95% protection levels' for metals in soil. *Oikos*, **66**, 137–41.

Huggett, R.J., Kimerle, R.A., Mehrle, P.M. and Bergma, H.L. (1992) *Biomarkers: Biochemical, Physiological and Histological Markers of Anthropogenic Stress*, Lewis Publishers, Boca Raton, FL, USA.

Kammenga, J.E., Van Gestel, C.A.M. and Bakker, J. (1994) Patterns of sensitivity to cadmium and pentachlorophenol among nematode species from different taxonomic and ecological groups. *Arch. Environ. Contam. Toxicol.*, **27**, 88–94.

Klerks, P.L. and Weis, J.S. (1987) Genetic adaptation to heavy metals in aquatic organisms: a review. *Environ. Pollut.*, **45**, 173–205.

Koehn, R.K. (1978) Physiology and biochemistry of enzyme variation: the interface of ecology and population genetics, in *Ecological Genetics: the Interface* (ed. P.F. Brussard), Springer, New York, pp. 51–72.

Kooijman, S.A.L.M. (1987) A safety factor for LC_{50} values allowing for differences in sensitivity among species. *Water Res.*, **21**, 269–76

Låg J., Hvatum, O.Ø. and Bølviken, B. (1970) An occurrence of naturally lead-poisoned soil at Kastad near Gjøvik, Norway. Årbok 1969, *Norges Geologiske Undersökelse*, **266**, 141–59.

Lavie, B. and Nevo, E. (1982) Heavy metal selection of phosphoglucose isomerase allozymes in marine gastropods. *Mar. Biol.*, **71**, 17–22.

Lavie, B. and Nevo, E. (1986) The interactive effects of cadmium and mercury pollution on allozyme polymorphisms in the marine gastropod *Cerithium scabridum. Mar. Pollut. Bull.*, **17**, 21–3.

Lower, W.R. (1975) Gene frequency differences in *Drosophila melanogaster* associated with lead smelting operations. *Mutat. Res.*, **31**, 315.

Luoma, S.N. (1977) Detection of trace contaminant effects in aquatic ecosystems. *J. Fish. Board Can.*, **34**, 436–9.

Maruyama, T. and Fuerst, P.A. (1985) Population bottlenecks and nonequilibrium models in population genetics. II. The number of alleles in a small population that was formed by a recent bottleneck. *Genetics*, **111**, 675–89

Michod, R.E. (1979) Evolution of life histories in response to age-specific mortality factors. *Am. Nat.*, **113**, 531–50.

Molander, S. and Blanck, H. (1992) Detection of pollution-induced community tolerance (PICT) in marine periphyton communities established under diuron exposure. *Aquat. Toxicol.*, **22**, 129–44.

Molander, S., Blanck, H. and Söderström M. (1990) Toxicity assessment by pollution induced community tolerance (PICT), and identification of metabolites in periphyton communities after exposure to 4,5,6-trichloroguaiacol. *Aquat. Toxicol.*, **18**, 115–36.

Molander, S., Dahl, B., Blanck, H., Josson, J. and Sjöström, M. (1992) Combined effects of Tri-n-butyl Tin (TBT) and diuron on marine periphyton communities detected as pollution-induced community tolerance. *Arch. Environ. Contam. Toxicol.*, **22**, 419–27.

Murdoch, M.H. and Hebert, P.D.N. (1994) Mitochondrial DNA diversity of brown bullhead from contaminated and relatively pristine sites in the great lakes. *Environ. Toxicol. Chem.*, **13**, 1281–9

Okkerman, P.C., Van de Plassche, E.J., Emans, H.J.B. and Canton, J.H. (1993) Validation of some extrapolation methods with toxicity data derived from multiple species experiments. *Ecotox. Environ. Safety*, **25**, 341–59

Peakall, D. B. (1992) *Animal Biomarkers as Pollution Indicators*, Chapman & Hall, London.

Pearson, T.H., Gray, J.S. and Johannessen, P.J. (1983) Objective selection of sensitive species indicative of pollution-induced change in benthic communities. 2. Data analyses. *Mar. Ecol. Progr. Ser.*, **12**, 237–55.

Posthuma, L. (1990) Genetic differentiation between populations of *Orchesella cincta* (Collembola) from heavy-metal contaminated sites. *J. Appl. Ecol.*, **27**, 609–22.

Posthuma, L. and Janssen, G.M. (1995) Genetic variation for life-history characteristics in *Orchesella cincta* (L.) in relation to evolutionary responses to metals in soils. *Acta Zool. Fenn.*, **196**, 301–6.

Posthuma, L. and Van Straalen N.M. (1993) Heavy-metal adaptation in terrestrial invertebrates: a review of occurrence, genetics, physiology and ecological consequences. *Comp. Biochem. Physiol.*, **106C**, 11–38.

Posthuma L., Hogervorst R.F. and Van Straalen N.M. (1992) Adaptation to soil pollution by cadmium excretion in natural populations of *Orchesella cincta* (L.) (Collembola). *Arch. Environ. Contam. Toxicol.*, **22**, 146–56.

Posthuma, L., Hogervorst R.F., Joosse E.N.G. and Van Straalen N.M. (1993a) Genetic variation and covariation for characteristics associated with cadmium tolerance in natural populations of the springtail, *Orchesella cincta* (L.). *Evolution*, **47**, 619–31.

Posthuma, L., Verweij R.A., Widianarko B. and Zonneveld C. (1993b) Life-history patterns in metal-adapted Collembola. *Oikos*, **67**, 235–49.

Posthuma, L., Boonman, H., Mogo, F.C. and Baerselman, R. (1994) *Heavy Metal Toxicity in Eisenia andrei Exposed in Soils From a Gradient Around a Zinc Smelter (Budel) and Comparison with Toxic Effects in OECD-Artificial Soil.* RIVM-report no. 719102033, Dutch National Institute for Public Health and Environmental Protection *RIVM*, Bilthoven, The Netherlands.

Posthuma, L., Weltje, L. and Notenboom J. (1995) *Mixtures of toxicants and their effects on soil animals.* RIVM-Annual Scientific Report 1994, RIVM, Bilthoven, The Netherlands. pp. 137–8.

Postma, J.F. and Davids, C. (1995) Tolerance induction and life-cycle changes in cadmium exposed *Chironomus riparius* (Diptera) during consecutive generations. *Ecotox. Environ. Safety*, **30**, 195–202.

Postma, J.F., Mol, S., Larsen, H. and Admiral, W. (1995) Life-cycle changes and zinc shortage in cadmium tolerant midges, *Chironomus riparius* (Diptera) reared in the absence of cadmium. *Environ. Toxicol. Chem.*, **14**, 117–22.

Reznick, D.N., Bryga, H. and Endler, J.A. (1990) Experimentally induced life-history evolution in a natural population. *Nature*, **346**, 357–9.

Shaw, J. (1988) Genetic variation for tolerance to copper and zinc within and among populations of the moss, *Funaria hygrometrica* Hedw. *New Phytol.*, **109**, 211–22.

Siepel, H. (1994) *Structure and Function of Soil Microarthropod Communities.* PhD Thesis, Landbouw Universiteit Wageningen, The Netherlands.

Slooff, W., Van Oers, J.A.M. and De Zwart, D. (1986) Margins of uncertainty in ecotoxicological hazard assessment. *Environ. Toxicol. Chem.*, **5**, 841–52.

Suter, G.W. (1992) *Ecological Risk Assessment*, Lewis Publishers, Boca Raton, FL, USA.

Swift, M.J., Heal, O.W. and Anderson J.M. (1979) *Decomposition in Terrestrial Ecosystems.* Studies in Ecology, no. 5. Blackwell Scientific Publications, Oxford, UK.

Tranvik, L., Sjögren M. and Bengtsson, G. (1994) Allozyme polymorphism and protein profile of *Orchesella bifasciata* (Collembola): indicative of extended metal pollution? *Biochem. Syst. Ecol.*, **22**, 13–23.

Tyler, G. (1975) Heavy metal pollution and mineralisation of nitrogen in forest soils. *Nature*, **255**, 701–2.

Tyler, G. (1984) The impact of heavy-metal pollution on forests: a case study of Gusum, Sweden. *Ambio*, **13**, 18–24.

Van Beelen, P. and Fleuren-Kemilä, A.K. (1993) Toxic effects of pentachlorophenol and other pollutants on the mineralization of acetate in several soils. *Ecotox. Environ. Safety*, **26**, 10–17.

Van Beelen, P. and Van Vlaardingen, P.L.A. (1994) A method for the ecotoxicological risk analysis of polluted sediments by the measurement of microbial activities, in *Ecotoxicology of Soil Organisms* (eds M.H. Donker, H. Eijsackers and F. Heimbach), Lewis Publishers, Boca Raton, FL, pp. 105–12.

Van Beelen, P., Fleuren-Kemilä, A.K., Huys, M.P.A., Van Mil, A.C.H.M. and Van Vlaardingen, P.L.A. (1990) Toxic effects of pollutants on the mineralization of substrates at low environmental concentrations in soils, subsoils and sediments, in *Contaminated Soil '90* (eds F. Arendt, M. Hinseveld, and W.J. van den Brink), Kluwer Academic Publishers, Dordrecht, The Netherlands, pp. 431–8.

Van Beelen, P., Fleuren-Kemilä, A.K., Huys, M.P.A., Van Montfort A.C.P. and Van Vlaardingen, P.L.A. (1991) The toxic effects of pollutants on the mineralization of acetate in subsoil microcosms. *Environ. Toxicol. Chem.*, **10**, 775–89.

Van Capelleveen, H.E. (1987) *Ecotoxicity of Heavy Metals for Terrestrial Isopods.* PhD Thesis, Vrije Universiteit, Amsterdam.

Van Ewijk, P.H. and Hoekstra, J.A. (1993) Calculation of the EC_{50} and its confidence interval when subtoxic stimulus is present. *Ecotox. Environ. Safety*, **25**, 25–32.

Van Straalen, N.M. (1993a) An ecotoxicologist in politics. *Oikos*, **66**, 142–3.

Van Straalen, N.M. (1993b) Soil and sediment criteria derived from invertebrate toxicity data, in *Ecotoxicology of Metals in Invertebrates* (eds R. Dallinger and P.S. Rainbow), Lewis Publishers, Boca Raton, FL, USA, pp. 427–41.

Van Straalen, N.M. (1994) Open problems in the derivation of soil quality criteria from ecotoxicity experiments, in *Contaminated Soil '93* (eds F. Arendt, G.J. Annokkée, R. Bosman and W.J. van den Brink), Kluwer Academic Publishers, Dordrecht, The Netherlands, pp 315–26.

Van Straalen, N.M. and Denneman, C.A.J. (1989) Ecotoxicological evaluation of soil quality criteria. *Ecotox. Environ. Safety*, **18**, 241–51.

Van Straalen, N.M., Groot, G.M. and Zoomer, H.R. (1986) Adaptation of soil Collembola to heavy-metal soil contamination, in *Proceedings of the International Conference on Environmental Contamination, Amsterdam*, CEP Consultants, Edinburgh, UK, pp. 16–20.

Van Straalen, N.M., Schobben, J.H.M. and De Goede, R.G.M. (1989) Population consequences of cadmium toxicity in soil microarthropods. *Ecotox. Environ. Safety*, **17**, 190–204.

Van Straalen, N.M., Schobben, J.H.M. and Traas, T.P. (1992) The use of ecotoxicological risk assessment in deriving maximum acceptable half-lives of pesticides. *Pest. Sci.*, **34**, 227–31.

Van Straalen, N.M., Leeuwangh, P. and Stortelder, P.B.M. (1994) Progressing limits for soil ecotoxicological risk assessment, in *Ecotoxicology of Soil Organisms* (eds M.H. Donker, H. Eijsackers and F. Heimbach), Lewis Publishers, Boca Raton, FL, USA, pp. 397–409.

Van Wensem, J., Vegter, J.J. and Van Straalen, N.M. (1994) Soil quality criteria derived from critical body concentrations of metals in invertebrates. *Appl. Soil. Ecol.*, **1**, 185–91.

Verkleij, J.A.C., Bast-Cramer, W.B. and Levering, H. (1985) Effects of heavy-metal stress on the genetic structure of populations of *Silene cucubalus*, in *Structure and Functioning of Plant Populations*, (eds. J. Haeck and J.W. Woldendorp) North Holland Publishing Company, Amsterdam, The Netherlands, pp. 355–65.

Vonk, J.W. and Matla, Y.A. (1993) *A Test for Effects of Chemicals on Glutamate Mineralisation in Soil.* TNO Institute of Environmental Sciences, Report no. IMW-R 93/097

Wagner, C. and Løkke, H. (1991) Estimation of ecotoxicological protection levels from NOEC toxicity data. *Water Res.*, **25**, 1237–42.

Wängberg, S.-Å. and Blanck, H. (1990) Arsenate sensitivity in marine periphyton communities established under various nutrient regimes. *J. Exp. Mar. Biol. Ecol.*, **139**, 119–34.

Wängberg, S.-Å., Heyman, U. and Blanck, H. (1991) Long-term and short-term arsenate toxicity to freshwater phytoplankton and periphyton in limnocorals. *Can. J. Fish. Aquat. Sci.*, **48**, 173–82.

Wilson, J.B. (1988) The cost of heavy-metal tolerance: an example. *Evolution*, **42**, 408–13.

PART THREE

The soil as an ecosystem

6   Linking structure and function in marine sedimentary and terrestrial soil ecosystems: implications for extrapolation from the laboratory to the field

THOMAS L. FORBES AND LIV K. KURE

6.1 ECOTOXICOLOGICAL EXTRAPOLATION AND THE STRUCTURE AND FUNCTION OF ECOSYSTEMS

A central goal of ecotoxicology is the development of sufficient ecological understanding to enable the accurate prediction of the behaviour and effects of contaminants in the environment. Progress toward this goal has been slow. We believe future advancement requires increased emphasis on interdisciplinary studies which comprehensively investigate chemical fate and effect at spatial and temporal scales relevant to the natural systems of interest (Forbes and Forbes, 1994). Building a bridge from the laboratory to the field will require improved extrapolation models incorporating a more complete understanding of the relationship between ecosystem structure and function than presently exists (Forbes and Forbes, 1993).

From the perspective of ecotoxicology, questions concerning ecosystem function must include the fate and effects of contaminants. For example, the degradation rates of many organic contaminants in both marine and terrestrial systems will to some degree be related to the processes of sediment or soil ingestion and transport due to the local fauna (Jones and Lawton, 1995). This perspective necessitates integrated study of the feedback relationships between contaminant fate and effect. We argue that by forming a more complete mechanistic understanding of the important structural and functional

Ecological Risk Assessment of Contaminants in Soil. Edited by Nico M. van Straalen and Hans Løkke. Published in 1997 by Chapman & Hall, London. ISBN 0 412 75900 4

interactions within ecosystems, a better understanding of the linkage between the fate and effect of pollutants can be achieved.

The goal of this contribution is to begin development of an explicit theoretical framework for linking structure and function in ecosystems containing anthropogenic contaminants. Below we examine potential linkages between marine benthic community structure and the fate of a representative organic contaminant in sediment. Our specific choice of models and parameters exemplifies only one of many possible approaches (e.g. Lubchenco, 1995). Nonetheless it contains two features we feel are essential to any successful effort to improve understanding. The first is an underlying mechanistic basis. Formalization of underlying mechanisms often necessitates the second feature of our particular model which is a significant degree of site or system specificity. We develop these ideas further below.

6.1.1 KEY CONCEPTS RELEVANT TO ECOTOXICOLOGICAL EXTRAPOLATION

Several definitions are required in order to engage in coherent discussion of linkages between the structure of ecological systems and their component functions or processes. The definitions below are by no means the only defensible ones that could be developed. The important point is that the terms of reference have been defined so that the discussion can proceed on common ground. The key definitions needed are those for the concepts of scale, function, structure and that for the ecosystem itself.

Scale refers to size in both time and space. Consideration of scale is central to organizing observation and essential in any attempt to extrapolate from laboratory or field studies to the much larger systems which are often the focus of management decisions (Schindler, 1995). An object is large-scale relative to another object if perceiving it requires relatively longer periods of time or larger regions in space.

Here we define an **ecosystem** as a set of interlinked, differently scaled pathways of processes and fluxes among organisms and between organisms and their environment. These pathways may be intangible and diffuse in space but easily defined in terms of turnover times, half-lives and mass balances (O'Neill *et al.*, 1986; Allen and Hoekstra, 1993; Jones and Lawton, 1995). Some degree of internal cycling is required for ecosystems to be coherent entities worthy of investigation and the perceived temporal and spatial scales are delimited by the process(es) under study.

We use the terms **function** and **process** synonymously. Unfortunately, these terms have been used in at least two ways in the past by different groups of ecologists. Ecosystems ecologists generally define process as a sequence of transformations of matter and/or energy which may be expressed as a rate (Jones and Lawton, 1995). Examples include denitrification, photosynthesis, sulphate reduction, the remineralization of organic matter, the rate at which sediment or soil is transported by biological activity, and so on. The second definition of process is more often employed by community ecologists and uses the terms to refer to dynamic interactions among whole organisms or between organisms and their environment. Examples include the processes of predation, competition, disturbance and parasitism (Paine, 1994). These differences in usage reflect important differences in outlook and have been a source of confusion in the past. These outlooks are qualitatively different and reflect fundamentally different organizational paradigms. These are the conservation of mass and energy on the one hand and natural selection and evolution on the other. Nevertheless, note that either of the above definitions can be used to express a rate. Part of our goal below is to begin to integrate these two perspectives into a larger, more inclusive framework.

The **structure** of an ecosystem is defined here as the species abundances of the organisms participating in ecosystem processes at any given point in time. This is a somewhat restricted definition of structure. This seems most appropriate for discussions of ecotoxicological extrapolation which attempts to relate laboratory toxicity data on single species to effects in natural environments. Previous more expansive definitions have, for example, defined structure as the trophic scaffolding resulting from species interactions in a food-web (Paine, 1980).

(a) The 'functional' ecosystem perspective

For processes such as nutrient cycling or energy flow, which are of primary concern to ecosystems ecologists, it has generally proven quite difficult to determine any simple or unambiguous mapping between process and structure (O'Neill *et al.*, 1986; Allen and Hoekstra, 1993 treat this topic in detail). This means that the investigation of ecosystem function as defined above can become intractable if the system is viewed solely as a nested hierarchy within which individual organisms are discrete parts. As alluded to above in our dichotomous definition of process, this intractability is reflected in two somewhat divergent views of ecosystems. With only a slight risk of oversimplification, we term these the **functional** and **structural** perspectives (Jones and Lawton, 1995). The functional perspective focuses primarily on questions related to the flux of matter and energy in ecosystems. This view would contend that the process of carbon cycling is inherently interesting and important in and of itself. When seen strictly in terms of individual species assembled within an abiotic context, investigation of processes such as carbon cycling can rapidly become mired in a morass of detail. Simplification is often achieved by adopting a strongly functional or process-oriented perspective. An example from terrestrial ecology involves the concept of the rhizosphere (Lynch, 1990), which is composed of both biotic and abiotic functional components in such a way as to allow a conceptually simplified model of the nutrient cycle.

(b) The 'structural' ecosystem perspective

In contrast, the structural perspective – also often referred to as population and community ecology – focuses on questions concerning the mechanisms which control the structure of communities and ecosystems. Here, the focus is primarily on the distribution and abundance of organisms with much less emphasis placed on fluxes of matter and energy (Paine, 1994; Jones and Lawton, 1995).

The fragmentation manifest in the above perspectives must be overcome for further significant progress to occur in linking ecosystem structure and function. Ecologists now realize that a complete science of ecology must struggle with the difficult process of combining these two seemingly disparate perspectives (Schulze and Mooney, 1993; Kareiva *et al.*, 1993; Jones and Lawton, 1995). In order adequately to protect natural ecological systems from anthropogenic contaminants, ecotoxicology must also develop a much deeper understanding of how ecosystem structure and function are related. Only then will there exist a rational basis for the protection of both essential services to humans and the species composition of ecosystems.

6.1.2 MASTER PROCESSES AND THEIR RELATION TO ECOSYSTEM STRUCTURE

We employ a modelling approach below to suggest that these two ecological perspectives, the interrelated but distinct views of ecosystem structure and function, can be brought more closely together by the identification of key processes that can be linked to ecosystem structure in a simplified way. The strategy involves expanding and generalizing a fundamental insight into the nature of community structure attributed to Watt (1947) but succinctly articulated by Paine (1980) that '...pattern is generated by process'. Here, a unidirectional causal flow from process to structure is invoked. This chain of causation has been demonstrated repeatedly in experiments by population and community ecologists (Paine, 1994). Process is seen as organizing community structure with local variations generating observable pattern (Paine, 1994). Here, Paine (1994) is referring to process in the restricted sense of dynamic interactions among whole organisms. We would extend the concept to include both definitions of ecosystem process above and add that considering only a linear, one-way causal chain is incomplete. Process generates pattern but process must depend in some way on structural elements (Jones and Lawton, 1995). Integration of ecosystem structure and function will come about only when the loop is closed by more comprehensive investigation of the relationship between pattern and process.

The mutual interdependence of process and structure is clearly illustrated by a classic series of manipulative field experiments in the marine rocky intertidal environment. Paine (1966, 1969) demonstrated that the relatively diverse community structure observed along the Pacific north-west coast of America was actively maintained by the process of predation by the starfish *Pisaster ochraceus* on the otherwise competitively dominant mussel *Mytilus californianus*. Paine's work led to the general concept of a **keystone predator** – a single species whose predation on a competitive dominant has a disproportionate effect on community structure.

This keystone predator system can be generalized and viewed from a slightly different perspective than that originally taken by Paine. For example, Paine's experiments also demonstrated that a single structural component (*P. ochraceus*) could, through control of a critical intermediate process (the rate of removal of *M. californianus* from its rocky substrate), exert a high degree of feedback control on the structure of the entire community. Removal of *Pisaster* breaks a critical link in the system and removes the feedback control (Figure 6.1(a)). Here the effect of a particular species of starfish on the process of mussel removal rate strongly modulates the competitive interactions among other species in the community. This effect occurs by way of the subsequent intermediate process of space monopolization by *M. californianus*. Note that it is the process – the number of mussels removed per square metre of habitat per day – that feeds back to control community structure. The same community structure will develop whether the remover is a starfish or a rubber-booted ecologist.

We suggest that by careful choice of a governing or **master process**, it may often be possible to successfully and mechanistically couple ecosystem function to structure. The master process may in turn be coupled to one or more additional processes or directly to the process of interest – in the present case the fate of an organic contaminant in sediment. A critical point is that the interaction strengths directly linking ecosystem structure to contaminant degradation be relatively strong relative to other possible interactions (Figure 6.1(b); Paine, 1980). Further simplification might be achieved by identifying critical species groups or guilds exerting primary control over the master process. Certainly for marine sedimentary systems, and probably also for terrestrial soil systems, particulate transport due to the burrowing and feeding activities of the resident fauna (i.e. bioturbation) is such a master process. Marine soft-substrate community structure affects nitrification and denitrification rates by controlling the nature of vertical sediment mixing and burrow irrigation. The nitrogen cycle is in turn closely linked to primary production and carbon loading in sediments and thus provides an additional feedback between benthic fauna and biogeochemical processes (Giblin *et al.*, 1995). Bioturbation is thus one example of a master process that can be used as a 'structure–function couple'. Keystone predation is an additional example from the marine rocky intertidal. By generalizing Paine's original conception such that master processes are employed as critical links between structure and function, we hope to improve our ability to understand the often complex interactions between these two important perspectives on ecosystems. The purpose of the remaining discussion in this contribution is to make the above abstract ideas concrete.

We will use the above perspective to develop one example of a theoretical framework for linking pollutant effect with chemical behaviour in the field.

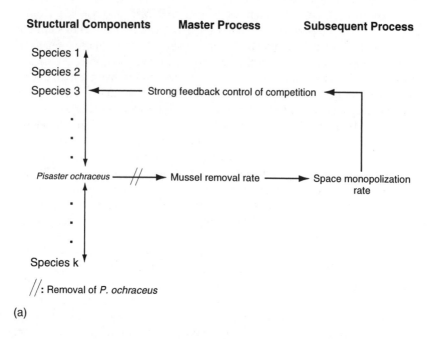

(a)

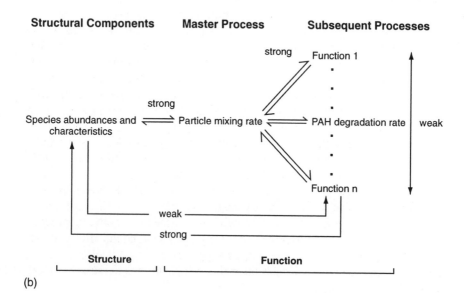

(b)

This process will involve drawing some important connections between the structure and function of aquatic soft-bottom benthic systems. Where possible we will attempt draw parallels with soil ecosystems. No attempt is made at comprehensive coverage of either the marine sedimentary or terrestrial soil environments. We simply wish to highlight promising directions for further research.

6.1.2 THE APPROACH

To provide a firm footing, we will focus on the important ecotoxicological problem area of organically contaminated sediments. We will demonstrate how knowledge of organism abundance, size and feeding rate can be used to model organism-driven particle transport rates in marine sediments. We then show that bioturbation can in turn influence contaminant fate – demonstrating the potential formation of feedback loops where fate is mediated by biological effect. Finally, we suggest ways in which this approach might be extended to terrestrial soil ecosystems. We conclude that by carefully considering variables that bridge scales and mechanistically link system structure and function, it should be possible to develop a much more predictive and quantitative understanding of contaminant fate and effect in complex sedimentary and soil ecosystems.

6.2 CLOSING THE LOOP: RELATING ECOSYSTEM STRUCTURE TO FUNCTION

6.2.1 THE INFLUENCE OF SEDIMENT AND SOIL-DWELLING ORGANISMS ON SUBSTRATE PHYSICAL AND CHEMICAL PROPERTIES

In order to link community structure and the ecosystem function defined as the degradation rate of a specific organic contaminant, we must know how benthic organisms can influence the physical and chemical properties of sediments. For reasons that will become clear below, we will focus on the effects of the macrobenthic infauna on the physical and chemical properties of sediments. Macrobenthos are those organisms living within the sediment that do not pass through a 1mm mesh sieve. This size class includes those organisms

Figure 6.1 (see facing page) Examples of feedback relationships between ecosystem structure and function. Structure is viewed as mediated by a critical or 'master' process. (a) Feedback control of marine rocky intertidal community structure by the predatory starfish, *Pisaster ochraceus*, through the process of mussel removal. Removal of *Pisaster* breaks the feedback loop. (b) Potential relationships among soft-bottom benthic community structure and function from an ecotoxicological perspective. The critical link in the structure–function couple is the process of bioturbation which is hypothesized to link structure and function through a feedback loop of strong interactions.

which are expected to most strongly influence biogenic particle transport in sediments (Wheatcroft *et al.*, 1990) and probably soils as well.

Studies of the effects of invertebrates on the properties of soils has a long history dating from at least the work of Charles Darwin (Darwin, 1881). Likewise, oceanographers and limnologists have found that the physical and chemical evolution of a deposit can be profoundly influenced by the activity of benthic organisms and the biogenic structures they produce. This means that organisms inhabiting sediments must exert some degree of control over the fate of contaminants. The fate in turn influences the bioavailability of the contaminants and thus the eventual effects on the organisms themselves (Rice and Whitlow, 1985). One of the most visually obvious physical effects of organisms is the construction of burrows which directly alter the sedimentary matrix (Rhoads and Boyer, 1982). Burrows can completely change sedimentary diffusion geometry which often results in an increase in the total anoxic–oxic boundary and changes the surface area available for diffusive exchange with the overlying water column (Aller, 1982). Organism activities change the relative dominance and distribution of important oxidation–reduction reactions and increase the overall biochemical heterogeneity of near-surface sediments. The irrigation activities (i.e. the advection of overlying water across the sediment–water interface) of populations of benthic infauna strongly influence the rate at which water and dissolved chemical species are exchanged across the sediment–water interface.

These physical and chemical alterations are necessarily tightly coupled to changes in microorganism activity and abundance. For example, the regions around burrow structures are often sites of greatly enhanced microbial activity and increased meiofaunal and microbial population densities (Hylleberg, 1975; Aller and Yingst, 1978; Kristensen *et al.*, 1985; Aller and Aller, 1986). With regard to the fate of organic pollutants, these factors will be most important for organic matter-rich hypoxic or anoxic coastal sediments where the impact from land-derived sources is greatest. The ecological role of the macrobenthos with regard to the chemical and physical properties of aquatic sediments is summarized in Figure 6.2.

The physicochemical properties of soils are likewise influenced by their epi- and endogeic inhabitants. Soil fauna influence ecosystem processes primarily through their interaction with the microbial community and alteration of soil physical properties (Parmelee, 1995). The many species of earthworms are probably of singular importance in this regard because they tend to be among the largest and deepest-dwelling of the earth movers and ingesters (Satchell, 1983; Lavelle *et al.*, 1989; Alban and Berry, 1994; Scheu and Parkinson, 1994).

Like those of their marine counterparts, the biogeochemical and ecological effects of earthworms are important and diverse. In addition to indirect effects on nutrient cycling through interactions with microbes, earthworms have been shown to have large direct effects on nitrogen cycling. Nitrogen flux through

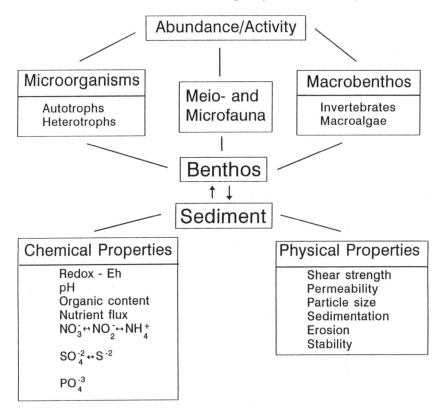

Figure 6.2 The sediment and benthos as an integrated system. (Adapted from Forbes and Forbes, 1994.)

the tissue of worms can exceed 100 kg ha^{-1} y^{-1} (Lee, 1985). Worms have also been shown to enrich heavy metals in both their faecal material (casts) and burrow linings (Protz *et al.*, 1993; Tomlin *et al.*, 1993) leading to, among other things, facilitated transfer of metals to aquifers. This is due to the predominantly vertical orientation of the burrows of many species which can extend to depths of 50 cm or more and function as preferential water flow conduits (Trojan and Linden, 1992; Edwards *et al.*, 1993; Ligthart *et al.*, 1993; Stehouwer *et al.*, 1993). Worms also effect the decomposition of organic matter in diverse ways, including incorporation of surface leaf litter directly into the soil at depth and activating both remineralization and humification of organic matter (Stout, 1983; Lavelle *et al.*, 1989; Anderson, 1995).

To achieve the goal of linking structure and function in a quantitative and predictive way, this chapter will focus on the role of biogenic particle transport in determining organic contaminant diagenesis in marine sediments. Contaminant diagenesis is the sum total of all physical, chemical and biological processes affecting the fate of a contaminant after it arrives at the sediment–water interface (Berner, 1980). The example is well suited to the goal of linking biological structure with ecosystem processes because marine sediments exhibit extremely tight biogeochemical coupling between the infaunal inhabitants and the physicochemical characteristics of sediments. This approach can be directly extended to soil systems through the process of soil and detrital feeding by terrestrial invertebrate populations.

6.2.2 PERSISTENT ORGANIC CONTAMINANTS AND THE SEDIMENTARY SINK

Most contaminants that do not rapidly degrade will eventually find their way to the sedimentary sink. Here, we will consider a polycyclic aromatic hydro-carbon (PAH; fluoranthene) as a model organic micro-contaminant. PAH are important because of their toxicity and persistence. Many PAH are known carcinogens and are now ubiquitous in both aquatic and terrestrial systems (Neff, 1985; Shiaris, 1989). A number of environmental factors are known to influence PAH degradation rates in sedimentary environments. The most important factors are probably oxygen concentration, organic matter content, temperature, nutrient concentration, the local hydrodynamic regime and the degree to which microbes are grazed by detritivores. Many of these factors are strongly influenced by benthic organisms (Figure 6.2) (Aller, 1988; Forbes and Forbes, 1994; Levinton, 1995).

Due to their extremely low solubilities, most PAH (and other persistent organic contaminants) entering the aquatic environment in solute form will be adsorbed rapidly onto particle surfaces where, being organic themselves, they associate strongly with particulate organic matter (Neff, 1979; Swartz *et al.*, 1990). In addition to sedimenting in regions of high particulate organic matter (POM) deposition, the slow degradation rates of most PAH often cause a high degree of accumulation in sediments. These characteristics lead to the frequently observed situation of low water column but high sediment and organism tissue concentrations.

6.2.3 BIOTURBATION AND THE FATE AND CYCLING OF PARTICLE-BOUND PAH

The strong particle-reactivity and intimate association with sedimentary organic matter suggest that one of the most important influences on the fate of PAH will be the displacement and mixing of sediment by benthic organisms (bioturbation). We focus on bioturbation as a master process linking benthic ecosystem structure and function by summing the effects of individuals or guilds of community members on the movement of sedimentary particulates – thus coupling the benthos and the physical and chemical properties of sediments (Guinasso and Schink, 1975; Aller and Cochran, 1976; Aller,

1982). These conceptualizations closely follow approaches taken previously by oceanographers and biogeochemists (Rice and Rhoads, 1989). The degradation of a pollutant will be assumed to be dependent on its own reaction kinetics as well as on the biological and physical mixing of the sedimentary matrix.

6.3 MODEL DEVELOPMENT

To see clearly the potential effects of infaunal benthos on PAH diagenesis, we will first examine the simplest possible case in which microbial degradation occurs in the complete absence of any bioturbation by benthic animals. This exercise highlights the need for a truly multidisciplinary approach incorporating the physical, chemical and biological aspects of the system of interest.

Once on the bottom, a particle-bound PAH may undergo physical resuspension and transport, incorporation into the deposit, redistribution due to the activities of the benthos, and chemical reactions within the sediment (Berner, 1980; Aller, 1982). This is true for any particle-bound, degradable pollutant reaching the aquatic environment. General site-specific information regarding the interactions among physicochemical and biological variables will probably be essential for the understanding of structure–function relationships in any particular ecosystem.

6.3.1 DIAGENESIS OF PAH IN THE ABSENCE OF BENTHIC ORGANISMS

After a large oil spill, a severe physical disturbance, or a period of organic enrichment and associated anoxia one often finds sediments in which there is a complete absence of benthic macrofauna. Without the presence of benthic organisms, any variability in the spatial distribution of particle-reactive contaminants is predominantly in the vertical direction. The pollutant concentration can then be fairly accurately described as a function of time and depth in the sediment by the following equation:

$$\frac{\partial C_v}{\partial t} = v \frac{\partial}{\partial x}(\omega C_v) - R \qquad (6.1)$$

where C_v (t,x) is the mass of pollutant per volume of sediment; t is time; x is depth within the sediment column (positive downward, origin at the sediment–water interface); ω is the sedimentation or burial rate in units of length per time; and R is the degradation rate of the pollutant in units of mass per time per volume of sediment (Berner, 1980).

To further simplify the situation we assume that compaction of sediment is negligible (ω is constant as a function of depth) and the system is at steady-state (i.e. the rate of change of C_v at any given depth is zero). In addition, biodegradation is modelled as a first-order rate process, independent of depth, and k is the first-order rate constant in units of time^{-1} such that $R = kC_v$. The pollutant concentration is then only a function of depth and Equation (6.1) simplifies to the ordinary differential equation:

$$0 = -\omega\frac{\mathrm{d}C_v}{\mathrm{d}x} - kC_v \tag{6.2}$$

The assumption of steady-state means that depositional inputs and transport within the sediment by such processes as bioturbation, resuspension, burial and physical mixing by currents and waves are balanced and tend to average out over time. The effect of this balance is to keep the pollutant depth profile constant over time. Rearranging Equation (6.2) slightly we have:

$$\frac{\mathrm{d}C_v}{\mathrm{d}x} = \frac{-kC_v}{\omega} \tag{6.3}$$

which says that under steady-state conditions the PAH concentration gradient is directly proportional to the decay constant (k) and inversely proportional to the sedimentation rate (ω). The solution to Equation (6.3) under the boundary condition $C_v(0) = C_{v0}$ demonstrates that the concentration of a PAH undergoing steady-state sedimentation and first order decay is an exponential function of depth:

$$C_v(x) = C_{v0}e^{\frac{-kx}{\omega}} \tag{6.4}$$

Thus, the greater the biodegradation rate constant (k), the steeper the decline in PAH with depth. The total **inventory** (I) of contaminant under a unit square of seafloor surface may be obtained by integrating Equation (6.4) from $x = 0$ to $x = \infty$. Then I (in g cm^{-2}) is equal to $C_{v0}\omega/k$. The inverse proportionality between the sedimentation and decay rates means that in eutrophic coastal environments with high sedimentation rates, PAH concentration will decrease less rapidly with depth. This will act to increase the total inventory and drive the contaminant deeper into the deposit. Though we do not do so here, the decay constant may also be modelled as a function of local environmental conditions, which might include the activities of the benthos (e.g. irrigation).

All else being equal, the exact PAH depth profile will depend on the surface concentration, C_{v0} (mass of pollutant per volume sediment). The surface concentration is in turn dependent on the local hydrodynamic, biological and chemical aspects of the ecosystem.

To incorporate the local hydrodynamic conditions into the model we must consider the local physical oceanography and the nature of the PAH depositional flux to the seafloor. Analogous physical processes in terrestrial soil systems would include local rainfall and wind conditions which would influence soil particulate sources and would in turn act to control contaminant inventories through input rate, soil hydration state and rapidity of solute transport to depth.

For marine systems the local hydrodynamic conditions can be easily incorporated as the boundary conditions of Equation (6.1). For example, if there is a constant PAH flux (J_{PAH}: mass of PAH per unit surface per time) to the sediment surface that is independent of the sedimentation rate of inorganic particulates, then the relevant boundary condition is:

$$x = 0, \ C_{v0} - \frac{J_{PAH}}{\omega} \qquad (6.5)$$

where C_{v0} is the initial volume concentration of PAH (mass per volume sediment at the interface). This boundary condition stipulates that the initial surface concentration of PAH is inversely proportional to sedimentation rate but directly proportional to PAH deposition rate. This uncoupling of contaminant deposition from sedimentation rate could occur, for example, in shallow coastal environments where there is a large amount of benthic algal production (and possibly PAH accumulation) at the sediment–water interface. In many coastal environments local oceanographic conditions can dominate the sedimentation of pollutants such that the surface concentration of pollutant is held at a constant value – by 'smearing over' any spatial heterogeneity – and Equation (6.4) applies with C_{v0} fixed by the local hydrodynamic conditions.

We can now examine the fate (i.e. steady-state depth profile) of PAH in sediments in the absence of any effects due to bioturbation by benthic animals. Surface sediments from polluted areas often exhibit concentrations of individual PAH of the order of 1–10 μg g^{-1} of dry sediment. For example, a value of 4.7 μg g^{-1} has been reported for fluoranthene contamination in Hudson/Raritan estuary surface sediment (National Oceanic and Atmospheric Administration, 1989). Note that mass of contaminant per volume of sediment (C_v) can be converted to mass per unit weight of sediment (C_m) if one knows sediment bulk density (ρ: g cm^{-3}) and porosity (φ: cm^{-3} pore space cm^{-3} solid) and uses the relation $C_m = C_v / [\rho (1-\varphi)]$.

We can then plot the expected profiles for a range of first-order degradation (k) and steady-state fluxes (J_{PAH}) by employing Equation (6.4) (Figure 6.3). Here, the sedimentation rate was held constant at 2 cm yr^{-1} (= 6.3 × 10^{-8} cm s^{-1}) and sediment bulk density (ρ) and porosity (φ) were taken to be 2.6 g cm^{-3} and 0.85 respectively. Note that the PAH inventory is positively related to J_{PAH} and inversely to k. Note also that the biodegradation rate constants used here may be notably higher than those of natural sediments. If PAH degrades at negligible rates under anaerobic conditions (Bauer and Capone, 1985), then field profiles will approach the vertical in the absence of fauna and time-dependent variations in input.

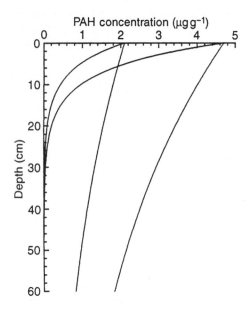

Figure 6.3 PAH concentration (μg PAH g^{-1} dry sediment) as a function of depth. Profiles are plotted for two flux values (J_{PAH}) with two values of the biodegradation rate constant ($k = 10^{-9}$, 10^{-8} s^{-1}) each. Flux values from right to left are: $J_{PAH} = 11.1 \times 10^{-8}$ and 5.0×10^{-8} cm s^{-1} and correspond to initial surface concentrations of 4.7 and 2.1 μg g^{-1} respectively.

In deeper environments the flux of contaminants to the seafloor is often directly proportional to the overall particulate sedimentation rate (ω). In such cases the depth profile of the contaminant will be dependent on both the degradation rate constant and the sedimentation rate. These dependencies are illustrated in Figure 6.4 for an initial individual PAH surface concentration of 4.7 μg g^{-1}. Holding degradation rate constant in the absence of bioturbation shows that sedimentation rate can strongly influence the overall concentration profile and inventory of contaminant (Figure 6.4(a)). Likewise, the effect of a decrease in degradation rate is qualitatively similar to that of an increase in sedimentation rate (Figure 6.4(b)). In general however, when the additional effects of bioturbation are considered, ω will be expected to play a role of importance only when sedimentation rates are exceptionally high.

The situation in which the local physical oceanographic conditions control the surface concentration of contaminant produces depth profiles that are qualitatively similar to those depicted in Figure 6.4(a) and (b). Here, decreasing degradability will act to increase penetration into the sediment in a manner analogous to that shown for increasing sedimentation rate in Figure 6.4(b). Increasing the sedimentation rate has the same qualitative effect as decreasing degradation rate (k).

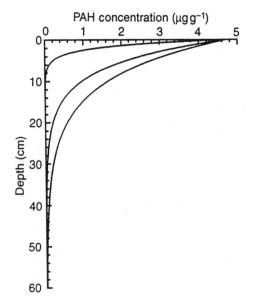

(a)

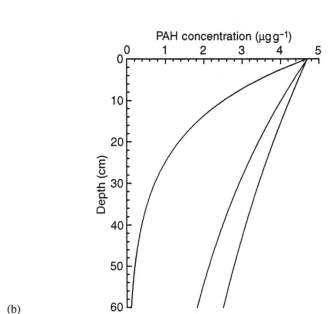

(b)

Figure 6.4 PAH concentration (μg PAH g^{-1} dry sediment) as a function of depth. (a) The three curves are model plots for sedimentation rates of 1.6, 6.3 and 9.5$\times$10^{-8} cm s^{-1}, where $k = 10^{-8}$ s^{-1}. (b) Curves are model plots for sedimentation rates of 1.6, 6.3 and 9.5$\times$10^{-8} cm s^{-1}, where $k = 10^{-9}$ s^{-1}. Both plots are for the situation in which PAH flux to the sediment surface is proportional to ω such that the initial surface concentration is 4.7 μg g^{-1}.

6.3.2 DIAGENESIS OF PAH IN THE PRESENCE OF BENTHIC ORGANISMS

The simplest effect of animals on the distribution of the particulate material of sediments is the homogeneous or non-selective mixing of particles due to the activities of burrowing, tube construction and deposit feeding. This is the only case we will consider here. Organisms are capable of causing particle movement in a manner entirely different from that caused by physical resuspension, sedimentation and compaction.

Nonetheless, this mixing process can be quantified by defining the parameter D_b to be a random mixing or **biodiffusivity** coefficient for sedimentary particulates (Berner, 1980). Recently, a 'random walk' conception of particle biodiffusivity has been developed which places D_b on a more solid mechanistic foundation and allows the possibility of linking benthic community structure and function through sediment mixing by infauna. This is accomplished by relating D_b directly to organism body length and the time a particle spends at rest (Wheatcroft et al., 1990). In most coastal and deep-sea sediments the value of this coefficient is primarily determined by biological mixing, although purely physical processes can also play an important role. For simplicity, we will not consider the contribution of physical or strongly directional biological mixing here. Under these conditions the change in PAH concentration as a function of depth and time in the presence of bioturbation will follow the relation (Berner, 1980):

$$\frac{\partial C_v}{\partial t} = \frac{\partial}{\partial x}\left[D_b\left(\frac{\partial C_v}{\partial x} \right)\right] - \omega\frac{\partial C_v}{\partial x} - kC_v \tag{6.6}$$

where D_b is a parameter expressing the contribution of biological mixing in 'diffusive' units of cm^2 yr^{-1}. We assume no compaction and that D_b is independent of depth in the sediment. Values for D_b can be determined using particle-following tracers. In field situations this is usually done by measurements of naturally occurring radionuclides (Guary et al. 1988), or additions and later quantification of glass beads or fluorescent particles (Wheatcroft, 1991, 1992).

Assuming steady-state conditions and depth and time independent values of ω, D_b and k; the solution for Equation (6.6) has the form:

$$C_v(x) = C_{v0}e^{-\lambda x} \tag{6.7}$$

where

$$\lambda = \frac{\omega}{2D_b} - \left[\frac{\omega^2}{4D_b^2} + \frac{k}{D_b} \right]^{1/2} \tag{6.8}$$

Model PAH depth profiles incorporating the effects of bioturbation are shown in Figure 6.5. Model parameters are given in the figure caption. The range of biodiffusivity values (D_bs) cover those measured over a broad spectrum of environments from coastal to deep sea (Matisoff, 1982). The effect of increasing animal activity on the depth profile is similar to that of decreasing degradability (k), that is, the PAH inventory increases. Note that when PAH flux is proportional to ω, the surface concentration will be constant if ω is constant. Then changes in bioturbation rate will cause proportional changes in pollutant inventory. Increased inventory means increased exposure. The danger of incompletely specifying the system of interest is also readily apparent, particularly if too small a scale is chosen to include all relevant processes related to the phenomena of interest. For example, comparison of Figures 6.4(a) and 6.5 reveals that nearly identical changes in PAH inventory can be brought about by changes in either sedimentation rate, degradation rate or macrofaunal activity. Only with some knowledge of both the hydrodynamic and biological context can accurate predictions of future impacts or changes in contaminant loading be made.

Thus, examination of the profiles alone cannot be used to infer which system processes are most important because identical profiles can be generated

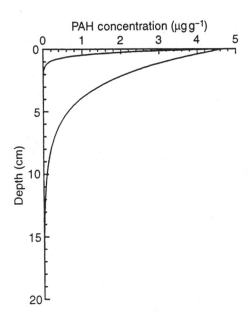

Figure 6.5 PAH concentration (μg PAH g^{-1} dry sediment) as a function of depth. Flux of PAH is proportional to sedimentation rate. Curves are for values of the biodiffusivity coefficient D_b of 6×10^{-9} cm^2 s^{-1} ($I_{PAH} = 1.4$ μg cm g^{-1}) and 600×10^{-9} cm^2 s^{-1} ($I_{PAH} = 11.8$ μg cm g^{-1}), where $k = 10^{-7}$ s^{-1} and $\omega = 10^{-8}$ cm s^{-1}.

by radically different processes. Enhanced sedimentation rates caused by increased eutrophication, degradation rate constant changes due to community structural changes and behavioural shifts involving bioturbation rate changes due to pollutant effects may all be capable of generating identical contaminant depth profiles. Emphasizing the overwhelming importance of scale effects in ecology, we conclude that **local** physical, chemical and biological aspects must all be incorporated in models seeking a predictive understanding of pollutant behaviour in ecosystems.

6.4 ECOTOXICOLOGICAL CONSEQUENCES OF BIOTURBATION

The increase or decrease in pollutant inventory as a function of increasing or decreasing organism activity has important consequences for the organisms themselves and illustrates the situation in which pollutant fate and effect can become strongly linked when whole communities or ecosystems are considered. Figure 6.5 demonstrated how infaunal organisms can influence their own exposure to contaminants. This feedback loop can be quantified by the functional relationship between pollutant inventory (and thus exposure) and bioturbation rate. Thus, a link between ecosystem structure and function can be made using knowledge of how the abundance and distribution of infaunal benthic species influences D_b. This 'biotic loop' or causal feedback relationship between the benthic fauna and contaminant is illustrated diagramatically in Figure 6.6. Approaches which assess only the effect of contaminated sediment on the biota (pathway A, Figure 6.6) are incomplete. Development of a greater predictive ability in sediments inhabited by macrobenthos will require an understanding of the functional relationships in pathways B and C, which must be coupled with specific knowledge of the abiotic context. We expect that these types of feedback relationships are a very general phenomenon. For example, even in the simplest of laboratory systems, the degradation rate of an organic toxicant will be heavily dependent on the effect of the chemical on the microbial community doing the degrading.

6.4.1 BIOTURBATION, CONTAMINANT INVENTORY AND EXPOSURE

Most infaunal organisms will be exposed to organic pollutants either from desorption into the pore water and subsequent uptake or through ingestion of contaminated sediment. In certain cases predation may also play a role. Regardless of the mechanism, contaminant inventory is an extremely important factor determining exposure. For example, if equilibrium partitioning theory holds and organisms accumulate contaminants primarily through contact with pore water, then the exposure concentration will be directly proportional to the sedimentary inventory of contaminant (Forbes and Forbes, 1994). Coupled with the fact that irrigation activity can increase sedimentary pore water content, actual contaminant exposures could either be greater (if

The 'Biotic Loop'

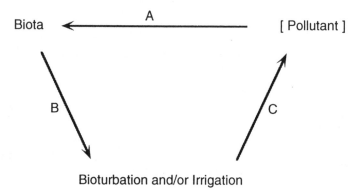

A : Traditional approaches focus here

Figure 6.6 Illustration of the 'biotic loop' or feedback coupling between the biota and contaminant concentration and distribution. A: relationship between the concentration of pollutant and its effect on the resident biota. The focus of sediment bioassays and effect assessments. B: relationship between the biota and their local effects relevant to pollutant fate. C: relationship between biotically driven environmental changes and pollutant fate.

partitioning theory holds) or less than (if irrigation flushes pore water on a time scale faster than that of equilibrium partitioning) that which would occur in the absence of significant macrofaunal activity. If exposure occurs by means of ingestion of contaminated sediment, then inventory is obviously also an important factor determining bioaccumulation, especially for deposit-feeding animals preferentially ingesting the organic fraction of sediments (Lopez *et al.*, 1989).

We can quantify the PAH inventories in the top 20 cm for the two curves in Figure 6.5 by integrating Equations (6.7) and (6.8) over the depth to which bioturbation occurs (e.g. 0–20 cm). Then for slowly reworked sediment I_{PAH} = 1.4 μg PAH cm g^{-1} and for the sediment undergoing more rapid bioturbation I_{PAH} = 11.8 μg PAH cm g^{-1}. Regardless of the degree to which pore water or ingestion serves as the route of uptake, exposure can be enhanced by more than a factor of eight through bioturbation activities alone. The relationship between PAH inventory and bioturbation can be calculated by integrating Equations (6.7) and (6.8) over a range of D_bs and is shown in Figure 6.7. Note that the greatest change in I_{PAH} per unit change in D_b is predicted to occur at the lowest reworking rates.

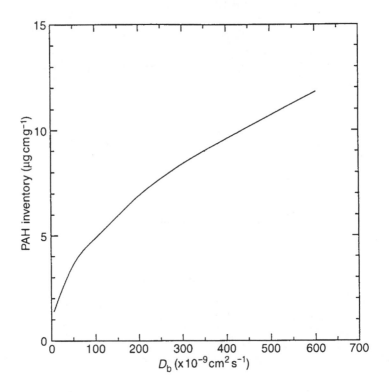

Figure 6.7 PAH inventory versus the bioturbation parameter (D_b). Flux of PAH to the seafloor is proportional to bulk particulate sedimentation rate.

6.4.2 FATE–EFFECT AND STRUCTURE–FUNCTION COUPLING

Regardless of the principal mode of exposure, the ultimate effect of any particle-reactive, biodegradable organic contaminant will be a function of hydrodynamic, chemical and biological factors. These relationships can be conceptualized by noting that exposure/effect will be some function of the contaminant inventory and bioturbation activity. From a biological perspective, D_b will be a function both of the biological system under consideration and the effect of the contaminant on that system. When viewed in terms of pollutant fate and behaviour at the ecosystem level, one of the most important effects of the contaminant will be its influence on organism activity or D_b. Thus, structure and process can be linked to the extent that the biological mixing coefficient (D_b) can be decomposed or modelled as a function of individual and population-level parameters. Recent theoretical work has succeeded in decomposing D_b into fundamental dimensions of length and time that can be related directly and mechanistically to animal activities (Wheatcroft *et al.*,

1990). This work has opened the door to direct coupling of soft-bottom benthic ecosystem structure and process.

(a) The mechanistic decomposition of D_b into parameters relevant to community structure

The random walk conception of particle biodiffusion allows sediment grains an equal probability of making a step in any direction. Particle motion is a function of the distance a particle moves in one step and the time it remains at rest between successive steps. Thus, the vertical component of biodiffusivity can be expressed as (Wheatcroft *et al.*, 1990):

$$D_b = \frac{l^2}{8R} \tag{6.9}$$

Here, l is mean particle step length (cm), which is related to organism size, and R is mean particle rest period (days). Because l is squared, D_b will be quite sensitive to step length. Most biological particle movement in sediment occurs through the process of deposit feeding and feeding activity will thus be strongly related to l (Wheatcroft *et al.*, 1990). This means D_b will be heavily influenced by animal size and/or how much movement occurs between ingestion and egestion of sediment.

Thus, the average distance a particle moves will be related to the size distribution of the community of deposit feeders. Furthermore, inspection of Equation (6.9) indicates that low densities of large deposit feeders (high l) may often have a greater positive effect on D_b than greater densities of small organisms (low R). The time a particle spends at rest will primarily be a function of deposit-feeder population density, particle size, horizontal sediment transport rate and sedimentation rate. This conception allows a mechanistic coupling between the properties of organisms and communities and particle motion and has been used to successfully predict the range of particle movement in microcosm experiments employing known densities and sizes of the polychaete, *Capitella* species 1 (T. Forbes, M. Holmer and V. Forbes, unpublished results).

We suggest that the manner in which species distributions, population densities, individual size distributions and so forth interact to control the tempo and mode of bioturbation will form the most important link between ecosystem structure and effects and behaviour of particle-reactive pollutants.

Recent microcosm experiments from our laboratory using sediments inhabited by different densities of *Arenicola marina* clearly demonstrate some of the theoretical processes and effects discussed above. Figure 6.8 shows depth profiles of ^{14}C activities (DPM g^{-1}) as a function of worm density. In this experiment, ^{14}C-fluoranthene-labelled sediment was added as a thin layer (< 0.5 cm) 3 weeks prior to the profile determinations shown here in

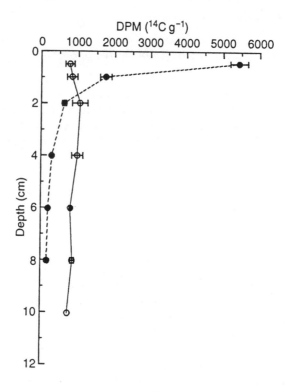

Figure 6.8 ^{14}C activity (DPM g[-1]) versus depth as a function of the number of *Arenicola marina* per core after a 4-week incubation period (±1 s.e.). [14]C originally added as ^{14}C-labelled fluoranthene bound to sediment and placed as a thin layer at the sediment surface. See text for experimental details.

concentrations far too low to have had an effect on the animals (<<1 pg g[-1]). Note that even though the assumptions differ from those of the model curves described above (e.g. $\omega = 0$), one can readily see the qualitative similarity to the model profiles in terms of the effect of the worms. The worms have greatly lowered the surface concentration and driven the [14]C (fluoranthene plus metabolites) deeper into the sediment. These are important effects both in terms of changes in pollutant distribution within the sediment and potential for exposure of the infauna. Figure 6.9 demonstrates that *Arenicola marina* density and presumably bioturbation and irrigation activity can also have a strong effect on the ultimate biodegradation of ^{14}C-labelled fluoranthene. Here, ultimate biodegradation was measured as cumulative $^{14}CO_2$ production in microcosms with different worm densities. Fluoranthene half-life increased as a function of increasing worm density. This occurred because the worms transported fluoranthene and its degradation products deeper into anoxic regions of the sediment and away from more rapid aerobic degradation at the sediment–water interface.

From the above discussion it is clear that natural communities or ecosystems exhibiting relatively high bioturbation rates can function as what Aller (1982) has termed **geochemical hotspots**. This has important implications for particle-reactive pollutant fate. One prediction is that organic contaminant inventories will increase (by an amount dependent on the local physical conditions) as a function of bioturbation rate (Figure 6.7). All other things being equal, systems with high bioturbation rates will exhibit a tendency to scavenge or accumulate pollutants. However, the actual fate of the pollutant will be dependent on the interactions among the key parameters discussed above. Among the most important parameters controlling pollutant fate will be biotic processes such as D_b, irrigation rate and biodegradation rate (including the effect of the contaminant on these processes), as well as abiotic processes such as sedimentation rate and the local hydrodynamic regime. We predict that the

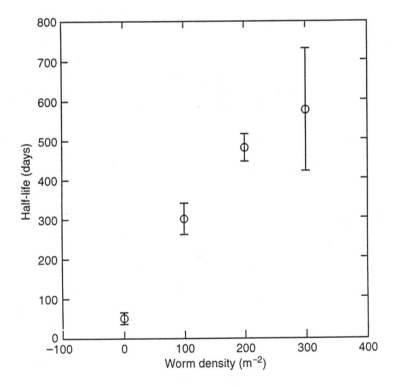

Figure 6.9 Fluoranthene half-life versus the number of *Arenicola marina* per core measured over a 4-week incubation period (±1 s.e.). Values calculated from cumulative CO_2 release as a percent of the activity added on day 0. See text for details.

natural outcome of this process will be a spatial and temporal mosaic of pollutant exposure and loading which is strongly correlated with local reworking rates. These rates in turn will be a function of ecosystem structure and dynamics.

The final link in the pollutant fate and effect dynamic comes about if the sediment and water-processing activities of the resident organisms are influenced by the exposure levels they have created. This is illustrated as a strong function to structure feedback in Figure 6.1(b). Thus in addition to knowledge of the local hydrodynamic regime, ecotoxicological studies of marine sedimentary systems will need to couple local biotic data with the system processes of interest (e.g. contaminant biodegradation). This theoretical development led Forbes and Forbes (1994) to propose a central role for biological mixing as a master process in coupling benthic community structure to the important ecosystem functions such as contaminant biodegradation and behaviour (Figure 6.10).

6.4.3 GUILDS OF BIOTURBATING SPECIES

We suggest that a promising first step in relating ecosystem structure to function would be to take advantage of any 'functional redundancy' with regard to suspected master processes to construct guilds of species with similar effects. This will require knowledge of the natural history and function of the system of interest and necessitates some degree of site specificity. For example, one might begin by grouping benthic species by the manner in which they bioturbate or irrigate sediments. In marine systems a rather clear distinction can be made between organisms that tend to be large, inhabit the deeper sediment zones, and move large amounts of particulate matter in an advective and very

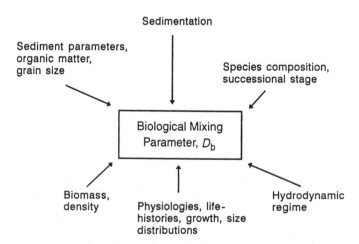

Figure 6.10 The central role of bioturbation in coupling the fate and effects of biodegradable contaminants in marine sediments. (Adapted from Forbes and Forbes, 1994.)

directional manner as opposed to those smaller organisms which live near the sediment–water interface and move smaller amounts of material in a more diffusive manner. Both theory (Wheatcroft, 1990) and experiment (Rice, 1986) indicate that large deposit-feeding animals (e.g. *Arenicola marina*) will in nearly all cases dominate biogenous particle mixing. Thus, one can construct guilds in terms of their effects on the tempo and mode of biological particle mixing. This approach has several important ecotoxicological consequences. It suggests that important guild members are the species that should be targeted in laboratory toxicity testing. It may be more important to determine pollutant effects on these species than to simply search for the 'most sensitive test species', especially since species which are generally sensitive to a wide range of toxicants may not exist (Forbes and Depledge, 1992).

6.5 CONCLUSIONS

We conclude that a productive way to link structure to function may be through the use of guilds of species which control master processes. For particle-reactive pollutants this will mean organisms that move particles around and transport water across the sediment–water interface. It may be productive to classify guilds by the 'tempo and mode' of their bioturbation and irrigation activities. In these cases it is possible to directly link structure to function by measuring the relationships between species functional type (advective–deep versus diffusive–shallow bioturbators), population density, body size and their relation to bioturbation rates. This requires a degree of site-specific knowledge or categorization in terms of the hydrodynamic regime and individual species in the system(s) of interest. One needs to know the system to identify the key processes. Improved prediction requires more site specificity than is presently practised.

In terms of the design of laboratory test systems, this means that it is critically important to understand the effects of contaminants on important functional groups of species. Which species are most important may depend on which process is being considered. Nevertheless, some generalization is possible. For particle-reactive pollutants, the species influencing D_b are the species of interest. In terms of subsequent effects due to fate/effect feedbacks, it may be more important to protect *Arenicola marina* than *Corophium volutator*. We should not just be searching only for the most sensitive species in terms of test system development.

It is easy to see why purely structural measurements fail to tell the whole story. Critical system variables are simply left out. Extrapolation from simple laboratory tests (which at best can make predictions about ecosystem structure) must often fail. The structural view is valid, but incomplete. In nature, the interactions within the biotic components and between the biotic and abiotic components mean that it will not be possible to make accurate predictions of the fate and effects of contaminants in the field based solely on laboratory tests of single species or suites of species – no matter what the end-point. However, the good news is that **if** it can be shown that laboratory testing can protect the structure of ecological systems and **if** protection of structure generally means protection of process or function, then it may be possible to use suites of carefully chosen laboratory test systems for the protection of communities or assemblages. Accurate assessment of whether or not systems are actually being protected using any extrapolation scheme will first require development of solid general principles incorporating the important abiotic parameters of the ecosystem of concern. A multidisciplinary approach is essential. The proper size or spatiotemporal scale of the ecosystem of interest will follow naturally from the questions asked.

6.6 SUMMARY

Improved prediction of the behaviour and effects of contaminants in the environment will require increased emphasis on strongly integrated interdisciplinary studies. An important goal of these studies must be to develop a more complete understanding of the relationship between ecosystem structure and function. Only by developing a mechanistic understanding of important ecosystem structural and functional interactions will it be possible to improve our ability to predict from necessarily smaller-scale investigations to the larger scales that are of interest to environmental managers.

As a step in this direction we define a **master process** as a process that is determined to be of over-riding importance in the coupling of system structure and function. We then use this concept to extend the ecological insight that 'process generates pattern' to include the feedback relationships that must be considered in any attempt to link ecosystem structure and function. As an illustration, we develop an explicit theoretical framework for linking key aspects of the structure and function of marine sedimentary ecosystems containing organic contaminants and undergoing biological mixing.

We conclude that successful linkages between structure and function will of necessity incorporate a great degree of site specificity. One needs to know the system to identify the key processes. Extrapolation from simple laboratory tests on individual species can at best allow predictions about effects on ecosystem structure. Purely structural measurements will fail to allow accurate prediction of contaminant effects in nature because they provide an incomplete picture of the system. In nature, strong interactions within the biotic components of an ecosystem and between the biotic and abiotic components necessitate a comprehensive multidisciplinary approach.

ACKNOWLEDGEMENTS

We thank Valery Forbes and Marianne Holmer for generously supplying unpublished data. Critical reviews by Valery Forbes and two anonymous reviewers greatly improved earlier drafts of this manuscript. Special thanks to Niels Bohse Hendriksen and Andreas Petersen for an introduction to the vast literature on the ecology and geochemical effects of earthworms. This work was supported by research grants from the Danish National Science Council (TLF, grant nos. SNF 11-0289-1 and SNF 11-1047-1) and PhD stipends from the Danish Research Academy and the National Environmental Research Institute (LKK).

REFERENCES

Alban, D.H. and Berry, E.C. (1994) Effects of earthworm invasion on morphology, carbon, and nitrogen of a forest soil. *Appl. Soil Ecol.*, **1**, 243–9.

Allen, T.F.H. and Hoekstra, T.W. (1993) *Toward a Unified Ecology*, Columbia University Press, New York.

Aller, R.C. (1982) The effects of macrobenthos in the chemical properties of marine sediment and overlying water, in *Animal–Sediment Relations: The Biogenic Alteration of Sediments* (eds P.L. McCall and M.J.S. Tevesz), 2nd vol., *Topics in Geobiology*, Plenum Press, New York, pp. 53–102.

Aller, R.C. (1988) Benthic fauna and biogeochemical processes in marine sediments: The role of burrow structures, in *Nitrogen Cycling in Coastal Marine Environments* (eds T.H. Blackburn and J. Sørensen), John Wiley and Sons Ltd, Chichester, pp. 301–38.

Aller, J.Y. and Aller, R.C. (1986) Evidence for localized enhancement of biological activity associated with tube and burrow structures in deep-sea sediments at the HEBBLE site, Western North Atlantic. *Deep Sea Res.* **33**, 755–90.

Aller, R.C. and Cochran, J.K. (1976) $^{234}Th/^{238}U$ disequilibrium in the near-shore sediment: particle reworking and diagenetic time scales. *Earth Planet. Sci. Lett.*, **29**, 37–50.

Aller, R.C. and Yingst, J.Y. (1978) Biogeochemistry of tube dwellings: a study of the sedentary polychaete *Amphitrite ornata* (Leidy). *J. Mar. Res.*, **36**, 201–54.

Anderson, J.M. (1995) Soil organisms as engineers: Microsite modulation of macroscale processes, in *Linking Species and Ecosystems* (eds C.G. Jones and J. H. Lawton), Chapman & Hall, London, pp. 94–106.

Bauer, J. and Capone, D.J. (1985) Effects of four aromatic pollutants on microbial glucose metabolism and thymidine incorporation in marine sediments. *App. Env. Microbiol.*, **49**, 828–35.

Berner, R.A. (1980) *Early Diagenesis: A Theoretical Approach*, Princeton University Press, Princeton.

Darwin, C. (1881) *The Formation of Vegetable Mould through the Action of Worms, with Observations on their Habits*. J. Murray, London.

Edwards, W.M., Shipitalo, M.J., Owens, L.B. and Dick, W.A. (1993) Factors affecting preferential flow of water and atrazine through earthworm burrows under continuous no-till corn. *J. Environ. Qual.*, **22**, 453–7.

Forbes, T.L. and Forbes, V.E. (1993) A critique of the use of distribution-based extrapolation models in ecotoxicology. *Funct. Ecol.*, **7**, 249–54.

Forbes, V.E. and Depledge, M.H. (1992) Predicting population response to pollutants: the significance of sex. *Funct. Ecol.*, **6**, 376–81.

Forbes, V.E. and Forbes, T.L. (1994) *Ecotoxicology in Theory and Practice*, Chapman & Hall, London.

Giblin, A.E., Foreman, K.H. and Banta, G.T. (1995) Biogeochemical processes and benthic community structure: Which follows which? in *Linking Species and Ecosystems* (eds C.G. Jones and J. H. Lawton), Chapman & Hall, London, pp. 37–44.

Guary, J.C., Guegueniat, P. and Pentreath, R.J. (eds) (1988) *Radionuclides: A Tool for Oceanography*, Elsevier Applied Science, London.

Guinasso, N.L., Jr and Schink, D.R. (1975) Quantitative estimates of biological mixing rates in abyssal sediments. *J. Geophys. Res.*, **80**, 3032–43.

Hylleberg, J. (1975) Selective feeding by *Abarenicola pacifica* with notes on *Abarenicola vagabunda* and the concept of gardening in lugworms. *Ophelia*, **14**, 113–37.

Jones, C.G. and Lawton, J.H. (eds) (1995) *Linking Species and Ecosystems*, Chapman & Hall, London.

Kareiva, P.M., Kingsolver J.G. and Huey, R.B. (eds) (1993) *Biotic Interactions and Global Change*, Sinauer Associates, Sunderland.

Kristensen, E., Jensen, M.H. and Andersen, T.K. (1985) The impact of polychaete (*Nereis virens* Sars) burrows on nitrification and nitrate reduction in estuarine sediments. *J. Exp. Mar. Biol. Ecol.*, **85**, 75–91.

Lavelle, P., Barois, I., Martin, A., Zaidi, Z. and Schaefer, R. (1989) Management of earthworm populations in agro-ecosystems: a possible way to maintain soil quality? in *Ecology of Arable Land* (eds M. Clarholm and L. Bergström), Kluwer Academic Publishers, Dordrecht, The Netherlands, pp. 109–22.

Lee, K.E. (1985) *Earthworms: Their Ecology and Relationships with Soils and Land Use*, Academic Press, London.

Levinton, J.S. (1995) Bioturbators as ecosystem engineers: control of the sediment fabric, inter-individual interactions, and material fluxes, in *Linking Species and Ecosystems* (eds C.G. Jones and J. H. Lawton), Chapman & Hall, London, pp. 29–36.

Ligthart, T.N., Peek, G.J.W.C. and Taber, E.J. (1993) A method for the three dimensional mapping of earthworm burrow systems. *Geoderma*, **57**, 129–41.

Lopez, G., Taghon, G. and Levinton, J. (eds) (1989) *Ecology of Marine Deposit Feeders*, Springer-Verlag, New York.

Lubchenco, J. (1995) The relevance of ecology: the societal context and disciplinary implications of linkages across levels of ecological organization, in *Linking Species and Ecosystems* (eds C.G. Jones and J. H. Lawton), Chapman & Hall, London, pp. 297–305.

Lynch, J.M. (ed.) (1990) *The Rhizosphere*, J. Wiley & Sons Ltd, Chichester.

Matisoff, G. (1982) Mathematical models of bioturbation, in *Animal–Sediment Relations: The Biogenic Alteration of Sediments* (eds P.L. McCall and M.J.S. Tevesz), 2nd vol., *Topics in Geobiology*, Plenum Press, New York, pp. 289–330.

National Oceanic and Atmospheric Administration (1989) An evaluation of candidate measures of biological effects for the National Status and Trends Program. *NOAA Technical Memorandum* NOS OMA 45 NOAA, Seattle, WA.

Neff, J.M. (1979) *Polycyclic Aromatic Hydrocarbons in the Aquatic Environment: Sources, Fates and Biological Effects.* Applied Science Publishers, London.

Neff, J.M. (1985) Polycyclic aromatic hydrocarbons, in *Fundamentals of Aquatic Toxicology* (eds G.M. Rand and S.R. Petrocelli), Hemisphere, New York, pp. 416–54.

O'Neill, R.V., DeAngelis, D.L., Waide, J.B. and Allen, T.F.H. (1986) *A Hierarchical Concept of Ecosystems*, Princeton University Press, Princeton, New Jersey.

Paine, R.T. (1966) Food-web complexity and species diversity. *Am. Nat.*, **100**, 65–75.

Paine, R.T. (1969) A note on trophic complexity and community stability. *Am. Nat.* **103**, 91–3.

Paine, R.T. (1980) Food-webs: linkage, interaction strength and community infrastructure. The third Tansley lecture. *J. Anim. Ecol.*, **49**, 667–85.

Paine, R.T. (1994) *Marine Rocky Shores and Community Ecology: An Experimentalist's Perspective*, Ecology Institute, Oldendorf/Luhe.

Parmelee, R.W. (1995) Soil fauna: Linking different levels of the ecological hierarchy, in *Linking Species and Ecosystems* (eds C.G. Jones and J. H. Lawton), Chapman & Hall, London, pp. 107–16.

Protz, R., Teesdale, W.J., Maxwell, J.A., Campbell, J.L. and Duke, C. (1993) Earthworm transport of heavy metals from sewage sludge: a micro-PIXE application in soil science. *Nuclear Instruments and Methods in Physics Research*, **B77**, 509–16.

Rhoads, D.C. and Boyer, L.F. (1982) The effects of marine benthos in the physical properties of sediments: a successional perspective, in *Animal–Sediment Relations: The Biogenic Alteration of Sediments* (eds P.L. McCall and M.J.S. Tevesz), 2nd vol., *Topics in Geobiology*, Plenum Press, New York, pp. 3–52.

Rice, D.L. (1986) Early diagenesis in bioadvective sediments: relationships between the diagenesis of beryllium-7, sediment reworking rates, and the abundance of conveyor-belt deposit-feeders. *J. Mar. Res.*, **44**, 149–84.

Rice, D.L. and Rhoads, D.C. (1989) Early diagenesis of organic matter and the nutritional value of sediment, in *Ecology of Marine Deposit Feeders* (eds G. Lopez, G. Taghon and J. Levinton), Springer-Verlag, New York, pp. 59–97.

Rice, D.L. and Whitlow, S.I. (1985) Diagenesis of transition metals in bioadvective sediments, in *Heavy Metals in the Environment*, 2nd vol., CEP Consultants, Edinburgh, pp. 353–5.

Satchell, J.E. (ed.) (1983) *Earthworm Ecology: From Darwin to Vermiculture*, Chapman & Hall, London.

Scheu, S. and Parkinson, D. (1994) Effects of earthworms on nutrient dynamics, carbon turnover and microorganisms in soils from cool temperate forests of the Canadian Rocky Mountains – laboratory studies. *Appl. Soil Ecol.*, **1**, 113–25.

Schindler, D.W. (1995) Linking species and communities to ecosystem management: a perspective from the experimental lakes experience, in *Linking Species and Ecosystems* (eds C.G. Jones and J.H. Lawton), Chapman & Hall, London, pp. 313–25.

Schulze, E.-D. and Mooney, H.A. (eds) (1993) *Biodiversity and Ecosystem Function*, Springer-Verlag, Berlin.

Shiaris, M.P. (1989) Seasonal transformation of naphthalene, phenanthrene, and benzo(a)pyrene in surficial estuarine sediments. *Appl. Environ. Microbiol.*, **55**, 1391–9.

Stehouwer, R.C., Dick, W.A. and Traina, S.J. (1993) Characteristics of earthworm burrow lining affecting atrazine sorption. *J. Environ. Qual.*, **22**, 181–5.

Stout, J.D. (1983) Organic matter turnover by earthworms, in *Earthworm Ecology: From Darwin to Vermiculture*, Chapman & Hall, London, pp. 35–48.

Swartz, R.C., Schults, D.W., DeWitt, T.H., Ditsworth, G.R. and Lamberson, J.O. (1990) Toxicity of fluoranthene in sediment to marine amphipods: a test of the equilibrium partitioning approach to sediment quality criteria. *Environ. Toxicol. Chem.*, **9**, 1071–80.

Tomlin, A.D., Protz, R., Martin, R.R., McCabe, D.C. and Lagace, R.J. (1993) Relationships among organic matter content, heavy metal concentrations, earthworm activity, and soil microfabric on a sewage sludge disposal site. *Geoderma*, **57**, 89–103.

Trojan, M.D. and Linden, D.R. (1992) Microrelief and rainfall effects on water and solute movement in earthworm burrows. *Soil Sci. Soc. Am. J.*, **56**, 727–33.

Watt, A.S. (1947) Pattern and process in the plant community. *J. Ecol.*, **35**, 1–22.

Wheatcroft, R.A. (1991) Conservative tracer study of horizontal sediment mixing rates in a bathyal basin, California borderland. *J. Mar. Res.*, **49**, 565–88.

Wheatcroft, R.A. (1992) Experimental tests for particle-size dependent bioturbation in the deep ocean. *Limnol. Oceanogr.*, **37**, 90–104.

Wheatcroft, R.A., Jumars, P.A., Smith, C.R. and Nowell, A.R.M. (1990) A mechanistic view of the particulate biodiffusion coefficient: step lengths, rest periods and transport directions. *J. Mar. Res.*, **48**, 177–207.

7   A food-web approach to assess the effects of disturbance on ecosystem structure, function and stability

JOHN C. MOORE AND PETER C. DE RUITER

7.1 DISTURBANCE AND STABILITY

Numerous field and laboratory studies have demonstrated that disturbances that alter the densities and physiologies of organisms or change the species make-up of communities affect ecosystems in ways that would be characterized as 'de-stabilizing.' Modelling studies have also focused on stability, but in a well-defined mathematical sense. Conspicuously absent from the scientific dialogue has been an operational definition of stability that has both empirical and mathematical meaning. Disturbance is central to the concept of stability. Research on the effects of disturbance on ecosystems has either focused on how the disturbance alters aspects of ecosystem structure (e.g. species richness and diversity, food-web connectance, and trophic structure) or ecosystem processes (e.g. nitrogen dynamics and decomposition). Remarkably, few studies have attempted to integrate structure and processes.

This lack of congruence between theory and observationally based definitions of stability, and the recognition that both ecosystem structure and processes are affected by disturbance, has influenced the field of ecotoxicology. Soil processes, soil community structure, and soil community functions are three aspects of soils viewed as end-points in risk assessment and biomonitoring efforts (Eijsackers and Løkke, 1992). Many tests have been recommended to study each aspect independent of the other. For example, primary production, the decomposition of organic matter and the mineralization of nitrogen are key ecosystem processes that are often monitored. Structural aspects of communities that are typically monitored are species richness and evenness of selected taxa (e.g. nematodes, *sensu* Bongers,

Ecological Risk Assessment of Contaminants in Soil. Edited by Nico M. van Straalen and Hans Løkke. Published in 1997 by Chapman & Hall, London. ISBN 0 412 75900 4

1990), while functional aspects of communities might include measurements related to important mutualisms (e.g. *Rhizobium*, *Frankia* or mycorrhizae). Integrating processes, community structure, and community function is critical to ecotoxicological studies.

There are two objectives to this chapter. The first is to link two modelling approaches that are used in food-web research (Moore *et al.*, 1996). Focusing on food-webs provides a means of integrating structure and processes. Food-webs are descriptions of the trophic interactions among organisms within a community and, as such, they depict both structural and functional aspects of communities. The second objective is to take the linked models and demonstrate that there is an interconnectedness among species that is often overlooked by theoreticians, experimentalists and environmental managers. The implications for ecotoxicology are clear, as a key subtheme to the chapter is the effect of disturbance on individual species, processes within communities and whole communities. The chapter will begin by contrasting how ecologists and the regulatory process use models. Next, a historical perspective on the development and use of models is provided by asking why energetics has been ignored by many ecologists. Basing food-web descriptions and models on ecological energetics is critical to the integration of structure and function. Finally, the chapter proposes a hypothesis that attempts to explain how disparate disturbances may operate to disrupt ecosystem stability through a common mechanism.

7.2 MODELLING DYNAMICS AND ENERGETICS

7.2.1 WHY USE MODELS?

A model of an ecosystem is a caricature or incomplete representation of nature. Models are judged by their accuracy in reproducing results and by their explanatory and predictive abilities. Ecologists use models as tools to integrate and analyse data, and as guides to future research. For example, the simple multi-species models used by May (1973) and Pimm (1982) were adaptations of the familiar Lotka–Volterra models. These models were used to explore general patterns in the ways that species interact, and the constraints imposed on the stability of an ecosystem by the interactions. An alternative to the models that describe population dynamics are the process oriented ecosystem models. Process-oriented ecosystem models (Lindeman, 1942; Odum, 1957; Hunt *et al.*, 1987) are used to explore energy flow and material cycles within and among ecosystems. In short, to ecologists models are a useful part of the **scientific process**.

Those involved in the **regulatory process** use models as risk assessment tools, i.e. they study the impacts of toxicants on organisms and on communities. Many of the same principles used by ecologists are applied when developing risk assessment models and many of the same models might be used by ecologists and regulators. The more obvious triumphs of modelling in risk assessment can be seen in the ubiquitous use of models in the routine testing of toxicants on organisms, the development of standards, the evaluations of chemicals, and in ecoepidemiological studies.

While the scientific and regulatory processes have much in common, they naturally conflict with one another. Many systems models are remarkably accurate in predicting the rates of carbon and/or nitrogen mineralization (Hunt *et al.*, 1987; De Ruiter *et al.*, 1994). Yet few systems ecologists would recommend that these estimates of mineralization rates be used as a trigger in a decision tree of a regulatory process. On the other hand, regulatory authorities cannot accept what they perceive as the vagaries of most systems models. An implicit expectation of the regulatory process is that the models will provide accurate estimates of a particular variable in order to meet the requirements that are mandated by policy makers.

7.2.2 AN AVERSION TO ENERGETICS

A schism between community and ecosystem ecology developed in the 1950s that persists today. Community ecologists study the factors that influence the abundance and distribution of species. For example, the effect of clear-cutting a forest would be assessed in terms of how the clear-cut affected the distribution of forest species. The succession that follows would be studied in terms of what plant species returned and the timing of their return. Ecosystem ecologists more typically focus on processes that result from the interplay of species and the abiotic component of the environment. The same clear-cut would be studied from the standpoint of how the decomposition of organic material and the cycling and retention of plant limiting nutrients were affected.

Food-web research has also been affected by the community and ecosystem perspectives, in that either the interactions among species and the effects that these interactions had on the abundances of species were studied, or, the energetics of species interactions were studied. DeAngelis (1992) traced the history of the different perspectives and found that both originated at about the same time and have shared common players (see Lotka, 1956; Hutchinson, 1959). The reason that all but a few ecologists (Hunt *et al.*, 1987; DeAngelis, 1992; Moore *et al.*, 1993; De Ruiter *et al.*, 1994) have not applied energetics to their studies of species interactions and dynamics can be traced back to the classic work of Paine (1966, 1980) and the modelling approaches adopted by Pimm and Lawton (1977) and Cohen (1978).

Paine (1966) demonstrated that the effects that species have on other species within a community was not a function of the amount of energy that the organisms utilized. In other words, the per capita effects of species and the impact that they have on the structure of communities do not necessarily correlate to energy flow (Paine, 1980). Taken to an extreme, calculating energy flow among organisms would not yield any information about system stability, since stability is gauged by per capita effects, and per capita effects are not related to energy flow. This reasoning has led some to conclude that efforts to link energetics with dynamics hold little promise or should be abandoned (Paine, 1988; Polis, 1994).

Pimm and Lawton (1977) and Cohen (1978) demonstrated the importance of the architecture of species interactions (length of food-chains, complexity, types of interactions) to the local stability of a community. Local stability assumes that all species within a community possess a stable equilibrium. When pushed a small distance away from equilibrium (locally) the species return to their original equilibrium values. Their work revealed regularities in topological features such as the average length of food-chains, the incidence of omnivory, and the ratio of predators to prey. Patterns that emerged from each approach were explained in terms of stability. Among their conclusions were the statements that food-chains were typically of length 3–4 due to dynamic constraints, that food-chains do not form trophic loops and that omnivores should be rare. The conclusion that dynamics limited food-chain length is important to this chapter, and we will return to it. The long-held tenet that food-chain length was limited by the amount of available energy at the last trophic level (Hutchinson, 1959) was discounted. Comparisons of the average food-chain lengths from descriptions of food-webs from ecosystems that differed in productivity by several orders of magnitude revealed that more productive ecosystems had similar average food-chain lengths as lesser productive ecosystems (Pimm, 1982; Briand and Cohen, 1987).

7.2.2 MODELS OF POPULATION DYNAMICS BASED ON ENERGETICS

The approach discussed above focused on structural aspects of communities and contrasts sharply with one based on energetic properties of communities. Trophic interactions within food-webs are defined with measurable parameters that are rooted in ecological energetics (O'Neill, 1969; DeAngelis, 1975; De Ruiter *et al.*, 1994; Moore *et al.*, 1996). Rather than describe the dynamics of populations in terms of the changes in the numbers of individuals over time (e.g. no. bacterial cells g^{-1} dry soil yr^{-1}), the approach based on energetics describes the dynamics of populations in terms of changes in the amount of matter or energy over time (e.g. g bacterial C g^{-1} dry soil yr^{-1}). Two types of models are developed using the energetics approach. The first model was designed to study the effects of minor perturbations on the local stability of the system and system resilience (May, 1973; Moore *et al.*, 1993). Food-chains based on primary producers are modelled after Lotka–Volterra:

$$\frac{dX_i}{dt} = r_i X_i - \sum_{j=1}^{n} c_{ij} X_i X_j \qquad (7.1)$$

where X_i and X_j represent the population densities of primary producers and herbivores respectively, r_i is the specific growth rate of the herbivore (birth minus death unrelated to herbivory), and c_{ij} the coefficient of consumption of the herbivores on the primary producers. Detritus can be modelled after DeAngelis *et al.* (1989) and Moore *et al.* (1993):

$$\frac{dX_d}{dt} = R_d + \sum_{i=1}^{n}\sum_{j=1}^{n}(1-a_i)c_{ji}X_jX_i + \sum_{i=1}^{n}d_iX_i - \sum_{j=1}^{n}c_{dj}X_dX_j \qquad (7.2)$$

Here, R_d is the constant input from an allochthonous source, e.g. terrestrial inputs into streams, leaf litter fall on to soils, carbon-rich exudates from plant roots to soil. In addition to the allochthonous inputs, detritus cycles autochthonously, in the form of the metabolic wastes of consumers and the unassimilated fractions of prey killed, i.e. $\Sigma\ (1-a_j)c_{ji}X_jX_i$, and the corpses of consumers that die from causes other than predation, $\Sigma\ d_iX_i$.

The equations for consumers (microbes, herbivores, microbivores, and predators), X_i, are similar for both primary producer and detritus food-chains:

$$\frac{dX_i}{dt} = -d_iX_i + \sum_{i=1}^{n}a_ip_ic_{ji}X_jX_i - \sum_{j=1}^{n}c_{ij}X_iX_j \qquad (7.3)$$

In the absence of prey, the consumer dies at a constant rate, d_i. A first approximation of the specific death rate d_i would be the inverse of the life span of the consumer. The consumer increases as a function of the amount of prey consumed, its assimilation efficiency and its production efficiency, $\Sigma\ a_ip_ic_{ji}X_jX_i$.

The second type of model calculates feeding rates among organisms and rates of nitrogen or carbon mineralization for each species. Feeding rates are calculated after Hunt *et al.* (1987) using the notation of De Ruiter *et al.* (1994):

$$F_j = \frac{d_jB_j + P_j}{a_jP_j} \qquad (7.4)$$

where F_j is the feeding rate of species j, P_j is the death rate due to predation, and B_j is the average population size of species j obtained from field data.

For predators that feed on more than one prey type the preference for the prey and the densities of the prey type are considered:

$$F_{ij} = \frac{w_{ij}B_i}{\sum_{k=1}^{n}w_{kj}B_k} \qquad (7.5)$$

where w_{ij} is the preference weighting factor of predator j for prey i. Calculations of feeding rates begin with the top predator(s) since only non-predatory death accounts for losses, and proceed backwards through the food-web to the basal resources.

The two types of models can be linked by assuming that at steady state the predation rates described in Equations (7.1–7.3), $c_{ij}X_i^* X_j^*$ (the * indicates that the variable is at steady-state), are equal to the feeding rates in Equations (7.4–7.5) (De Ruiter *et al.*, 1994). Next, assume that the field population averages approximate the theoretical steady-state population densities, i.e. $B_i \approx X_i^*$. By substituting the field population averages, B_i, for the theoretical steady state population densities, X_i^*, the consumption coefficient c_{ij} is estimated as:

$$c_{ij} = \frac{F_{ij}}{B_i B_j} \qquad (7.6)$$

7.2.3 ENERGETICS AND DYNAMICS ARE INEXTRICABLY INTER-RELATED

Equations (7.1–7.3) are energetic analogues to the Lotka–Volterra based equations used by Pimm and Lawton (1977). Moore *et al.* (1993) repeated the exercise conducted by Pimm and Lawton (1977) using Equations (7.1–7.3) and demonstrated that the feasibility and dynamics of simple food-chains were a function of productivity. A food-chain was feasible if all species could maintain a positive equilibrium at the level of production provided. The dynamics of the food-chains were gauged by their return-times (time required to return to equilibrium following a disturbance). Longer food-chains were less likely to be feasible than shorter chains at low levels of productivity. This result was consistent with the earlier assertions of Hutchinson (1959) that as prey are consumed less potential energy resides at the higher trophic level than at the lower levels. The return-times of food-chains decreased with increased productivity and each chain possessed return-times with a unique limit. While consistent with Pimm and Lawton (1977) at lower levels of productivity, the results presented a different scenario at higher levels – longer food-chains possessed shorter return times.

The key to the results of Moore *et al.* (1993) can be demonstrated with a simple two-species food-chain and using it to work through a local stability analysis. The chain might represent plants and root-feeding nematodes. Using the parameters presented in Equations (7.1–7.3):

$$\frac{dX_1}{dt} = r_1 X_1 - c_{11}X_1^2 - c_{12}X_1 X_2 \qquad (7.7)$$

describes the dynamics of the plant population, X_1, and

$$\frac{dX_2}{dt} = -d_2 X_2 + a_2 p_2 c_{12} X_1 X_2 \qquad (7.8)$$

depicts the dynamics of the nematode population, X_2. The steady-state solutions to dX_1/dt and dX_2/dt are obtained by setting each equal to zero and solving for $X_1{}^*$ and $X_2{}^*$. As above, the * indicates that the variable is at steady-state. Hence:

$$X_1{}^* = \frac{d_2}{a_2 p_2 c_{12}} \tag{7.9}$$

$$X_2{}^* = \frac{r_1 a_2 p_2 c_{12} - c_{11} d_2}{a_2 p_2 c_{12}^2} \tag{7.10}$$

Equations (7.9) and (7.10) illustrate how energetics affects the feasibility of the food-chain. The steady states of both plant ($X_1{}^*$) and root-feeding nematode ($X_2{}^*$) are dependent on the characteristics of consumption and ecological efficiencies of the nematode ($\sim a_2 p_2 c_{12}$). For the root-feeding nematode, the steady-state is dependent on the rate of plant production, the characteristics of consumption and the ecological efficiency of the nematode, $r_1 a_2 p_2 c_{12}$, and the degree of intra-specific competition (self-limitation) within the plant community and the specific death rate of the nematode, $c_{12} d_2$. The steady-state of the plant is dependent on the specific death rate, d_2.

If the two-species model is locally stable, then the populations will return to their steady state values following a minor disturbance. For this particular two-species example, all feasible food-chains are locally stable. Nonetheless, the procedures described below apply to more complex systems where this conclusion may not be the case. To assess the local stability of the two species system the non-linear Equations (7.7) and (7.8) must be linearized. The linearized equations are expressed in terms of deviations of the populations from their steady-states. Hence, $x_1 = X_1 - X_1{}^*$, and $x_2 = X_2 - X_2{}^*$, represent these deviations or small disturbances. Next, substitute X_1 and X_2 with ($x_1 + X_1{}^*$) and ($x_2 + X_2{}^*$). The resulting equations are simplified to dx_1/dt and dx_2/dt since $X_1{}^*$ and $X_2{}^*$ are constants. The new equations represents the dynamics of the deviations of the populations from their steady-states.

The per capita effects, α_{ij}, of each species on the dynamics of the deviations of all species from their steady-states are calculated by taking the partial derivatives of the equations with respect to all species. These partial derivatives represent the interaction strengths among species. If a species has a net negative effect on the dynamics of another, then α_{ij} is negative. If the effect on dynamics is positive, then α_{ij} is positive. The interaction strength, α_{ij}, is zero if a species has no effect on the dynamics of another. The matrix composed of interaction strengths is called the **community matrix**. The elements of the community matrix for the two-species food-chain are:

$$
\begin{aligned}
\alpha_{11} &= -c_{11} X_1{}^* & \alpha_{12} &= -c_{12} X_1{}^* \\
\alpha_{21} &= a_2 p_2 c_{12} X_2{}^* & \alpha_{22} &= 0
\end{aligned}
\tag{7.11}
$$

The community is stable (x_1 and x_2 return to zero) if the eigenvalues, λ_i, of the community matrix are less than zero. If any eigenvalue is greater than or equal to zero, then x_1 and x_2 increase in size over time and the community is unstable. If the community is stable, the real part of the largest eigenvalue (closest to zero), λ_{max}, is used to estimate the return-time, RT:

$$RT = -\frac{1}{\lambda_{max}} \tag{7.12}$$

where

$$\lambda_{max} = \frac{\alpha_{11} + \sqrt{\alpha_{11}^2 + 4\alpha_{12}\alpha_{21}}}{2} \tag{7.13}$$

From Equation (7.13), it can be shown that the real part of λ_{max} decreases towards $c_{11}/2$ as r_1 increases. As r_1 increases the contribution of the discriminant to the solution decreases since the discriminant of the quadratic approaches zero and eventually becomes negative. Hence, the real part of RT decreases towards $c_{11}/2$ as r_1 increases.

This exercise demonstrates two important points. First, the dynamics of a species and the energetic properties of communities are intimately tied to one another. It is an intellectual false dichotomy to refer to either dynamics (*sensu* Pimm and Lawton, 1977) or the energetic properties (*sensu* Hutchinson, 1959) as being the sole or dominant factors that constrain the structure, function or stability of communities. Second, with regard to toxicants and their effects on species and communities, the scientific community investigating the effects and regulatory process must consider this inter-relationship. As illustrated in the two-species food-chain, a change in the density, birth rate, fecundity, death rate, feeding activity, the digestibility of prey (assimilation efficiency) or the physiology (production efficiency) of one species affects the other species and the dynamics of the chain as a whole.

Real communities are more complex than simple food-chains. Can the results presented above be applied to complex multi-species communities? The limiting factor in applying the techniques outlined above is not in describing the dynamics, but rather in solving for the steady state equilibria of all species in the community. De Ruiter *et al.* (1994) used Equation (7.6) to link Equations (7.1–7.3) with Equations (7.4–7.5). Equations (7.4–7.5) are at the heart of the models developed by Hunt *et al.* (1987), who developed a model of the below-ground food-web for the Central Plains Experimental Range (CPER) of the North American Shortgrass Steppe. It was one of the first models to adapt the food-web descriptions championed by Cohen (1978) to the ecosystem approach. De Ruiter *et al.* (1994) developed a means to take the energetic analogues to Lotka–Volterra equations [Equations (7.1–7.3)] and couple them with the equations developed by Hunt *et al.* (1987) in a fashion that allowed the stability of food-webs whose descriptions were based on field and laboratory data to be gauged in the mathematical sense (i.e. one could assess the local stability of real systems just as theoreticians had assessed the local stability of model systems). The procedure is similar to that applied to the two-species example presented above. The difference is that instead of solving for the steady-state values as shown in Equations (7.9) and (7.10), estimates of the steady-state from long-term averages obtained from field data are used. The coefficients of predation, c_{ij}, are estimated from the flux rates using Equation (7.6).

De Ruiter *et al.* (1995) used the approach outlined above to demonstrate that the local stability of real systems depends on the patterning of per capita effects within communities (Figure 7.1). Food-webs of native and agricultural soils from across the globe consistently exhibit a pattern in the per capita effects (the α_{ij} terms of the community matrix, see Equation 7.11) among species with the trophic position of the interaction. There is a clear asymmetry in the alignment and magnitude of the interaction strengths with increased trophic position. Interactions at the base of the food-web possess disproportionately large negative effects of predators on prey when compared with the positive effect of prey on predators for the same interaction. At higher trophic positions the converse is true, as the relative sizes of the positive effects of prey are disproportionately large relative to the negative effect of a predator on its prey. Monte Carlo simulations that compared food-webs that possessed the pattern (patterned webs) in interaction strength to webs with the same predator–prey pairings but with the pattern of interaction strength removed (random webs), demonstrate that the patterned webs were more likely to be stable than the random webs (Figure 7.2).

The analysis of the two-species model and multi-species communities demonstrates one of the principal tenets of systems theory. Systems possess emergent properties that are not readily detected (if at all) by studying subcomponents of the system in isolation, or collectively but not in an integrative manner. For ecosystems, an important emergent property is that the energetic organization of communities forms the basis of ecosystem stability. A stated goal within ecotoxicology is to integrate the structural and functional components of ecosystems. These results suggest that guidelines must provide identifiable attributes ecosystems that can be assessed through integrating structure and function.

7.3 CONCLUSIONS

A common feature of disturbed systems is a change or disruption in the patterning of interaction strength within the community. When species are removed from a system, added to a system (pests), when the physiologies and life-histories of species are altered, or when some aspect of environmental conditions changes and whole assemblages become more or less active, the common response is a change in the distribution of the per capita effects within the community. Changes in the patterning of per capita effects from a stable

Trophic Position

consumer

resource

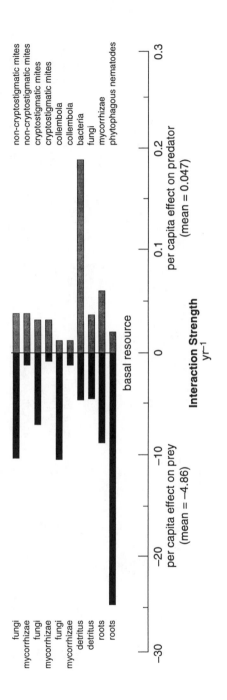

Figure 7.1 A plot of interaction strengths for each pairwise predator–prey interaction by the trophic position of the interaction for the soil food web of the Central Plains Experimental Range (CPER), Long Term Ecological Research Site, Nunn, CO, USA. Recall that the interaction strengths are the per capita effects represented as the α_{ij} or elements of the community matrix (see the example of the two-species model in the text). The interactions should be read from left to right (prey to predator) with trophic positions starting at the bottom (basal resources) and ending at the top (top predators). The negative effects of predators on prey (α_{ij}) are disproportionately large compared with the positive effects of prey on predators (α_{ji}) at low trophic positions, with the contrary true at higher trophic positions ($P < 0.05$). In addition to the CPER, De Ruiter et al. (1995) found this pattern in soil webs from Horseshoe Bend, GA, USA; the Lovinkhoeve, the Netherlands; and Kjettslinge, Sweden.

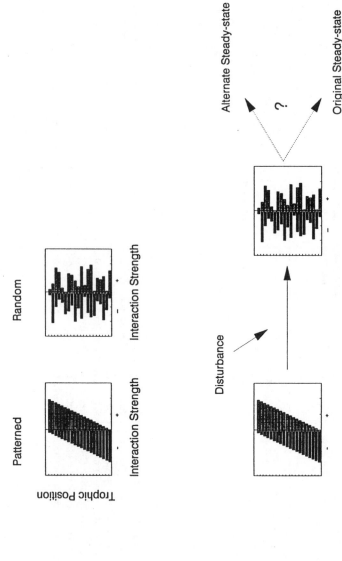

Figure 7.2 The pattern of interaction strength presented in Figure 7.1 is presented in a stylized form and compared with a hypothetical disturbed counterpart. The pattern was disrupted by taking a random sample from a uniform distribution of the mean ± 1 s.d. for the positive and negative interaction strengths (De Ruiter *et al.*, 1995). Monte Carlo simulations of food webs with (patterned webs) and without (random webs) the patterning of interaction strength indicated that the real webs were more likely to be stable than the random webs. It is hypothesized that disturbance operates to disrupt the stable patterning of interaction strength within an ecosystem. The fate of the community following the disturbance is less clear. If the ecosystem is locally stable it will return to its original steady state. If it is locally unstable, the ecosystem will change to an alternate steady state with the same species or form a new community.

to an unstable configuration and the response of the system to these changes are an aspect of disturbance that have not been considered in risk assessment and biomonitoring efforts (Figure 7.2). Proximate reasons for declines in populations or productivity, or changes in community structure, can be made from current biomonitoring practices, e.g. '...fungal populations increased because the toxicant reduced mycophagous Collembola densities by increasing mean instar duration and reducing fecundity'. What is missing is consideration of the emergent systems properties that all communities possess. Without adopting a systems perspective in risk assessment and biomonitoring efforts, the importance of these emergent properties will be overlooked.

There were two objectives to this chapter. First, models used to describe population dynamics and structural characteristics of ecosystems were linked with models used to describe ecosystem processes. This was achieved by describing dynamics in terms of units that were used in the process-oriented models, i.e. biomass rather than individuals were used in the models. The second objective was to demonstrate that ecosystems possess a high degree of interconnectedness which is important to the stability of the system as a whole. Meeting these objectives reveals two conclusions:

1. Energetics and dynamics are inextricably inter-related.
2. The energetic organization of communities forms the basis of ecosystem stability.

The implications of the results presented here and the conclusions drawn are both daunting and challenging. They are daunting in the sense that if we continue to focus our risk assessment efforts of single species, groups of species, or important ecosystem processes we are not likely to gain much insight into the nature of how toxicants affect systems. The challenge will be to incorporate the approaches presented here into the risk assessment and biomonitoring efforts that try to understand how disturbance operates within an ecosystem. The clear implications are that studying single species or groups of species will not yield the insight we seek, but rather a whole systems perspective should be adopted.

7.4 SUMMARY

An important goal of ecotoxicology is to augment current risk assessment and biomonitoring practices that study the effects of contamination (disturbance) on single species, whole communities and processes with a more integrative approach. Models of food-webs are presented that integrate the population dynamics of species with the structural and functional aspects of communities that could serve in risk assessment and biomonitoring efforts. A unique feature of the models is that they are described in terms of parameters that are currently being used in risk assessment and biomonitoring. The models demonstrate that: (i) population dynamics and ecosystem productivity are inter-related in ways that influence the stability of ecosystems; and (ii) that the energetic organization of the ecosystem forms the basis of ecosystem stability. It is hypothesized that a common feature of disturbance is to change the energetic organization of communities from one that is stable to one that is less likely to be stable. The importance of the energetic organization of an ecosystem to its stability is an emergent system property that is currently not considered in the field of ecotoxicology. It is proposed that integrating structural and functional attributes of ecosystems could best be achieved if ecotoxicology adopts a broader systems approach.

REFERENCES

Bongers, T. (1990) The maturity index: an ecological measure of environmental disturbance based on nematode species composition. *Oecologia*, **83**, 14–19.

Briand, F. and Cohen, J. (1987) Environmental correlates to food-chain length. *Science*, **238**, 956–8.

Cohen, J.E. (1978) *Food-webs in Niche Space. Monographs in Population Biology*, Vol. 11. Princeton University Press, Princeton, New Jersey.

DeAngelis, D.L. (1975) Stability and connectance in food-web models. *Ecology*, **56**, 238–43.

DeAngelis, D.L. (1992) *Dynamics of Nutrient Cycling and Food-webs*, Chapman & Hall, London.

DeAngelis, D.L., Bartell, S.M. and Brentkert, A.L. (1989) Effects of nutrient recycling and food-chain length on resilience. *Nature*, **134**, 778–805.

De Ruiter, P.C., Neutel, A.-J. and Moore, J.C. (1994) Modelling food-webs and nutrient cycling in agro-ecosystems. *Trends Ecol. Evol.*, **9**, 378–83.

De Ruiter, P.C., Neutel, A.-J. and Moore, J.C. (1995) Energetics, patterns of interaction strengths and stability in real ecosystems. *Science*, **269**, 1257–60.

Eijsackers, H. and Løkke, H. (eds) (1992) *SERAS – Soil Ecotoxicological Risk Assessment System. A European Scientific Programme to Promote the Protection of the Health of the Soil Environment.* Report from a workshop held in Silkeborg, Denmark, 13–16 January 1992. National Environmental Research Institute.

Hunt, H.W., Coleman, D.C., Ingham, E.R., Ingham, R.E., Elliott, E.T., Moore, J.C., Reid, C.P.P., Rose, S.L. and Morley, C.R. (1987) The detrital food-web in a shortgrass prairie. *Biol. Fert. Soils*, **3**, 57–68.

Hutchinson, G.E. (1959) Homage to Santa Rosalia or why are there so many species? *Am. Nat.*, **93**, 155–60.

Lindeman, R.L. (1942) The trophic–dynamic aspect of ecology. *Ecology*, **23**, 399–418.

Lotka, A.J. (1956) *Elements of Mathematical Biology*, Dover Publications, New York.

May, R.M. (1973) *Stability and Complexity in Model Ecosystems. Monographs in Population Biology*, vol. 6. Princeton University Press, Princeton, New Jersey.

Moore, J.C., De Ruiter, P.C. and Hunt, H.W. (1993) The influence of productivity on the stability of real and model ecosystems. *Science*, **261**, 906–8.

Moore, J.C., De Ruiter, P.C., Hunt, H.W., Coleman, D.C. and Freckman, D.W. (1996) Microcosms and soil ecology: critical linkages between field research and modelling food-webs. *Ecology*, **77**, 694–705.

Odum, H.T. (1957) Primary production measurements in eleven Florida springs and a marine turtle grass community. *Limn. Oceanogr.*, **2**, 85–97.

O'Neill, R.V. (1969) Indirect estimates of energy fluxes in animal food-webs. *J. Theor. Biol.*, **22**, 284–90.

Paine, R.T. (1966) Food-web complexity and species diversity. *Am. Nat.*, **100**, 65–7.

Paine, R.T. (1980) Food-webs: linkage strength and community infrastructure. *J. Anim. Ecol.*, **49**, 667–85.

Paine, R.T. (1988) Food-webs – road maps of interactions or grist for theoretical development? *Ecology*, **69**, 1648–54.

Pimm, S.L. (1982) *Food-webs*, Chapman & Hall, London.

Pimm, S.L., and Lawton, J. (1977) The number of trophic levels in ecological communities. *Nature*, **268**, 329–31.

Polis, G.A. (1994) Food-webs, trophic cascades and community structure. *Aust. J. Ecol.*, **19**,121–36.

PART FOUR

The spatial component of soil communities

8   Scale dependency in the ecological risks posed by pollutants: is there a role for ecological theory in risk assessment?

PAUL C. JEPSON

8.1 THE ROLE OF ECOLOGY WITHIN RISK ASSESSMENT SYSTEMS

This chapter attempts to shed light on an area of tension in ecotoxicological risk assessment; the trade-off between experimental control in laboratory toxicology and the realism of field-based measurement. This trade-off is brought most sharply into focus through the issue of extrapolation of risk estimates between the laboratory and the field. It is a paradigm of ecotoxicology that it is possible to build tiers of test procedures, rooted in highly controlled, often elaborate, laboratory-based test systems from which we can make forecasts of potential harm to organisms in the polluted environment. These predictions then form the basis for regulatory decisions concerning pollutant release and control. This paradigm became established through the unification of toxicology (because of the need to understand the relationship between dose and effect) and environmental chemistry (because of the need to determine environmental concentration and hence, potential exposure). The massive inputs of science and technology involved in this process have been fuelled and motivated by the need for regulation, legislation and the setting of environmental quality standards. The output from risk assessment procedures fulfils the need in these formalized systems, for repeatability, reliability and low cost. This chapter examines whether or not the paradigm of extrapolating risks between the laboratory and the field has been rigorously challenged, and explores some of the alternative criteria that ecotoxicologists with a more ecological perspective might select. It argues that a truly unified and scientifically defendable

Ecological Risk Assessment of Contaminants in Soil. Edited by Nico M. van Straalen and Hans Løkke. Published in 1997 by Chapman & Hall, London. ISBN 0 412 75900 4

approach to risk assessment can only be achieved if the third conceptual arm of ecology is added to toxicology and environmental chemistry.

The role of ecology as a discipline within risk assessment systems is still uncertain. Ecological ideas play a minimal role in the initial construction of these systems, largely because the currency in which risk is expressed is normally toxicological. Unacceptable risks are often expressed as a high probability of suffering a toxic effect without determining how important that effect might be for the persistence of the affected population. By focusing upon acute toxicity, the danger of such a philosophy is that it may underplay the importance of chronic or low toxicity effects that could be damaging to populations that are sensitive to small changes in survival rate. If such occurrences were widespread in the real world, risk assessments based on laboratory toxicological criteria alone could make falsely negative predictions, with potentially hazardous consequences. The conventional approach also fails to recognize that even acutely toxic chemicals may have a relatively benign environmental impact in some circumstances, although this is part of the experience of any biologist who works in polluted environments.

It is a responsibility for ecologists to make a convincing case that risk assessment procedures require elaboration and further complication. These procedures have, after all, provided many important benefits, without necessarily being realistic. They are also designed to be conservative, and a proportion of false positives may be intrinsically acceptable. Arguing from the standpoint that a correct balance should be struck, and that the system should correctly be challenged to maintain rigour, how then should ecologists develop their argument? Two key steps are probably required before the case can be made: firstly, risk predictions should be examined to detect evidence for falsely positive or, more importantly, falsely negative predictions. Such evidence would be required if large changes to current procedures were to be envisaged. Secondly, a practical methodology, exploiting ecological ideas, should be developed which reduces the likelihood that these types of error will occur. The former step is complicated by a lack of validation of much risk assessment: it is clearly impossible to follow up every case of the thousands of materials subjected to regulatory scrutiny. In any case, we lack criteria of harmfulness with which to evaluate such data: changes in population density may be detectable, but are they of any significance for the populations that experience them? The latter step may be pursued through gradual research and development with case studies, some examples of which are reviewed below, and by seeking support from ecological theory for the development of criteria that provide more appropriate end-points for risk assessment.

8.2 WHY IS SCALE SO IMPORTANT TO ECOTOXICOLOGY?

It may be simply argued from first principles that as time elapses, following a pollution incident, the importance of the initial toxic effect for populations of exposed organisms may be reduced and ecological factors like potential for population recovery take precedence in determining the long-term impact of the toxic effect. This is particularly true for short-term pollutants like pesticides, which are patchily distributed and for macro-invertebrates that colonize agricultural systems, which tend to have powers of dispersal that exceed the scale of pesticide treatment. The potential for these invertebrates to recolonize the treated area is a critical factor in judging the long-term risk posed by a given chemical exposure. It is obvious that recolonization can only be investigated in an appropriately designed field study and that extrapolation from laboratory toxicological testing to an outcome that depends upon population dynamics is impossible (Figure 8.1). To what extent though, can generalizations be made from this specific case to all risk assessment? The relative scaling of pollutant distribution and persistence, relative to the dispersal range and increase rate of exposed organisms, seems to have a critical effect on our perceptions about the role of ecological criteria in determining risk. Thus for micro-invertebrates exposed to persistent, large-scale, heavy metal pollution under the soil, or in aquatic test systems, questions of scale and recovery rarely arise in risk assessment (Calow, 1993) and risk assessment procedures may not progress beyond controlled test systems or mesocosms in such a way as to permit recovery processes to take place.

Why has an apparent distinction arisen between pesticide risk assessment for beneficial invertebrates and risk assessment for other classes of pollutant and organisms? The answer may lie in the uniquely detailed knowledge we have of adverse consequences of pesticide side effects on beneficial invertebrates, namely pest resurgence and secondary pest outbreaks, that follow local extinctions of natural enemies. Intensive and/or large-scale applications of pesticides are widely known to exacerbate pesticide impact on beneficial invertebrates and to increase the risks of these adverse effects taking place. The short time-scales involved, and the high powers of dispersal and reproduction of the organisms concerned, greatly assisted the process of discovery, and the economic importance of pest attack meant that much research has been undertaken in the field by ecologists. It is now widely accepted that the nature of the sprayed system, the distribution of spraying, and the powers of recovery of the exposed organism all contribute to the ultimate end-point of local extinction, and risk assessment procedures are evolving to incorporate these factors. Is there evidence that a more general application of these ideas throughout ecotoxicology is needed or justified, however?

8.3 RELATIVE SCALING OF CHEMICAL PERSISTENCE, SPATIAL DISTRIBUTION AND ORGANISM CHARACTERISTICS: GENERATING A NEW PARADIGM FOR ECOTOXICOLOGY?

Conclusions from verbal reasoning in this complex argument may rest on small changes of emphasis and perspective. Some formality to the reasoning process

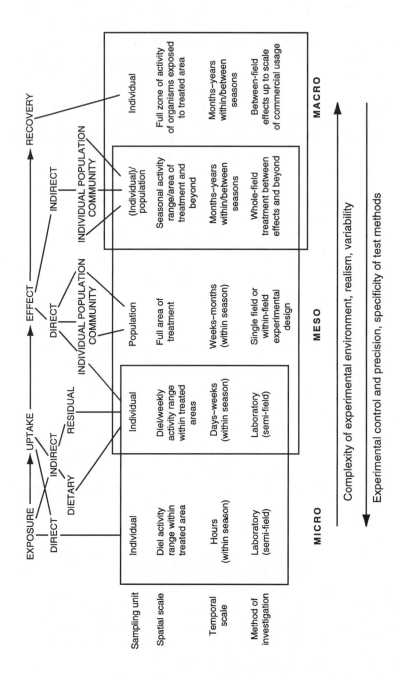

Figure 8.1 The evolving process of exposure, poisoning and recovery for beneficial invertebrates exposed within an arable crop. The overlapping boxes indicate the appropriate scale and methodology for investigating each stage of the process. The overlap of the boxes indicates the limits to which extrapolations may be made between the micro, meso and macro scales. (Modified from Jepson, 1989.)

may be introduced by exploiting mathematical logic to investigate the interrelationships between pollutant distribution, the rate of decay in toxicity, and the mobility and reproductive rate of the organism. Arguably, it is the immensely variable scaling of these factors, which are common to all possible ecotoxicological scenarios, which has prevented some generality about the importance of ecological factors in ecotoxicological risk assessment from emerging.

Simple models (Box 8.1: Jepson and Sherratt, 1996) reveal that widely differing population trends emerge when pollutant persistence and the movement rate of the organism are varied, even by small amounts, holding toxicity and the area polluted constant. We are not in a position yet to predict with confidence what, for example, the relative effects of a doubling of toxicity or a doubling of persistence will be. We can argue, however, that the importance of ecological processes, such as dispersal, has been underestimated in risk assessment, and that factors such as toxicity, persistence, distribution and the movement rate of organisms, are likely to trade-off against each other in such a way that even mildly toxic materials could prove hazardous in some situations, and highly toxic materials may be relatively benign in others. Risk assessment systems could theoretically therefore generate errors, by not accounting for the ecological mechanisms that underlie some responses to pollutant exposure.

What evidence is there from ecotoxicology in the real world to support this argument? We should seek for example, evidence that the spatial scaling or heterogeneity of pollution matches the scaling of movement by organisms in such a way that ecotoxicological effects are modified. We should also identify evidence from the real world for adverse effects in situations where risk assessment would guide us to expect otherwise and vice versa.

8.4 EVIDENCE FOR SCALING DEPENDENCY IN ECOTOXICOLOGICAL EFFECTS

The following examples form a spectrum of scales from hundreds of square kilometres down to cubic metres and smaller. All except one employ some form of modelling to elucidate the processes in question, although there is qualitative support for the conclusions reached from field-collected data. In each case, interpretation of the ecotoxicological outcome depends upon a quantitative understanding of pollutant distribution, pollutant decay rate and the rate of movement of the organism. Toxicological data may have identified the potential for harm; however, this assessment might have led to falsely positive conclusions in all cases (i.e. predictions of harm that may not occur in the real world), with opportunities for falsely negative predictions in some. Most of the studies go beyond rediscovering that exposure is both complex and important in risk assessment, they reveal that risk is a function of temporal and spatial scaling for components of habitat, pollutant and organism, and that these need to be considered before risk can be quantified.

Box 8.1 Summary of basic features of a simple model that explores the relationship between the proportion of a habitat that is contaminated by a toxic pollutant, the toxicity and persistence of the pollutant and the rates of population increase and dispersal of an exposed organism (Jepson and Sherratt, 1996)

The system being modelled is an area of habitat that contains a proportion P (where $0 \leq P \leq 1$) which is contaminated by a pollutant. Area P is inhabited by x individuals of the exposed organism, and the uncontaminated part of the habitat contains y individuals. There is movement by diffusion at rate α by individuals between these zones.

Area P has a high incidence of mortality, Z where $(0 < Z < 1)$, which declines over time according to a decay constant θ, thus:

proportion killed per unit time $= Ze^{-\theta t}$

The rate equations governing populations in the contaminated zone x and the uncontaminated zone y are as follows:

$$\frac{dx}{dt} = rx\left(1 - \frac{x}{PK}\right) - \alpha x + \alpha y - xZe^{-\theta t}$$

and

$$\frac{dy}{dt} = ry\left(1 - \frac{y}{(1-P)K}\right) + \alpha x - \alpha y$$

Where r is the intrinsic increase rate and K is the carrying capacity of the habitat. In the worst case scenario where the pollutant in area P is completely toxic and permanent, then the organism can still persist in the system, as long as:

$$r > \alpha$$

This solution may be obtained analytically. In addition, several numerical solutions are shown below in examples. The results are rather predictable and intuitive in such a simple system, they do however reveal the interplay between these factors and the potential for small changes to lead to markedly differing outcomes.

Parameters	**Outcome**
High toxicity ($Z = 1.0$)	Global extinction is possible if
Very high persistence (decay, $\theta = 0$)	the source area is exhausted
Low diffusion rate (exchange, $\alpha = 0.3$)	
High toxicity ($Z = 1.0$)	The system may be rescued by an
High persistence (decay, $\theta = 0.001$)	element of pollutant decay
Low diffusion rate (exchange, $\alpha = 0.3$)	
High toxicity ($Z = 1.0$)	The outcome is highly dependent
High persistence (decay, $\theta = 0.001$)	upon colonization rate
Low diffusion rate (exchange, $\alpha = 0.1$)	

Although this framework is very simple, it does enable the underlying relationships between some of the interacting phenomena that affect risk to be explored.

8.4.1 LINYPHIID SPIDERS IN SPRAYED FARMING SYSTEMS

Linyphiidae are highly susceptible to pesticides (Everts *et al.*, 1989), but they are also highly dispersive and are among the first invertebrates to reinvade agricultural fields after spraying (Thomas *et al.*, 1990). Although high risk is predicted from laboratory-based or small-scale field testing (Everts *et al.*, 1991; Jagers op Akkerhuis and Van der Voet, 1992) it is apparent that reinvasion may reduce the impact of this on a local scale (Thacker and Jepson, 1993). Linyphiid populations may become more sensitive to pesticide impacts as the scale and intensity of spraying increases, and it would be predicted from ecological reasoning that population persistence might be sensitive to rate of dispersal, among other factors. Both of these predictions have been confirmed by spatial modelling taking into account the key features of habitat, pesticide toxicity and persistence, and linyphiid dispersal, that are likely to mediate the impact of spraying (Halley *et al.*, 1996; Figure 8.2). The level of risk posed by pesticides to Linyphiidae is therefore highly dependent upon the temporal and spatial scale being considered. Expressed toxicologi-

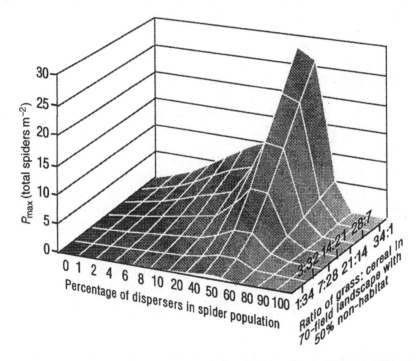

Figure 8.2 Maximum linyphiid spider population density estimates, from modelling, in one sprayed cereal fields in an 80-field landscape. The ratio of grass to sprayed cereals is varied in a landscape consisting of 50% non-habitat, where 75% of spiders dispersing into such areas do not survive. The percentage of dispersing spiders in the population is varied between 0 and 100% to examine the sensitivity of the population in the sprayed field to the intensity of spraying and the ability of Linyphiidae to recognize. Population persistence is highly sensitive to small changes in dispersal rate. (From Halley *et al.*, 1996.)

cally, Linyphiidae are subject to high risk from use of organophosphate and pyrethroid insecticides. The importance of these effects varies from being trivial in many realistic sprayed systems, because recolonization is rapid, to being potentially damaging in other systems, where reinvasion rate is restricted through more intensive spraying or differing landscape layout. It is only by investigating the ecological dimension of risk to spiders that the combination of factors leading to local extinction can be identified.

8.4.2 MIGRATORY LOCUSTS AND NON-TARGET INVERTEBRATES EXPOSED TO PERSISTENT PESTICIDE BARRIERS

Nymphal hopper bands of migratory locusts disperse up to several kilometres by walking before the final moult to the flying adult. They are highly susceptible to many pesticides, and large areas are treated to ensure effective population suppression. These insects form an interesting test case for the arguments developed in the simple modelling exercise in Box 8.1, because pesticides of widely differing environmental persistence have been employed in control. In theory, if the line of reasoning is continued, there should be a trade-off between the proportion of the area that is treated and the persistence of the pesticide, to achieve the same level of effect on a population of given mobility. This has been confirmed in practice (Bouaichi *et al.*, 1994; Cooper *et al.*, 1995), with 50 m wide barriers of persistent pesticide (active for up to 60 days), separated by untreated areas 600 m wide, being just as effective as conventional blanket treatments to the whole area with less persistent chemicals (effective for 3–5 days only). In the case of Cooper *et al.*'s (1995) study, ecotoxicological surveys on non-target invertebrates have demonstrated low levels of harm to susceptible, but less dispersive organisms, in the barrier sprayed zone (C. Tingle, personal communication): a finding that is again consistent with a line of reasoning that extends from the simple model in Box 8.1. The message for ecotoxicology is clear: risk in the field depends critically upon habitat and organismal characteristics (i.e. patchiness of contamination and organism dispersal rate) and their interplay with chemical persistence. Even very persistent chemicals may confer limited harm if organisms are sedentary or chemical distribution is restricted. In this case again, risk assessment based upon small-scale or laboratory measurements would not be capable of partitioning potential harm among the community of exposed organisms or of identifying which situations are potentially the most damaging.

8.4.3 CARABID BEETLES EXPOSED TO DIFFERENT PATTERNS AND INTENSITIES OF PESTICIDE SPRAYING

Carabid beetles are susceptible to local extinction in sprayed farming systems (Burn, 1989), despite their relative tolerance to pesticides (Wiles and Jepson, 1992). Modelling (Sherratt and Jepson, 1993) suggests that extinction may occur, even at realistic rates of spraying in UK arable crops, through an inter-action between the limited dispersal powers of some species, the toxicity, dis-tribution and frequency of spraying and habitat features such as field size and the penetrability of field boundaries. The outcome of spraying in terms of risks to particular field populations is highly unpredictable: risk estimated on the basis of toxicity and exposure alone is a poor guide to the eventual out-come. A risk assessment model incorporating some of the essential features of this system (i.e. characteristics of dispersal, population dynamics and habi-tat) could again, however, identify the circumstances in which toxic effects could lead to local extinctions. Importantly, the analysis of Sherratt and Jepson (1993) reveals that even mildly toxic pesticides, yielding low risk from conventional laboratory-based procedures, could trigger local extinctions if used frequently enough or on a large enough scale. This potential for falsely negative results adds to the evidence that ecological criteria should be incor-porated within risk assessment procedures.

8.4.4 SOIL-INHABITING ORGANISMS EXPOSED TO HEAVY METALS

Soil is a highly heterogeneous medium and pollutants within it are patchily distributed. Arguably, risk should be dependent upon the relative scaling of pollutant distribution and the pattern of movement of the organism, even on the small spatial scales that are appropriate to soil invertebrates. Marinussen and Van der Zee (1994) employed aspects of toxicology, soil science, geosta-tistics and ecology in a theoretical investigation of this issue for a group of taxa, exposed to cadmium contamination in soil. They demonstrated that properties of the organism including mobility and area occupied had impor-tant consequences for risk in the field. They concluded that the distribution of bioavailable pollutant concentrations, organism residence time, size of living area and mobility must be quantified in order to provide realistic estimates of exposure and risk. Adapted to explore the consequences of chemical persis-tence and habitat heterogeneity for risk, or parametrized with more detailed and precise behavioural and ecological parameters, theoretical exercises of this form could be used to identify the optimum requirements for risk assess-ment across a wide range of taxa.

8.5 A ROLE FOR ECOLOGICAL THEORY IN RISK ASSESSMENT: COULD METAPOPULATION THEORY PROVIDE UNIFICATION IN ECOTOXICOLOGY?

Ecological risk is clearly affected by the temporal and spatial scaling of pol-lutant distribution and persistence and of organism distribution and move-ment. In order to incorporate this complexity within risk assessment prac-tices, however, some general rules or unifying principles need to emerge. In ecology, metapopulation theory (Hanski and Gilpin, 1991) has provided new insights into spatial processes and provided explanations for patterns of population persistence in patchily distributed organisms. Essentially, metapopulations consist of subpopulations which are distributed among a mosaic of habitat patches. Local populations may become extinct from time to time, however, metapopulation persistence is maintained by a dynamic balance between the rates of colonization and extinction. In their simplest form, metapopulations consist of clusters of equivalently sized subpopulations that are sufficiently distanced to be separate, but not so widely distributed that colonization is prevented. The subpopulations should not be synchronized to such a degree that one catastrophe, i.e. pollution episode, affects all groups equally. The main contribution made by developments in metapopulation theory has been the refocusing by ecologists on to colonization and extinction rates as factors that may mediate persistence.

Superficially, metapopulation theory seems to satisfy some of the requirements for a unifying theory in ecotoxicology. It incorporates patchiness and movement as key features and can be used to determine the likelihood of a population persisting when either colonization or extinction rates are altered. If classical metapopulations were widely distributed in nature, then ecotoxicologists would have a powerful tool, in all the theory that surrounds this topic, for elaborating risk assessment procedures to take into account some of the spatial and temporal factors that mediate risk. Some case studies seem to lend themselves to this approach. For example, Carabidae in sprayed farm systems (Jepson, 1993; Sherratt and Jepson, 1993) seem to have metapopulation structure imposed upon them by the subdivision of arable landscapes into fields and the asynchronous nature of hazardous farming practices which take place in different crop types. The limited powers of dispersal of many species also conveniently seems to fit with the requirements of theory for a degree of connection that is not so all-embracing as to effectively synchronize the whole system. In this case, laboratory toxicology and classical toxicological risk assessment procedures may be used to determine death rates from spraying which are then inserted into appropriate metapopulation models to determine those combinations of spraying and landscape type that confer a serious risk of local extinction taking place. In this case, a clear and distinct role for ecological data is found, preserving the best aspects of current risk assessment practices, but exploiting the findings in a new way to produce a more refined approach to ecological risk assessment. There is less scope for exploiting metapopulation theory in other areas of ecotoxicology, however.

Classical metapopulations as defined above are probably rare in nature (Harrison, 1991). Three alternative scenarios for population structure have been identified:

1. Firstly, mainland – island dynamics, where persistence of the global population depends upon survival of an extinction-resistant source population (e.g. Schoener, 1991; Harrison *et al.*, 1988). For ecotoxicologists, the key issue here would be preservation of the source population. As the

example in Box 8.1 illustrates, there will be trade-offs between factors such as toxicity, chemical persistence and organism movement rate where sink populations (i.e. populations that depend upon the main source) are exposed to pollutant risk. If the main source population is affected then survival of the global system depends upon toxicity, persistence and the likelihood that the population will fall below density thresholds for viability: exchange of individuals with other habitat patches may only delay the eventual demise of the population. In both scenarios insights would be needed to determine the ecological risk posed by a given pollutant type; the questions asked and the way the data are interpreted would differ from a classical metapopulation approach, however, because population persistence is not solely dependent upon the dynamic balance between colonization and extinction rates.

2. Secondly, many populations are subdivided over a patchily distributed habitat. Dispersal between patches may, however, be so rapid that the (meta)population is effectively a single, patchy population with synchronized dynamics (e.g. Harrison and Thomas, 1991). This may apply to the case of Linyphiidae in sprayed farming systems and for some of the soil organisms, especially micro-invertebrates, in the case studies above. Although the patchiness of the habitat may contribute greatly to population stability by the influence that it has upon predator–prey interactions (Hassell, 1978), the ecotoxicologist would need to be wary of focusing upon between-patch dynamics as being the critical factor in population persistence. The survival of the population is determined by the long-term potential of the whole patchy habitat for population growth and survival.

3. Finally, population patches may become so separated from one another that they are effectively isolated and persistence then becomes a function of the general loss of habitat quality rather than the probability that recolonization may take place. This may commonly be the case for many soil invertebrates; however, there is a dearth of data concerning dispersal and colonization rates and the question must remain an open one. In this case again, ecotoxicologists would need to be sufficiently aware of the prevailing dynamics to conclude that the population in question was isolated: this case, however, would provide the best application for the conventional approach of modern ecotoxicology, which treats species as if their populations were isolated in space and time.

8.6 LESSONS TO LEARN FROM CONSERVATION BIOLOGY

Conservation biology similarly considers the persistence of populations that are patchily distributed, and is forced to deal with questions about reserve size and connectedness which impinge upon the same aspects of ecological theory as those that might be used by ecotoxicologists. The evolution of conservation biology as a discipline has not been punctuated, as in ecotoxicology, by the integration of other disciplines, such as toxicology and chemistry. Conservationists have not had to wrestle with the technology of testing and extrapolation of test results from laboratory systems to the field. Perhaps for this reason, conservationists have a head start in the application of ecological theory to the solution of their problems. Can ecotoxicologists learn from this?

One message from the experience of conservation biology with theory is that the relationship can be a tempestuous one, with many blind alleys (Simberloff, 1988). Although conservationists have sought a theoretical construct that may be applied generally across all problems, they have not to date been successful. In particular, metapopulation biology, which was seen as capturing many key issues, represents the real world too poorly to be generally applied (Harrison, 1993). Ecotoxicology should not fall into the same trap.

In both ecotoxicology and conservation biology, there is, however, a need to incorporate theory and ecological ideas if we are to manage and protect habitats effectively. The patchy nature of habitats, and the distribution and movement patterns of the organisms that occupy them, have a profound influence on the outcome of episodes of pollution. This chapter has argued that spatial processes should be explicitly incorporated with risk assessment systems in the future. The way ahead is probably to align with the rapid development in the spatial sciences, and apply geostatistics and spatial modelling to a far greater degree, as has been the case with conservation biology. The price of following this, however, is the large amount of data required to generate effective models (Halley *et al.*, 1996). Ecological systems are also highly variable, and the outcome of risk assessment models that exploit ecological data, will by their nature be much more variable than ecotoxicologists and regulators are used to dealing with. Pragmatic regulators should perhaps, seek guidance from models in the future, and recognize that risk is not solely determined by toxicity and bioavailability. Much more intensive research is, however, required to demonstrate that there are advantages to following this costly path.

8.7 THE NEXT STEP FOR ECOTOXICOLOGY

8.7.1 RULES OF THUMB AND THE INCORPORATION OF ECOLOGICAL THEORY

A practical and elegantly argued procedure for exploiting spatial dynamics and landscape ecology, to be followed by ecotoxicologists, has been outlined by Fahrig and Freemark (1995). They suggest that simple models should be developed that incorporate the spatiotemporal dynamics of the population and of the toxic event, and that predictions from these should be validated by field observation. Fahrig and Freemark (1995) provide abundant examples from the general ecological literature to support this approach, and the case studies for Linyphiidae, and Carabidae above, which incorporate toxic events, reinforce the potential value of taking this pathway, although the models in question, especially for spiders, may be far more elaborate than Fahrig and Freemark intend.

Although the carabid and linyphiid case studies provide general predictions about pesticide risks, neither incorporates the structure of real landscapes, and neither is species specific. This reinforces the main limitation to this approach: the detail needed in order to be able to correctly assign risk in a real system. Field experience tells us that Carabidae in sprayed farm systems may exhibit extinction, superabundance or neutrality in the face of heavy spraying regimes (Burn, 1989). Although their underlying dynamics are likely to be explained by models such as those developed by Sherratt and Jepson (1993), the biological, chemical and toxicological detail required to be able to predict exactly which population trajectory a given species will follow, makes it unlikely that models alone will suffice in risk assessment. Ecotoxicologists must more often ask: 'what level of detail in understanding was required to explain the outcome of a given toxic episode?' Only by experience can the appropriate trade-off between model and experiment be struck.

This summarizes the problem facing an ecologist in ecotoxicology. Ecological theory has made enormous progress in its ability to explain trends in population persistence and extinction. The ecologist will always recognize, however, that there are many possible outcomes from a given set of starting conditions. The firm predictions of laboratory-based risk assessment do not fit readily with this view of the world: they can, however, be placed in an ecological context, and greatly enrich the value of the data obtained. The cost of this approach is an apparent loss of precision. The gain may be that we will get closer to predicting the outcome that is most likely. Most importantly, if ecologists accept that current methodology carries an unpredictable risk of false negatives and false positives, it can be argued that these risks are minimized by the incorporation of more detail of the spatiotemporal dynamics of pollutant, habitat and organism.

8.8 SUMMARY

By seeking support from ecological theory criteria may be developed that provide more appropriate end-points for risk assessment, compared with current methodology. The scaling of pollutant distribution and persistence, relative to the dispersal range and capacity for population increase of the exposed organisms plays a crucial role in determining the ultimate end-point of local extinction. Simple models are able to demonstrate that the ecological risk of a pesticide spray to surface-active invertebrates is determined by trade-offs between toxicity, persistence and the rate of movement. Four case studies are reviewed in this chapter that provide evidence for scale-dependency in ecotoxicological effects. Metapopulation theory seems to satisfy some of the requirements for a unifying theory in ecotoxicology, but true metapopulations are rare. Present uncertainties in the current, test system-based methodology for risk assessment may be minimized by the incorporation of more details on the spatiotemporal dynamics of pollutant, habitat and organism.

REFERENCES

Bouaichi, A., Coppen, G.D.A. and Jepson, P.C. (1994) Barrier spray treatment with diflubenzuron (ULV) against gregarious hopper bands of the Moroccan Locust (*Dociostaurus maroccanus* Tunberg) (Orthoptera: Acrididae) in N.E. Morocco. *Crop Protect.*, **13**, 60–72.

Burn, A.J. (1989) Interactions between cereal pests and their predators and parasites, in *Pesticides and the Environment: the Boxworth study* (eds P. Greig-Smith, G.H. Frampton and A. Hardy), HMSO, London, pp. 110–31.

Calow, P. (ed) (1993) *Handbook of Ecotoxicology*, Blackwell Scientific Publications, Oxford, UK.

Cooper, J.F., Coppen, G.D.A., Dobson, H.M., Rakotonandrasana, A. and Scherer, R. (1995) Sprayed barriers of diflubenzuron (ULV) as a control technique against marching hopper bands of migratory locust *Locusta migratoria capito* (Sauss) (Orthoptera: Acrididae) in Southern Madagascar. *Crop Protect.*, **14**, 137–43.

Dunning, J.B., Danielson, B.J. and Pulliam, R.H. (1992) Ecological processes that affect populations in complex landscapes. *Oikos*, **65**, 169–75.

Everts, J.W., Aukema, B., Hengeveld, R. and Koeman, J.H. (1989) Side-effects of pesticides on ground-dwelling predatory arthropods in arable ecosystems. *Environ. Poll.*, **59**, 203–25.

Everts, J.W., Willemsen, M., Stulp, M., Simons, B., Aukema, B. and Kammenga, J. (1991) The toxic effect of deltamethrin on linyphiid and erigonid spiders in connection with ambient temperature, humidity, and predation. *Arch. Environ. Contam. Toxicol.*, **20**, 20–4.

Fahrig, L. and Freemark, K. (1995) Landscape scale effects of toxic events for ecological risk assessment, in *Ecological Toxicity Testing: Complexity and Relevance* (eds J. Cairns Jr. and B. Niederlehner), Lewis Publishers, Boca Raton, FL, pp. 193–208.

Halley, J.M., Thomas, C.F.G. and Jepson, P.C. (1996) A model of the spatial dynamics of linyphiid spiders in farmland. *J. Appl. Ecol.*, **33**, 471–92.

Hanski, I. and Gilpin, M. (1991). Metapopulation dynamics: brief history and conceptual domain. *Biol. J. Linnean Soc.*, **42**, 3–16.

Harrison, S. (1991) Local extinction in a metapopulation context: an empirical evaluation. *Biol. J. Linnean Soc.*, **42**, 73–88.

Harrison, S. (1993) Metapopulations and conservation, in *Large-scale* Ecology and Conservation Biology (eds P.J. Edwards, R.M. May and N.R. Webb). Blackwell Science Oxford, pp. 111–28.

Harrison, S. and Thomas, C.D. (1991) Patchiness and dispersal in the insect community on ragwort (*Senecio jacobaea*). *Oikos*, **62**, 5–12.

Harrison, S., Murphy, D.D. and Erlich, P.R. (1988) Distribution of the Bay Checkerspot butterfly, *Euphydryas editha bayensis*: evidence for a metapopulation model. *Am. Nat.*, **132**, 360–82.

Hassell, M.P. (1978) *The Dynamics of Arthropod Predator–Prey Systems*, Princeton University Press, Princeton.

Jagers op Akkerhuis, G.A.J.M. and Van der Voet, H. (1992) A dose–effect relationship for the effect of deltamethrin on a linyphiid spider population in winter wheat. *Arch. Environ. Contam. Toxicol.*, **22**, 114–21.

Jepson, P.C. (1989) The temporal and spatial dynamics of pesticide side-effects on non-target invertebrates, in *Pesticides and Non-target Invertebrates* (ed P.C. Jepson), Intercept, Wimborne, Dorset, UK, pp. 95–128.

Jepson, P.C. (1993) Insects, spiders and mites, in *Handbook of Ecotoxicology* Vol. 1 (ed. P. Calow). Blackwell Science, Oxford, pp. 299–325.

Jepson, P.C. and Sherratt, T.N. (1996) The dimensions of space and time in the assessment of ecotoxicological risks, in *Ecotoxicology: Ecological Dimensions* (eds D.J. Baird *et al.*). Chapman & Hall, London, pp. 43–54.

Marinussen, M.P.J.C. and Van der Zee, S.E.A.T.M. (1994) Spatial variability, risk, and extent of soil pollution: conceptual approach of estimating the exposure of organisms to soil contamination, in *Groundwater Quality Management.* Proceedings of the GQM 93 conference held at Talin, September, 1993. IAHS Publication no. 220.

Schoener, T.W. (1991) Extinction and the nature of the metapopulation. *Acta Oecol.*, **12**, 53–75.

Sherratt, T.N. and Jepson, P.C. (1993) A metapopulation approach to modelling the long-term impact of pesticides on invertebrates. *J. Appl. Ecol.*, **30**, 696–705.

Simberloff, D.S. (1988). The contribution of population and community biology to conservation science. *Annu. Rev. Ecol. Syst.*, **19**, 473–511.

Thacker, J.R.M. and Jepson, P.C. (1993) Pesticide risk assessment and non-target invertebrates: integrating population depletion, population recovery, and experimental design. *Bull. Environ. Contam. Toxicol.*, **51**, 523–31.

Thomas, C.F.G. Hol, E.H.A. and Everts, J.W. (1990) Modelling the diffusion component of dispersal during the recovery of a population of linyphiid spiders from exposure to an insecticide. *Funct. Ecol.*, **4**, 357–68.

Wiles, J.A. and Jepson, P.C. (1992) The susceptibility of a cereal aphid and its natural enemies to deltamethrin. *Pest. Sci.*, **36**, 263–72.

9   *Dispersal, heterogeneity and resistance: challenging soil quality assessment*

GÖRAN BENGTSSON

9.1 STATISTICAL ANALYSIS OF SPATIAL HETEROGENEITY

One of the most common, fundamental and intuitively attractive methods to assess environmental impact of pollutants is a survey of species density at a number of sites in a gradient around a known or expected source of pollution. The numbers derived are gross estimates of the site-specific response in mortality, reproduction, and immigration/emigration taken together and would normally represent the cumulated population performance, usually integrated over more than one generation. For certain groups that are more susceptible or exposed than others to the pollutants, a pattern may appear that relates the density variation in the gradient to the exposure data. One such example is given by Bengtsson *et al.* (1983) from their survey of earthworms in forests around a brass mill in south-east Sweden. It seems justifiable to suggest from their Figure 2 that the density of earthworms be inversely related to the concentration of metals in the soil. The predictive power of these data is, however, very weak because of a great variability in numbers between replicate samples. This is especially true if individual species are considered (coefficient of variation, CV, ranging from 100 to 500% for $n = 15$, Table 1 in Bengtsson *et al.*, 1983). The confidence interval for the correlation between the soil metal concentration and the earthworm density is such that a huge number of replicates would be required from a randomly selected site to tell whether earthworm density was influenced by the soil metal concentration or not (Figure 9.1).

Such great spatial variability, considered as indicative of a patchy distribution of individuals, is not limited to earthworms among soil invertebrates. Collembolans are also highly variable in their distribution, and their total

Ecological Risk Assessment of Contaminants in Soil. Edited by Nico M. van Straalen and Hans Løkke. Published in 1997 by Chapman & Hall, London. ISBN 0 412 75900 4

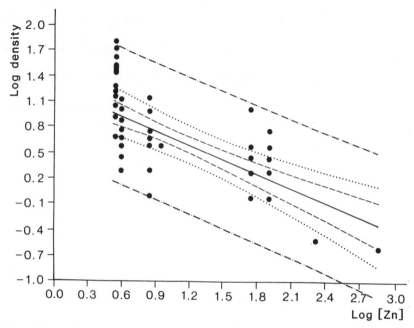

Figure 9.1 Regression of log density (individuals 0.25 m^{-2}) of earthworms on log concentration of extractable zinc (μg g^{-1}) in the A_0 horizon of soils around the brass mill at Gusum, Sweden. The 95% confidence interval is depicted for the average density of earthworms sampled (---), density in one new randomly selected 0.25-m^2 area (-----), and the mean density of 15 randomly sampled 0.25-m^2 areas (·····). (From Bengtsson *et al.*, 1983.)

abundance may vary by more than one order of magnitude in samples taken a couple of centimetres apart from each other in the mor horizon (Figure 9.2). The total number of collembolans at a site is often given with a CV greater than 1.0, and when individual species are considered, CVs may be even greater. If the density distribution of collembolans is assumed to be normal, the number of replicates (n) required to test for the probability of a certain relative difference between density means for two sites can be calculated from the standard deviation (Figure 9.3). A huge number of replicates would clearly have to be collected from a site if, let us say, a 20% difference between the means was the desired sensitivity. The patchy distribution is not always a consequence of the distribution of environmental parameters known to affect the performance of collembolans in laboratory experiments, such as soil moisture (Figure 9.4), organic carbon content, microbial respiration or density of fungal mycelium, that might have been helpful as correlates. Rather, one may have to consider other approaches to account for the variability.

One approach would be to use spatially and temporally oriented statistical analyses. One of the most common approaches to minimize the effect of heterogeneity on the data analysis – the randomized block design – has two

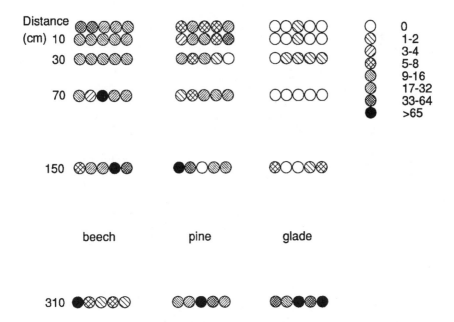

Figure 9.2 Patchiness in total abundance of collembolans in 10.2 cm² × 2 cm soil samples collected in PVC rings positioned close to each other in rows of five with increasing distance between the rows. Eight ranges of abundances are given in different symbols and three different biotopes are represented.

limitations in this respect. First, unless the heterogeneity present in the field is assessed in detail in advance, the block size employed does not necessarily correspond with the inherent heterogeneity pattern. Second, the usefulness of the information on spatial and temporal heterogeneity for the statistical data analysis and for the interpretation of the underlying ecological processes is overlooked. More powerful methods of analysis have been developed that consider the distribution of data in the field because of a stochastic process influenced by autocorrelation (Wilkinson *et al.*, 1983; Burgman, 1987; Zimmerman and Harville, 1991; Ver Hoef and Cressie, 1993). Autocorrelation refers to a tendency towards dependence in values of sampling units that are close to each other (e.g. Legendre, 1993 and references therein). Positive spatial autocorrelation of the data for example of the abundance and activity of soil organisms or metal concentrations in soil invertebrates may result from similarities in microenvironments or from random dispersal of propagules from the natal site. Negative spatial autocorrelation may result from, for example, competition for resources. Whenever an autocorrelation is evident, the basic assumption of independence of data of the classical ANOVA is violated. ANOVA is rather robust to small deviations

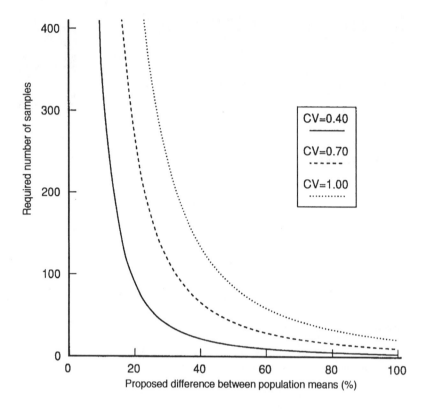

Figure 9.3 The number of samples required to detect a certain relative difference between two population means for different coefficients of variation (CV). The calculation assumes that densities are normally distributed and that the probability that differences are significant is 0.80. It is apparently not worthwhile to aim at detecting a 20% difference in population densities if the coefficient of variation is 0.70 or above.

from normality and homogeneity of variances, but it is sensitive to non-independence of observations, since these will not contribute a full degree of freedom. As a result, insignificant effects revealed by an ANOVA may imply that an effect of pollution, for example, is cancelled by spatial responses. Significant effects, on the other hand, may be reported as an artefact of auto-correlation producing a low within-group variance.

The stochastic process is reflected in the probability distribution of data from a single sampling unit sampled repeatedly over time. This stochastic pattern in nature introduces another source of randomness besides the one assigned to the sampling programme or experiment, so randomization of the design adds more variability to the analysis. Randomization yields a statistical

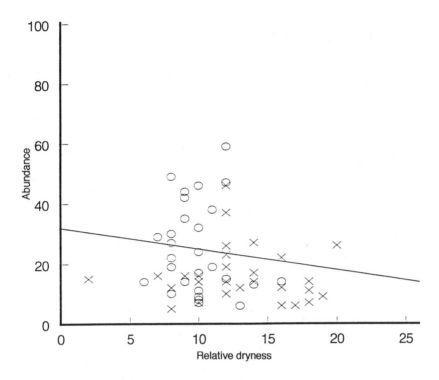

Figure 9.4 Correlation between the abundance of collembolans in 20.5 cm³ mor samples and soil moisture. The soil is water-saturated at a relative dryness of 0. The symbols show the distribution in a beech forest in May (×) and August (○) of the same year. The correlation coefficient is 0.01.

distribution of random errors, δ_{ij}, that can be calculated from the model Y_{ijk} = $\mu + \tau_k + \delta_{ij}$, where Y_{ijk} is the random variable, μ is the mean of the variable, τ_k is the kth treatment effect, i is the ith row and j is the jth column of the grid of plots in the field. A classical ANOVA can be used to estimate μ and τ_k, autocorrelation between sampling units is then modelled from the calculated δ_{ijk} by an iterative procedure, and a better estimate of μ and τ_k is finally derived from the procedure to detect spatial autocorrelation. The choice of model for the autocorrelation may not be overly critical, but the non-parametric Mantel test for spatial effects (Mantel, 1967; Burgman, 1987) does not require any assumptions about the form of autocorrelation or normality of the variables. Spatial methods that account for autocorrelation reduce variances by 20–80% compared with a classical ANOVA (Cullis and Gleeson, 1989; Grondona and Cressie, 1991; Zimmerman and Harville, 1991) and thus allow for the detection of effects on ecological variables that would otherwise escape as 'noise' in the data.

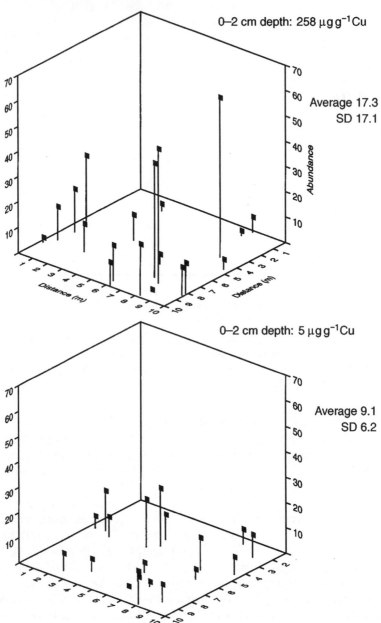

I have re-analysed the data on the distribution of collembolans near the brass mill at Gusum in south-east Sweden reported by Bengtsson and Rundgren (1988). The paper gives only means and variances, and the relationship between density and soil metal concentration is indicated by the bell-shaped distribution of collembolans with distance from the mill. When the total abundance of replicates is plotted against the sample coordinates, the patchiness becomes clearly visible (Figure 9.5). The abundance from one replicate to another is obviously more variable in the polluted soil than in the control soil, possibly in response to heterogeneity in the metal distribution, and most replicates close to each other have the same or similar abundances, that is, they are positively autocorrelated. The pattern of variability and auto-correlation at a site is not constant but varies from one year to another (Figure 9.6), suggesting that populations are very dynamic in their spatial distribution. I used three different statistical methods to re-analyse the data for the whole gradient reported in the Gusum study. When a classical ANOVA is used, the abundance of collembolans in the gradient is statistically invariable with the metal concentration, but if autocorrelation in their distribution is addressed, the metal effect is significant (Table 9.1).

9.2 LIFE-HISTORY MODEL FOR SPATIAL DISTRIBUTIONS

The second approach to account for the variability emphasizes the processes potentially responsible for the patchiness. We may consider two adjacent patches of collembolans and some processes that may contribute to their population dynamics and the distribution of specimens among the patches. The basic assumption for this approach is that the abundance distribution of collembolans at any time is the result of their abundance and performance in

Table 9.1 The data for the abundance of collembolans at sites in a metal concentration gradient in the vicinity of a brass mill, originally reported by Bengtsson and Rundgren (1988), re-analysed by three different statistical methods, two of which (variogram and Mantel test) address spatial heterogeneity. All three analyses were used to detect the effect of metals on the abundance. F is F-statistics, P is the significance level for wrongly rejecting the null hypothesis that there is no effect of metal concentration, τ is the treatment effect of the spatial model, and r is the Mantel statistic

ANOVA		Variogram		Mantel tests	
F	P	τ	P	r	P
2.48	0.21	−4.3	<0.05	0.136	0.001

Figure 9.5 The abundance distribution of collembolans in the A_0 horizon of mor soils in 10×10 m plots at two sites around the brass smelter at Gusum, Sweden. The lower diagram shows the distribution in 20.5 cm³ samples for a control site; the upper diagram shows the distribution of a polluted site.

TEMPORAL VARIABILITY

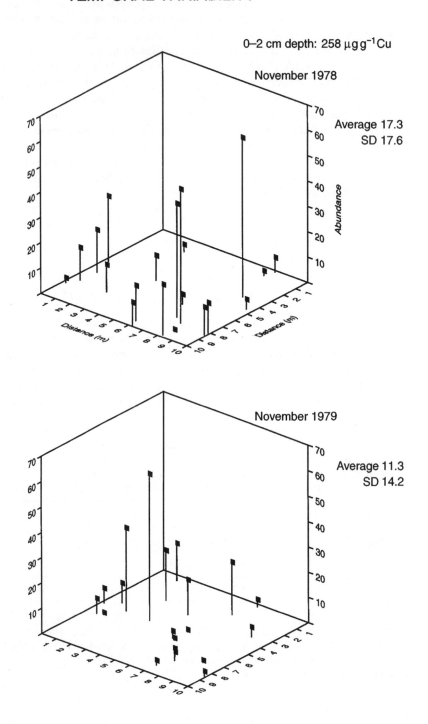

0–2 cm depth: 258 μg g^{-1}Cu

November 1978

Average 17.3
SD 17.6

November 1979

Average 11.3
SD 14.2

the patches during a preceding period such that the abundance in a patch any-time can be calculated and understood only by considering the past structure of the population and the variables affecting that structure. Abundance and demography are influenced by survival, reproduction and dispersal. Both survival and egg production are age-specific and related to soil moisture and food availability. Egg survival is known to depend on soil moisture and temperature, and dispersal tendencies depend on soil moisture, age, food density and population density, among several factors. A set of equations can be written to predict the change in distribution among patches, given that data are available on starting abundances, demography and patch-specific soil moisture and food conditions. Experimental data are needed for the functional relationships between survival, reproduction and dispersal on one hand, and soil moisture, food density and age structure on the other to run the model. Some of this is available in the literature but most of the data still needs to be collected.

Common to both approaches is the importance implicitly ascribed to dispersal, which may be considered as a life-history trait triggered by environmental factors (Southwood, 1962). There is evidence from soil collembolans for a density-dependent dispersal rate, but dispersal is also influenced by food density and moulting intensity (Bengtsson *et al.*, 1994a). Young adults seem to disperse more readily than both older ones and juveniles (M. Sjögren *et al.*, unpublished results), perhaps reflecting a risk-spreading behaviour in the matured collembolans, who lay several separate egg batches during their life time (Bengtsson *et al.*, 1985). This pattern of age variable dispersal and repeated production of dispersed egg batches may help to explain patchiness in distribution and autocorrelation of samples of collembolans; batches of ten eggs or more hatch in a habitat in which a majority of the juveniles may stay until they mature. The subsequent dispersal is slow in a homogeneous soil (some few cm^2 per day for *Onychiurus armatus* (Bengtsson *et al.*, 1994a; M. Sjögren *et al.*, unpublished results) and probably even more slowly in a patchy one, resulting in a fairly uniform density of individuals at short dispersal distances from the natal site.

The movement of collembolans in soil has been modelled by a deterministic compartmental analysis, in which individuals move in a density-dependent fashion from one patch to another. An alternative approach would be to model the distribution of distances from a natal patch, given a certain probability to stay in the natal patch and all others visited. A third approach would be a random walk, in which individuals move in a series of small random steps that can be characterized by a diffusion coefficient. The different modelling approaches are more or less appropriate to describe dispersal under certain experimental conditions. The compartmental analysis has been used to estimate dispersal rates of *O. armatus* between series of discrete and connected habitat patches

Figure 9.6 (see facing page) The abundance distribution of collembolans at a polluted site around the brass smelter at Gusum during two consecutive years. Upper diagram, 1978 data; lower diagram, 1979 data.

(Bengtsson *et al.*, 1994a). It revealed a density-dependent dispersal rate: individuals moved more than twice as fast between patches at a density of 90 000 m^{-2} compared with one of 30 000 m^{-2}. It also showed a food resource-dependent dispersal rate: the residence time in a sandy soil patch increased from about 2 days in the presence of 15 m fungal hyphae g^{-1} soil to 50 days when the soil had been enriched to support 35 m hyphae g^{-1}. It was also clear from the experiments that *O. armatus* can perceive the odour of fungal mycelium from a patch at a distance of at least 400 times their own body length – the compartmental model suggested that specimens moved five times faster from a patch at that distance to another patch enriched with hyphae of a preferred fungal species than they did when the preferred species was not inoculated to the soil.

9.3 DISPERSAL MECHANISMS FOR SINGLE CELLS

Density dependence is also obvious in the dispersal of soil bacteria which are present as free-living cells in the aqueous phase, as adsorbed cells on the solid, abiotic or biotic, phase, and in the interfaces between the gas, aqueous and solid phases. The movement of bacteria is primarily controlled by the flow of water through the pores, but like colloidal particles overall, cells are also diffusing so that some of them move faster and others slower than the average cells. Most interesting from an ecological and ecotoxicological point of view is the distribution of cells between the aqueous and solid phases. Cells move back and forth between the two phases at a rate that depends on a number of factors: water velocity and the physical and chemical characteristics of the two phases and of the cell surface. The adsorption of cells to a mineral particle in a homogeneous soil is limited both by diffusion of cells through the hydrodynamic boundary layer surrounding the particle and by chemical interaction between cell surface components and the inorganic and organic surface characteristics of the particle. Therefore, more cells tend to adsorb at higher flow velocities than at lower because the diffusion limitation is relaxed in a thinner boundary layer, and cells with a high degree of overall surface hydrophobicity adsorb more readily to organic coatings of the mineral particles than cells that are overall hydrophilic. The distribution of cells between the phases can be expressed by a single distribution coefficient reflecting conditions when densities in the two phases are equilibrated or as rate coefficients for the adsorption and desorption processes reflecting kinetics involved in the processes.

Adsorption of the cells results in a retardation of their dispersal compared with water. The breakthrough of cell populations in homogeneous soil columns has been successfully predicted by various models combining expressions for the advective flow, dispersion, and equilibrium- and kinetic-based adsorption (Corapciouglu and Haridas, 1984; Lindqvist and Bengtsson, 1991). The assumptions of these models are essentially analogous to those of the random walk models, i.e. all individuals are assumed to be alike and move independently of each other. Survival and growth rates can be included in the models and may be calculated from dispersal data if the system studied is well known. Density dependence has also been addressed in these models (Lindqvist and Enfield, 1992a), and the empirical evidence from column studies (Figure 9.7) suggests that soil particles are characterized by a limited sorption surface for which cells are competing. Growth conditions, such as energy and nutrient supply, influence the dispersal rate mainly through the behavioural flexibility of the cells. They tend to become more hydrophobic and less negatively charged when growth conditions in the aqueous phase become worse so that the cells separate themselves from the water package carrying it and more readily adsorb to a solid surface (Lindqvist and Bengtsson, 1991). Consequently, cell densities are often higher in the pore water than on the solid particles in nutrient- and carbon-rich water and lower when pore water conditions are oligotrophic.

The classical environmental problem that needs more thorough knowledge of the dispersal process is that of the fate of pathogenic bacteria. Also, other problems are associated with the ignorance of the dispersal behaviour of soil bacteria. One such problem is whether trace organics are degraded by bacteria in the aqueous or solid phase. The answer apparently depends on the distribution of bacteria as well as trace organics between the two phases, but there is still a long way to go before solid predictions can be made. Models including a degradation coefficient have been developed to describe the transport of trace organics in soil columns and successfully fitted to breakthrough data (Van Genuchten and Wagenet, 1989; Gamerdinger *et al.*, 1990). The present data from batch mineralization studies suggest that degradation occurs primarily in the aqueous phase (Subba-Rao and Alexander, 1982; Ogram *et al.*, 1985; Smith *et al.*, 1992) and is desorption rate-limited. Then, the degradation rate of trace organics in a soil is associated with the dispersal rate of bacteria, and those that move faster and spend a larger proportion of their life cycles in the aqueous phase are more likely to control the degradation rate than those that move slower.

Another problem is that of predicting the movement and fate of genetically modified organisms (GMOs). GMOs are commercially developed to accomplish, for example, pollutant degradation, pesticide control, wastewater treatment, polymer production, bioleaching of ores, biomonitoring and crop enhancement. To accomplish these tasks, GMOs have to be released into the soil, move, survive and possibly reproduce. The modified genes need to be stable and their expression alert as the cell moves through the soil. GMOs are known to survive for up to 1 month in unsterile soil (Beringer and Barth, 1988; Van Elsas *et al.*, 1989; Ramos *et al.*, 1991), but the mortality rate may vary from one strain to another and from one habitat to another. Even if the cell survives, the genes transferred or modified may be lost, especially if they are plasmid-bound, or subject to structural rearrangements, that modifies or restrains the

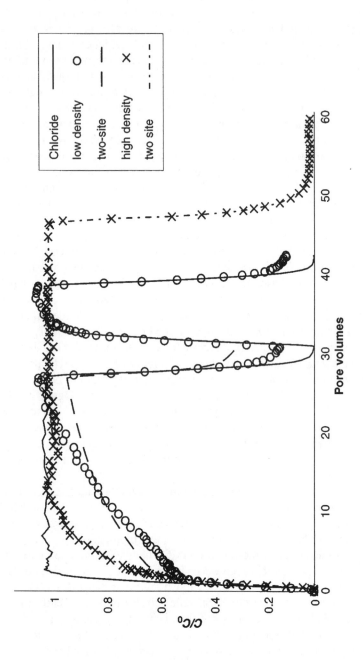

Figure 9.7 Dispersal of two densities of bacteria through a saturated soil column. The conservative tracer (chloride) indicates the breakthrough of water. The dispersal of the population at the low density is repeated after about 30 pore volumes of the first batch of the same population has moved through the column and sorbed to the particles. The y-axis gives the relative density of cells collected at the effluent of the column compared with the density continuously entering the column with the water flow.

gene expression (Schwab *et al.*, 1983; Hardman *et al.*, 1986; Roy and Chakravorty, 1986; Böttcher and Stelzer, 1989). Thus, the release and dispersal of GMOs in the soil environment also call for a description and estimate of the probability that genes are transferred between GMOs and indigenous bacteria. Not much is known about gene transfer frequency between strains in soil in general, but it is assumed that at least plasmid DNA is transferred between indigenous strains. Transfer of plasmid DNA between GMOs and indigenous bacteria has been demonstrated in a number of microcosm experiments (Trevors *et al.*, 1987; Henschke and Schmidt, 1989; Smit *et al.*, 1991), but the transfer frequency is generally low, implying that the probability of a successful transfer of plasmid genes largely depends on the population density. Gene transfer occurs at least in an aqueous solution and may be more likely in bacteria with a high propensity to disperse than in those that are more residential.

9.4 SOIL HETEROGENEITY REFLECTED IN DISPERSAL MODELLING

Much of what is known about dispersal of bacteria relies on data from controlled microcosm experiments with homogeneous soils and a few observations of released bacteria in the field. The translation of dispersal data from the microscopic or laboratory level to the macroscopic or field-scale level involves a number of assumptions and simplifications. Most of them are associated with the expression of local point estimates of flow and sorption into solute transport through field-scale physical heterogeneity. The scale dependency observed for the hydraulic conductivity has been addressed by a stochastic approach, in which the hydraulic conductivity is allowed to vary randomly in space, controlled by a limited number of statistical parameters (Dagan, 1989). As opposed to a deterministic approach, no detailed description of the geometry of the porous media is required. Similarly, the large spatial variability exhibited by sorption rate coefficients at the larger scales can be analysed by considering the interaction between the spatial distribution of sorption kinetics and the fluctuations of the flow velocity (Cvetkovic and Shapiro, 1990). A common modelling approach is to assume a steady-state flow in an ensemble of homogeneous vertical stream tubes with flow properties varying randomly between the tubes (Jury, 1982). It is generally agreed that the most important field-scale variability of solute transport is due to spatial variation in hydraulic conductivity (Gelhar, 1986; Sposito *et al.*, 1986).

With the flow velocities randomly varying, the expected density $\overline{N}(z,t)$ of cells at a given depth z can be expressed as

$$\overline{N} = \int_0^\infty N(z,t,D,v)P(v)\,\mathrm{d}v \qquad (9.1)$$

where D is the above-mentioned diffusion coefficient and $P(v)$ is the probability density function of the flow velocities. Under conditions of steady-state flow, the pore water velocity can be expressed as the saturated hydraulic conductivity, K_s, divided by the water content at saturation. K_s is usually regarded as a random variable with a log-normal distribution. The macro-scale transport process can be described by using either the resident concentration, that is, the mass of solute per volume of aqueous solution, or the solute flux concentration, that is, the average mass of solute per unit time crossing a unit area. Solute flux in a heterogeneous soil will result in a travel time distribution of solute reaching a certain depth in the soil at different times. The stochastic travel time analysis can be combined with an expression for non-equilibrium sorption kinetics, assuming negligible local dispersion and molecular diffusion. The expected mass flux of the sorbing solute is defined by a probability density function of the travel time. This can be obtained either analytically by assuming some correlation structure between the sorption coefficients and the hydraulic conductivity (Cvetkovic and Shapiro, 1990), or numerically by Monte Carlo simulations.

Whether the dispersal behaviour or dispersal rate of bacteria reflects the degree of contamination of a soil is not known. One may anticipate that hydrophobic cells are more retained in a soil that has become severely polluted by organic compounds because of increased hydrophobic interactions between the cells and the solid phase and that the density of dispersing populations is lower in a soil contaminated with, for example mobile metal ions because the ions taken up by the cells reduce the survival rate. It has long been known that trace organics may adsorb to cells (Baughman and Paris, 1981) and recent work has shown that soil bacteria can enhance the mobility of trace organics such as DDT by one to two orders of magnitude, depending on population density, strain characteristics and organic carbon content of the soil (Lindqvist and Enfield, 1992b). Two mechanisms have been suggested to explain this facilitated transport phenomenon. One refers to cells as colloidal particles that are considered as a separate, mobile organic phase to which trace organics can bind in preference to the solid phase (Enfield and Bengtsson, 1988). The other considers cells as modifiers of the solid phase properties and of the aqueous–solid interphase, which may make the solid phase less accessible for the trace organics or influence their speciation by changing pH of the interface (Bellin and Rao, 1993). Whatever the mechanisms, normal densities of bacteria in soils, 10^6–10^8 per gram, would have a significant impact on the mobility of trace organics with a log K_{ow} of 4–5 and above. This may be of special concern in areas with shallow aquifers or in soils subject to bioremediation by adding GMOs or stimulating the growth of indigenous bacteria.

9.5 DISPERSAL OF COLLEMBOLA IN METAL-POLLUTED SOIL

Further data on the influence of soil pollution on dispersal are available from experiments with soil invertebrates. In the above-mentioned experiments on the movement of *O. armatus* between discrete patches of soil, a geometric probability distribution was also appropriate as a description of dispersal distances. The resulting measurement is a propensity to stay or move rather than a rate, which may be a limitation when observations are to be translated between laboratory and field experiments. The probability to leave a patch varied between 0.1 and 0.5 depending on habitat conditions. M. Sjögren *et al.* (unpublished results) used the same species and model to calculate the leaving probability in experimental soils obtained by mixing a highly polluted soil with an unpolluted one from the same brass mill area as re-analysed for autocorrelation of collembolan densities. Specimens had a higher tendency to leave metal-polluted patches compared with less-polluted ones (Figure 9.8). If the variability is constant and reproducible and if 20 replicates are used, the experimental system would allow an enhanced dispersal rate to be detected, with an 80% probability, at concentrations of zinc and copper that are three- to four-fold higher than background levels in a forest soil. That is the same threshold level at which a soil exhibits a slightly reduced respiration and nitrogen mineralization rate (Tyler *et al.*, 1989).

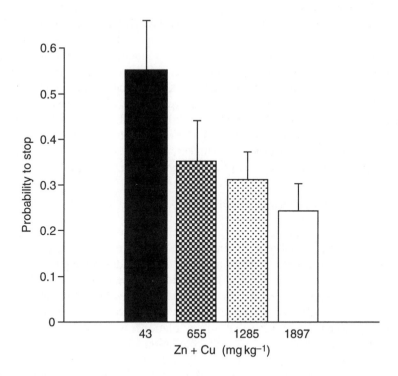

Figure 9.8 The probability to stop for populations of the collembolan species *Onychiurus armatus* while moving through mor soils containing different total concentrations of zinc and copper.

A similar approach was used to describe dispersal tendencies of two other collembolan species, *Folsomia fimetarioides* and *Isotomiella minor*, in an experimental gradient of metal-polluted soil (Bengtsson *et al.*, 1994b). Their dispersal tendencies were generally greater than that of *O. armatus*, with a probability to leave a patch above 0.5. They responded to increasing metal concentrations in the gradient by an exponential decrease of the probability to settle in a habitat. The probability model was developed to account also for metal resistance, which was demonstrated in a separate experiment by contrasting the survival of two populations of *F. fimetarioides* in the unpolluted and most heavily polluted sections of the gradient. The model also addressed the capability of collembolans to perceive changes in metal concentrations of a soil as they move from one habitat to another in the gradient. Comparisons of model simulations and observations of the distribution of collembolans in the gradient suggested that the dispersal pattern in the gradient was largely influenced by the size of metal resistance in the studied population and by their perception of the concentration differences of metals in the gradient. These findings have at least two implications for the assessment of toxicity of pollutants in a soil. First, deciding what populations are used in assessment procedures is important. Those that are resistant to the pollutants will give an under-estimated bias and those that are not will give an over-estimated bias. Second, knowing that most elements and compounds have a heterogeneous distribution in soil (Boekhold *et al.*, 1991; Bonmati *et al.*, 1991), one may anticipate that collembolans, and also other soil animals, experience a mosaic of fine-grained patches with respect to the distribution of metals and distribute themselves among the less-polluted patches. The reason that specimens collected in a polluted soil survive in far less than 100% when they are exposed to the same, homogenized soil in laboratory experiments may simply be the loss of the mosaic of refuges. Thus, it would pay to learn how resistance and dispersal have co-evolved in a gradient of pollution.

9.6 PERSISTENCE BY RESISTANCE AND DISPERSAL

Peace *et al.* (1989) have attempted to decide the conditions under which a population will persist in a moving climatic gradient by the dual processes of evolution and dispersal. Their approach can be adopted and developed to reveal how natural selection and pollution influence population persistence (Figure 9.9). Consider a species distributed across the pollutant gradient, with density $N(x, t)$ at location x at time t, where we measure time in units of generations. We assume that subpopulations are connected across the gradient by dispersal and that the population in a constant environment grows exponentially with some per capita growth rate equal to $\ln\overline{W}$, where $\overline{W}(x, \overline{z}, t)$ is the mean fitness of a subpopulation at any time t and $\overline{z}$ is the mean phenotype. The population density changes through time can be expressed as a standard reaction–diffusion equation with a density independent population growth:

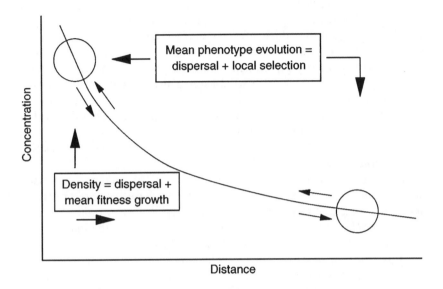

Figure 9.9 A model for the dual effect of evolution of resistance and dispersal for the distribution of a species in a pollutant gradient. Individuals are assumed to disperse in both directions in the gradient and dilute adaptations to high pollutant concentrations and redistribute resistant genotypes along the gradient.

$$\frac{\partial N}{\partial t} = D\frac{\partial^2 N}{\partial x^2} + N\ln\overline{W} \qquad (9.2)$$

where $N(x,t)$ is the density and D is the diffusion coefficient. The first term on the right-hand side quantifies how random dispersal causes population density at each location along the gradient to depend on the density at nearby locations. As a result, polluted environments will have higher population densities and unpolluted environments lower densities than would occur without dispersal.

The rate of evolution of the mean phenotype can be expressed as a function of dispersal and local selection:

$$\frac{\partial \overline{z}}{\partial t} = G\frac{\partial \ln\overline{W}}{\partial \overline{z}} + D\frac{\partial^2 \overline{z}}{\partial x^2} + 2D\frac{\partial \ln N}{\partial x}\frac{\partial \overline{z}}{\partial x} \qquad (9.3)$$

Where G is a constant additive genetic variance for the character, and G $\partial\ln\overline{W}/\partial\overline{z}$ quantifies the change of the mean phenotype caused by weak selection during one generation (Lande, 1976, 1979). The second term on the right-hand side of the equation gives the changes of the mean phenotype due to dispersal by simple mixing, and the last term arises because individuals immigrating to a particular locality are more likely to come from contiguous areas with relatively high population density. The fitness function, $\overline{W}(x, \bar{z}, t)$ has to be specified and three of the specified variables can be examined empirically. The *in situ* per capita per generation growth rate, $\ln\overline{W}$, can be determined in laboratory experiments with each population raised in its native soil. Net population dispersal can be estimated as described in laboratory experiments with native soils or in fenced field plots where unbounded random dispersal as well as preferential dispersal in pollutant-stratified soil can be observed. By experimental variation of population density, N, the density variation, if any, of the *in situ* growth rate can be examined. The additive genetic variance can be obtained from heritability analyses.

Dispersal itself may also be subject to selection. Although life-history traits overall show low heritability – possibly because they are affected by more environmental variation than the underlying metric traits (Price and Schluter, 1991) – there are some data in the literature on the inheritance of dispersal (Sakai, 1958; Roff, 1986; Johnson and Gaines, 1990). Several classes of theoretical models have been used to derive the evolutionary background for inheritance of dispersal. One class of models uses the idea of dispersal as an evolutionarily stable strategy (ESS) (Hamilton and May, 1977), and another large class of models assumes that dispersal has evolved as a mechanism for the avoidance of inbreeding. I have seen no attempts in that direction but measuring the survival and reproduction of dispersing compared with resident animals as a means to assess cost and benefits of dispersal in a gradient of pollution would be interesting. It is not an easy experimental task but it would be helpful as a test of the evolutionary basis of dispersal and as a means to establish soil quality criteria for the long-term influence of pollution.

9.7 CONCLUSIONS

Much of the present efforts to improve the protocol for soil quality assessment with respect to contamination are based on standardization of sensitive laboratory tests, even including standardized soils, and on monitoring certain key processes in the field. Standardizations are certainly helpful for intercalibration of test procedures and for rapid screening of the toxicity of chemicals but their present popularity tends to take researchers far away from the problems facing assessment approaches in real field soils. I have used this space to argue for the movement of attention among soil ecotoxicologists from standardization tests to the challenges represented by spatial and temporal heterogeneity, dispersal, and resistance, because those are in my opinion among the most important and under-rated phenomena to include in soil quality assessment.

9.8 SUMMARY

Soil quality assessment used to determine the effects and fate of contaminants needs to address heterogeneity in the distribution of organisms and contaminants and the rates of their movements, as well as the evolution of resistance in exposed populations. Being largely overseen in the present approaches to quality assessment, heterogeneity, movement and resistance are common phenomena in soils. The chapter gives as an example of heterogeneity the patchy distribution of collembolans among adjacent 10 cm² soil samples which were collected. The variability around the mean density of collembolans was so large that a huge number of replicates would be required to detect even a 50% change in density. The patchiness of collembolans did not correlate with any common environmental variables, such as soil moisture or food density, and it is suggested that a life-history modelling approach, including movement between patches, may help in understanding and predicting patchiness. Data on the distribution of collembolans in soils contaminated with metals from a brass mill are re-examined to demonstrate that statistical models accounting for autocorrelation are also useful to reveal effects of contaminants when the spatial and temporal variability in population density is large. Different approaches to describe and predict the movement pattern of collembolans and bacteria in soils are compared. Special attention is given to the dispersal of GMOs, to the enhanced transport of hydrophobic pollutants associated with dispersing bacteria and to dispersal of collembolans in metal contaminated soils. The quantification of the latter phenomenon is helpful in trying to understand the persistence of collembolans in contamination gradients as a result of dispersal and local adaptations to contaminants.

REFERENCES

Baughman, G.L. and Paris, D.F. (1981) Microbial bioconcentration of organic pollutants from aquatic systems – a critical review. *Crit. Rev. Microbiol.*, **8**, 205–28.

Bellin, C.A. and Rao, P.S.C. (1993) Impact of bacterial biomass on contaminant sorption and transport in a subsurface soil. *Appl. Environ. Microbiol.*, **59**, 1813–20.

Bengtsson, G. and Rundgren, S. (1988) The Gusum case: a brass mill and the distribution of soil Collembola. *Can. J. Zool.*, **66**, 1518–26.

Bengtsson, G., Nordström, S. and Rundgren, S. (1983) Population density and tissue metal concentration of lumbricids in forest soil near a brass mill. *Environ. Pollut. Ser. A*, **30**, 87–108.

Bengtsson, G., Gunnarsson, T. and Rundgren, S. (1985) Influence of metals on reproduction, mortality and population growth in *Onychiurus armatus* (Collembola). *J. Appl. Ecol.*, **22**, 967–78.

Bengtsson, G., Hedlund, K. and Rundgren, S. (1994a) Food- and density-dependent dispersal: evidence from a soil collembolan. *J. Anim. Ecol.*, **65**, 513–20.

Bengtsson, G., Rundgren, S. and Sjögren, M. (1994b) Modelling dispersal distances in a soil gradient: the influence of metal resistance, competition, and experience. *Oikos*, **71**, 13–23.

Beringer, J.E. and Barth, M.J. (1988) The survival and persistence of genetically-engineered microorganisms, in *The Release of Genetically-engineered*

Microorganisms (eds M. Sussman, C.H. Collins, F.A. Skinner and D.E. Stewart-Tull), Academic Press Ltd., London, pp. 29–46.

Boekhold, A.E., Van der Zee, S.E.A.T.M. and De Haan, F.A.M. (1991) Spatial patterns of cadmium contents related to soil heterogeneity. *Water Air Soil Pollut.*, **57–58,** 479–88.

Bonmati, M., Ceccanti, B. and Nannieri, P. (1991) Spatial variability of phosphate, urease, protease, organic carbon and total nitrogen in soil. *Soil Biol. Biochem.* **23,** 391–6.

Böttcher, I. and Stelzer, W. (1989) *In vitro* studies on long-term stability of R plasmids in *Escherichia coli* K12. *J. Basic Microbiol.*, **29,** 643–53.

Burgman, M.A. (1987) An analysis of the distribution of plants on granite outcrops in southern Western Australia using Mantel tests. *Vegetatio*, **71,** 79–86.

Corapciouglu, M.Y. and Haridas, A. (1984) Transport and fate of microorganisms in porous media: A theoretical investigation. *J. Hydrol.*, **72,** 149–69.

Cullis, B.R. and Gleeson, A.C. (1989) The efficiency of neighbour analysis for replicated variety trials in Australia. *J. Agric. Science, Cambridge*, **113,** 233–9.

Cvetkovic, V. and Shapiro, A. (1990) Mass arrival of sorptive solutes in heterogeneous porous media. *Water Resour. Res.*, **26,** 2057–67.

Dagan, G. (1989) *Flow and Transport in Heterogeneous Formations*, Springer-Verlag, New York.

Enfield, C.G. and Bengtsson, G. (1988) Macromolecular transport of hydrophobic contaminants in aqueous environments. *Ground Water*, **26,** 64–70.

Gamerdinger, A.P., Wagenet, R.J. and Van Genuchten, M.Th. (1990) Application of two-site/two-region models for studying simultaneous nonequilibrium transport and degradation of pesticides. *Soil Sci. Soc. Am. J.*, **54,** 957–63.

Gelhar, L.W. (1986) Stochastic subsurface hydrology. From theory to applications. *Water Resour. Res.*, **22,** S135–45.

Grondona, M.O. and Cressie, N. (1991) Using spatial considerations in the analysis of experiments. *Technometrics*, **33,** 381–92.

Hamilton, W.D. and May, R.M. (1977) Dispersal in stable habitats. *Nature*, **269,** 578–81.

Hardman, D.J., Gowland, P.C. and Slater, J.H. (1986) Large plasmids from soil bacteria enriched on halogenated alkanoic acids. *Appl. Environ. Microbiol.*, **51,** 44–51.

Henschke, R.B. and Schmidt, F.R.J. (1989) Survival, distribution, and gene transfer of bacteria in a compact soil microcosm system. *Biol. Fertil. Soils*, **8,** 19–24.

Johnson, M.L. and Gaines, M.S. (1990) Evolution of dispersal: theoretical models and empirical tests using birds and mammals. *Ann. Rev. Ecol. Syst.*, **21,** 449–80.

Jury, W.A. (1982) Simulation of solute transport using a transfer function model. *Water Resour. Res.*, **18,** 363–8.

Lande, R. (1976) Natural selection and random genetic drift in phenotypic evolution. *Evolution*, **30,** 314–34.

Lande, R. (1979) Quantitative genetic analysis of multivariate evolution, applied to brain: body size allometry. *Evolution*, **33,** 402–16.

Legendre, P. (1993) Spatial autocorrelation: trouble or new paradigm? *Ecology*, **74,** 1659–73.

Lindqvist, R. and Bengtsson, G. (1991) Dispersal dynamics of groundwater bacteria. *Microbial Ecol.*, **21,** 49–72.

Lindqvist, R. and Enfield, C.G. (1992a) Cell density and non-equilibrium sorption effects on bacterial dispersal in groundwater microcosms. *Microbial Ecol.*, **24,** 25–41.

Lindqvist, R. and Enfield, C.G. (1992b) Biosorption of dichlorodiphenyl-trichloroethane and hexachlorobenzene in groundwater and its implications for facilitated transport. *Appl. Environ. Microbiol.*, **58**, 2211–18.

Mantel, N. (1967) The detection of disease clustering and a generalized regression approach. *Cancer Res.*, **27**, 209–20.

Ogram, A.V., Jessup, R.E., Ou, L.T. and Rao, P.S.C. (1985) Effects of sorption on biological degradation rates of (2,4-dichlorophenoxy)acetic acid in soils. *Appl. Environ. Microbiol.*, **49**, 582–7.

Peace C.M., Lande R. and Bull J.J. (1989) A model of population growth, dispersal and evolution in a changing environment. *Ecology*, **70**, 1657–64.

Price, T. and Schluter, D. (1991) On the low heritability of life-history traits. *Evolution*, **45**, 853–61.

Ramos, J.L., Duque, E. and Ramos-Gonzalez, M-I. (1991) Survival in soils of an herbicide-resistant *Pseudomonas putida* strain bearing a recombinant TOL plasmid. *Appl. Environ. Microbiol.*, **57**, 260–6.

Roff, D.A. (1986) The genetic basis of wing dimorphism in the sand cricket, *Gryllus firmus* and its relevance to the evolution of wing dimorphism in insects. *Heredity*, **57**, 221–31.

Roy, S. and Chakravorty, M. (1986) Spontaneous deletions of drug-resistance determinants from *Salmonella typhimurium* in *Escherichia coli*. *J. Med. Microbiol.*, **22**, 119–23.

Sakai, K.I. (1958) Studies on competition in plants and animals. IX. Experimental studies on migration in *Drosophila melanogaster*. *Evolution*, **12**, 93–101.

Schwab, H., Saurugger, P.N. and Lafferty, R.M. (1983) Occurrence of deletion plasmids at high rates after conjugative transfer of the plasmids RP4 and RK2 from *Escherichia coli* to *Alcaligenes eutrophus* H16. *Arch. Microbiol.*, **136**, 140–6.

Smit, E., Van Elsas, J.D., Van Veen, J.A. and De Vos, W.M. (1991) Detection of plasmid transfer from *Pseudomonas fluorescens* to indigenous bacteria in soil by using bacteriophage ϕRf2 for donor counterselection. *Appl. Environ. Microbiol.*, **57**, 3482–8.

Smith, S.C., Ainsworth, C.C., Traina, S.J. and Hicks, R.J. (1992) Effect of sorption on the biodegradation of quinoline. *Soil Sci. Soc. Am. J.*, **56**, 737–46.

Southwood, T.R.E. (1962) Migration of terrestrial arthropods in relation to habitat. *Biol. Rev.*, **37**, 171–214.

Sposito, G., Jury, W.A. and Gupta, V.K. (1986) Fundamental problems in the stochastic convection–dispersion model of solute transport in aquifers and field soils. *Water Resour. Res.*, **22**, 77–88.

Subba-Rao, R.V. and Alexander, M. (1982) Effect of sorption on mineralization of low concentrations of aromatic compounds in lake water samples. *Appl. Environ. Microbiol.*, **44**, 659–68.

Trevors, J.T., Barkay, T. and Bourquin, A.W. (1987) Gene transfer among bacteria in soil and aquatic environments: a review. *Can. J. Microbiol.*, **33**, 191–8.

Tyler, G., Balsberg Påhlsson, A-M., Bengtsson, G., Bååth, E. and Tranvik, L. (1989) Heavy-metal ecology of terrestrial plants, microorganisms and invertebrates. *Water Air Soil Pollut.*, **47**, 189–215.

Van Elsas, J.D., Trevors, J.T., Van Overbeek, L.S. and Starodub, M.E. (1989) Survival of *Pseudomonas fluorescens* containing plasmids RP4 or pRK2501 and plasmid stability after introduction into two soils of different texture. *Can. J. Microbiol.*, **35**, 951–9.

Van Genuchten, M. Th. and Wagenet, R.J. (1989) Two-site/two-region models for pesticide transport and degradation: theoretical development and analytical solutions. *Soil Sci. Soc. Am. J.*, **53**, 1303–10.

Ver Hoef, J.M. and Cressie, N. (1993) Spatial statistics: analysis of field experiments, in *Design and Analysis of Ecological Experiments* (eds S.M. Scheiner and J. Gurevitch), Chapman & Hall, New York, London, pp. 319–41.

Wilkinson, G.N., Eckert, S.R., Hancock, T.W. and Mayo, O. (1983) Nearest neighbor (NN) analysis of field experiments. *J. Roy. Statist. Soc. Ser. B Meth.*, **45**, 152–212.

Zimmerman, D.L. and Harville, D.A. (1991) A random field approach to the analysis of field-plot experiments and other spatial experiments. *Biometrics*, **47**, 223–39.

PART FIVE

The role of ecological modelling

10 The use of models in ecological risk assessment

JOKE VAN WENSEM

10.1 DIFFERENT TYPES OF MODELS

For a long time, risk assessment for chemicals has been based almost exclusively on single-species laboratory tests. Methods have been developed to extrapolate results from single-species tests to so-called 'safe levels' in the environment. Despite the usefulness of some of these methods, it is felt that single-species tests alone do not provide sufficient basis for ecological risk assessment, and the use of micro- and mesocosms, and mathematical models is advocated to acquire additional data.

Models contribute to our understanding of the system that is studied, and can be used as a tool to discover research needs. This way of using models may be considered as purely scientific. Models can also be applied in risk assessment for predicting the risk of different (new) chemicals, for different systems, over longer time scales, for different environmental conditions. In this way models can be used as screening tools for chemicals, classifying chemicals in the range: no risk–high risk. This may save time and money during regulation procedures.

Models and calculation methods for risk assessment have been grouped into six different categories by the Health Council of the Netherlands (1994):

1. models for the distribution of chemicals in the various abiotic compartments of ecosystems;
2. toxicokinetic models for uptake and elimination of chemicals by organisms;
3. population models (responses of population variables to toxic stress);
4. models for biotic distribution (including models of food-chains and secondary poisoning);
5. food-web models;
6. ecosystem models (among which are so-called mass flow models).

Ecological Risk Assessment of Contaminants in Soil. Edited by Nico M. van Straalen and Hans Løkke. Published in 1997 by Chapman & Hall, London. ISBN 0 412 75900 4

However, not all models for ecological risk assessment can be placed in one of these categories, as they consist of a mixture of the different model types.

This chapter will deal with the question of which models are presently available to support ecological risk assessment for chemicals. Some recent modelling approaches that are promising and relevant for ecological risk assessment will be discussed at the 'state-of-the-art' level. Examples for two groups of models will be given: models that predict the influence of chemicals on processes in ecosystems (food-web and ecosystem models) and models (calculation methods) for the risk of exposure of higher organisms to chemicals (secondary poisoning). The first group consists of complicated models which are strongly based on ecological theories. It is a relatively new development to incorporate the effects of chemicals into this type of models; hence this group will be emphasized.

10.2 FOOD-WEB AND ECOSYSTEM MODELS

10.2.1 LITTORAL ECOSYSTEM RISK ASSESSMENT MODEL (LERAM)

LERAM is a bioenergetics ecosystem effects model (Hanratty and Stay, 1994). Flows of energy and biomass are modelled through 10 compartments in a littoral ecosystem. The compartments consist of (inorganic) nutrients, detritus and aggregations of species at the same taxonomic level with a similar function in the ecosystem (Figure 10.1). The model calculates changes in biomass in the compartments as a result of effects of chemicals on growth rates.

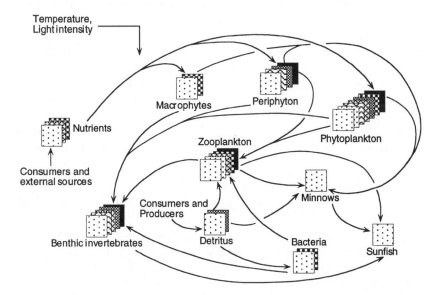

Figure 10.1 The littoral ecosystem risk assessment model (LERAM). Arrows represent flow of energy and biomass. Boxes represent aggregates of species with similar taxa and function. (From Hanratty and Stay, 1994.)

The biomass growth rate of primary producers (macrophytes, periphyton and phytoplankton; see Figure 10.1) is determined by the difference between photosynthetic rate and the rate of biomass loss. The photosynthetic rate depends on the maximum photosynthetic rate, light intensity (including shading by other groups), nutrient concentrations (N, P, Si) and temperature. Biomass loss rates of primary producers consist of rates of grazing, sinking, natural death and respiration. The biomass growth rate of consumers (bacteria, zooplankton, benthic invertebrates, minnows and sunfish; see Figure 10.1) is determined by the difference between consumption rate minus respiration, natural death and predation rates. The function for biomass growth rate of consumers is based on feeding functions, and allows for influences of temperature, prey density, feeding preferences, excretion and assimilation efficiencies.

The effects of chemicals on growth rates of the biomass are introduced using dimensionless scaling constants. For the estimation of these scaling constants, it is assumed that biomass loss due to a chemical is proportional to the LC_{50} (or EC_{50}) for a chemical. For example: if the concentration of a chemical is $0.5 \times LC_{50}$ for 96 hours, then for the model calculations it is assumed that after 4 days the biomass is decreased by 25% compared with the initial biomass. The scaling constants for the physiological parameters that determine biomass changes are calculated in such a way that the biomass changes, predicted by the LC_{50} values, are achieved. The scaling constants may range between 0 and 3: if a scaling constant of 3 is not large enough to simulate the required change in biomass, then a direct mortality rate, due to the chemical, is introduced. Thus, the effects of chemicals are not modelled, but introduced by scaling constants based on existing LC_{50} values for the chemical. This procedure assumes a linear exposure–response relationship, which overestimates the toxicity. The authors stated that this conservative estimate of the toxicity fits in LERAM as it was designed to determine the lowest observable effect concentration.

The means and standard deviations of the parameters in the model were calibrated using data from untreated field enclosures. The model was used to predict effects on a littoral ecosystem based on LC_{50} data for chlorpyrifos. The predictions were compared with data from field enclosures treated with chlorpyrifos. Both deterministic and probabilistic (using Monte Carlo iterations) predictions were made in the calibration and validation procedure. Figure 10.2 gives the LERAM predictions of zooplankton biomass compared with the measured biomass in enclosures treated with 0.5 µg l^{-1} chlorpyrifos. At this concentration the highest deviations between prediction and measurements were observed, up to 400%. At other concentrations the deviation ranged between 0 and 100%. The authors agree, that for a further development of the model, a more precise simulation of some of the compartments, and an improvement of the modelling of toxic effects are needed initially.

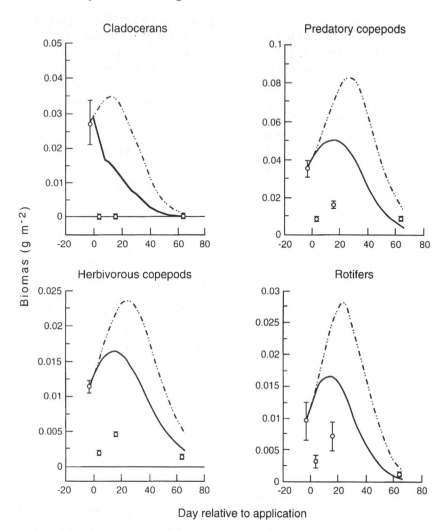

Figure 10.2 The LERAM predictions of zooplankton biomass in low (0.5 µg l⁻¹) chlor-pyrifos treatment enclosures (solid line) and in the same enclosures without treatment (dashed line). The open circles with bars are biomass estimates for abundances measured in the low treatment enclosures. Bars are the error associated with the biomass estimation technique. (From Hanratty and Stay, 1994.)

10.2.2 FOOD-WEB MODELS

This group of models describe below-ground food-webs in different terrestrial systems, including arable farming systems and shortgrass prairie (De Ruiter *et al.*, 1993a,b; Moore and De Ruiter, 1993; Moore *et al.*, 1993). The food-webs are represented by functional groups at different trophic levels

(Figure 10.3). The models calculate feeding rates among the functional groups and predict ecosystem functioning in terms of carbon and nitrogen mineralization. The models need parameter values for: biomass, specific death rates, prey preference weighing factors, assimilation efficiencies, production efficiencies and C : N ratios in all functional groups. Calculations are based on the assumption that the annual average biomass production rate of a functional group balances the rate of material loss through natural death and predation (i.e. assuming steady-state on a time scale of years).

The calculated carbon and nitrogen fluxes were compared with field data from different sites and from different arable farming systems. The results demonstrate that the food-web models were able to simulate nitrogen mineralization rates close to the rates obtained from field measurements (De Ruiter *et al.*, 1993b). The first trial to simulate effects of disturbance in the food-webs on flows of carbon and nitrogen was a rather theoretical: complete functional groups were eliminated from the food-web model, while the population size of the other groups remained the same. The conclusion of this exercise was that removal of a functional group led to a higher reduction in the nitrogen mineralization rate compared with the direct contribution of this group to the mineralization rate (De Ruiter *et al.*, 1993a). Recently, disturbances were simulated more realistically using a dynamic version of the food-web model. Disturbances were modelled as effects on consumption rates and the model calculate new equilibrium population sizes and equilibrium flow rates. Based on these new equations mineralization rates are calculated and compared with the rates of the undisturbed food-webs (De Ruiter *et al.*, 1994). The conclusion from this study was that effects of disturbances are often disproportionate. For example, the direct contribution of bacteria on the nitrogen mineralization rate is about 40 kg ha^{-1} yr^{-1}; increasing the predation rate on bacteria affects the overall mineralization rate by approximately 2 kg ha^{-1} yr^{-1}. The direct contribution of bacterivorous nematodes to nitrogen mineralization is less than 1 kg ha^{-1} yr^{-1}, but increasing the predation rate increases the overall nitrogen mineralization to more than 5 kg ha^{-1} yr^{-1}. These results are qualitatively similar to the outcome of the theoretical analysis of Yodzis (1988) and the outcome of controlled manipulation experiments (Paine, 1980, 1992).

10.2.3 CONTAMINANTS IN AQUATIC AND TERRESTRIAL ECOSYSTEMS

A family of models has been developed to calculate the accumulation of contaminants in ecosystems such as grassland, ruderal, shrubland and forest. The models are designated by CATS (contaminants in aquatic and terrestrial ecosystems). In this chapter the accumulation of cadmium in a meadow ecosystem is discussed (Traas and Aldenberg, 1992). This version of the CATS models predicts cadmium accumulation under different loading scenarios. The model simulates biomass fluxes through a food-web, consisting of functional groups and, separately, cadmium fluxes through a food-web and

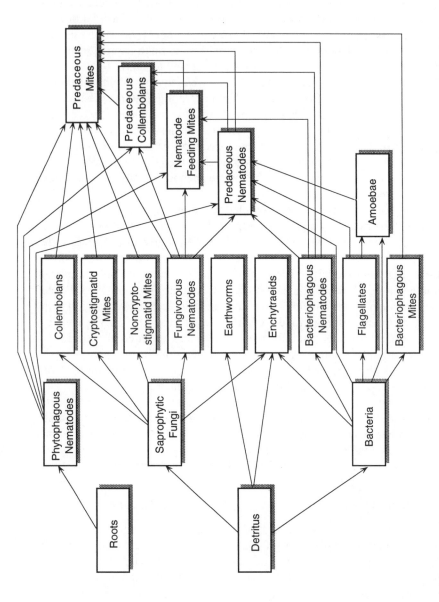

Figure 10.3 Diagram of a soil food-web for an arable field with an integrated management practice (low levels of N fertilizer and pesticide application and reduction of soil tillage) at the Lovinkhoeve experimental farm, Marknesse, Noordoostpolder, the Netherlands. (From De Ruiter *et al.*, 1993a.)

the environment. The (a)biotic compartments and biomass fluxes are shown in Figure 10.4 ; the related diagram of cadmium compartments and fluxes is not shown. The biotic compartments are named using the most dominant member of the functional group.

The increase in biomass of the functional groups is modelled by a set of independent logistic population growth equations. This means that the growth of biotic compartments is limited by food density and undefined environmental conditions. The use of such a generic growth model is, according to the authors, justified since functional groups are modelled (and not single populations). Another argument for this growth model is that the CATS models were developed to study the impact of emission reduction of chemicals and not to predict true population field dynamics.

Toxic effects of cadmium are not modelled. It is possible, however, to predict toxic effects when additional information about the relationship between accumulated levels of cadmium in biota and the effects of these internal levels is available.

The parameters in the model are given by ranges or distributions; using Monte Carlo simulations, distributions of accumulated levels in the compartment are calculated. Calibration of the model was carried out using simultaneous sampling of the parameters. The resulting model output was compared with ranges of acceptable output, based on measurements of the present situation in the field. The successful parameter sets were used to predict cadmium accumulation in the future under different scenarios. A representative result is shown in Figure 10.5.

10.2.4 OTHER ECOSYSTEM MODELS

Other modelling efforts concerning mass flows in ecosystems in relation to chemicals can be found in Mathes and Schulz-Berendt (1988), who have developed a simulation model for the nitrogen cycle in soils; in Bosatta (1982), who has modelled decomposition of organic matter in relation to acidification; and in Van Wensem (1992), who gives a simulation model for isopod-mediated decomposition in terrestrial microcosms, influenced by chemicals.

10.3 MODELS AND CALCULATION METHODS FOR SECONDARY POISONING

10.3.1 SECONDARY POISONING IN TERRESTRIAL FOOD CHAINS

This calculation method has been developed in the Netherlands as an addition to the direct effects assessment (DEA) method. The extrapolation method of Aldenberg and Slob (1993) is used for the DEA, and in case of a low number of data a modification of the EPA method is used, with assessment factors of 10, 100 or 1000 (Slooff, 1992). Both methods do not take the risk of secondary poisoning into account; therefore calculation methods for secondary

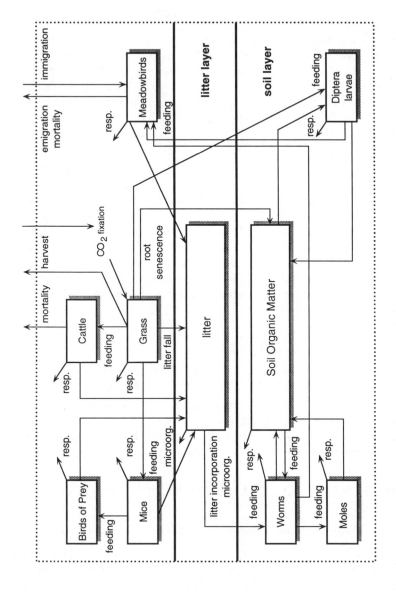

Figure 10.4 Diagram of biomass (boxes) and biomass fluxes (arrows) in a peat-land meadow. The biomass compartments represent functional groups and are called after the (locally) most dominant species in the group. (From Traas and Aldenberg, 1992.)

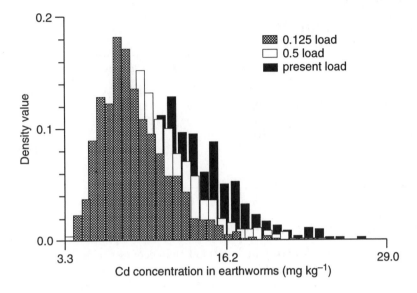

Figure 10.5 Distribution of cadmium (mg kg⁻¹ dry weight) in earthworms in the year 2015 under different loading scenarios, simulated by the CATS model. (From Traas and Aldenberg, 1992.)

poisoning were developed (Romijn *et al.*, 1993, 1994). The goal of these methods is to calculate maximum permissible concentrations for secondary poisoning in water and soil.

The food-chain soil → earthworm → bird, or mammal feeding on earthworms, was used as a model for soil (Romijn *et al.*, 1994). A maximum permissible concentration (MPC) in soil (mg chemical kg⁻¹ soil) is calculated by dividing the 'extrapolated' no observed effect concentration (NOEC, mg chemical kg⁻¹ food) for birds or mammals feeding on earthworms by the bioconcentration factor (BCF) between soil and earthworms. The 'extrapolated' NOEC is derived by the method of Aldenberg and Slob or the EPA method, based on NOECs of a chemical for mortality, reproduction or growth in birds or mammals. NOECs and BCFs are derived from the literature. In this way MPCs were calculated for lindane, dieldrin, DDT, pentachlorophenol (PCP), cadmium, inorganic mercury and methyl mercury, and compared with the results from the DEA method. The results (Table 10.1) indicate that in the case of cadmium and methyl mercury, higher organisms (represented by birds and mammals feeding on earthworms) might not be protected by the MPCs derived from the DEA.

An analogue of this method has been presented by Romijn *et al.* (1993) for water. In this case the food-chain water → fish → bird, or mammal feeding on fish, has been used as a model for secondary poisoning.

Table 10.1 Maximum permissible concentrations (MPCs) (mg kg⁻¹) for birds and mammals, based on secondary poisoning and MPCs based on the Dutch direct effects assessment (DEA) method. The MPCs for secondary poisoning that are lower than the MPCs for DEA are given in bold. (From Romijn *et al.*, 1994.)

Compound	MPC birds	MPC mammals	MPC DEA
Lindane	0.35	5.53	0.005
Dieldrin	0.88	1.06	0.05
DDT	0.78	27.22	0.01
PCP	16.5	3.72	0.17
Cadmium	**0.015**	0.867	0.17
Inorganic mercury	1.11	5.55	0.20
Methyl mercury	**0.011**	**0.012**	0.1

10.3.2 SECONDARY POISONING: A PROBABILISTIC APPROACH

The risk of secondary poisoning in a flood plain ecosystem has been estimated using a food-web model (Noppert *et al.*, 1993). The food-web contains various trophic levels, including plants (grass), soil invertebrates (including earthworms), mammals (three different mice species, mole, badger), and birds of prey (little owl and kestrel). For the uptake of chemicals, bioconcentration factors (BCFs), pH-dependent regression formulas or toxicokinetic one-compartment models were used, depending on the trophic level of the organism and the type of chemical. The values of the parameters: concentration in the soil, age, body weight and daily food intake were given as frequency distributions. The model calculations resulted in frequency distributions for concentrations of chemicals in the biota and exposure concentrations for the higher trophic levels. Comparison with critical levels (for example NOECs) gives an indication of the risk of secondary poisoning.

10.3.3 HERRING GULL BIOENERGETICS AND PHARMACODYNAMICS MODEL

This model combines the energy budget of the adult herring gull (including energy expenditures due to existence metabolism and foraging activities) and contaminant pharmacokinetics to predict changes in the body burden of the adult gull with time (Norstrom *et al.*, 1991). At the same time the concentration of the contaminant in the adult gull and the eggs is calculated. Feeding, overwintering, chick feeding, egg deposition and other gull activities are take into account in the energy budget. The environmental parameters, photoperiod and ambient temperature are taken into account and used to define energy intake and proportion of fat in the bird. The uptake of the chemical is modelled by using total food uptake, the chemical concentration in the aquatic portion of the food, and a chemical absorption efficiency factor. The fate and distribution of the chemical is described by a two-compartment model. The chemical is supposed to be partitioned between body fat and plasma and removed from the plasma at a certain clearance rate.

The model is based on equations for a generic, non-passerine, bird and the pharmacokinetic parameters take size into account. Although this makes the model, in principle, suitable for a more general application, the authors state that it has only been shown to be valid for herring gulls, and that most default values in the model apply to Lake Ontario in Canada.

10.3.4 EXPOSURE IN MODELS FOR SECONDARY POISONING

The models of Romijn *et al.* (1993, 1994) and Noppert *et al.* (1993) use BCF values to calculate the uptake of chemicals by organisms. BCF values are derived from the literature and vary strongly, as BCFs are dependent (among other things) on the concentration of the chemical in soil or water and the local soil or water conditions. To illustrate this situation the distribution of BCFs (on log-scale) for cadmium in mussels is given in Figure 10.6 (Haenen *et al.*, 1993). The BCFs were collected over a period of approximately 10 years at seven different locations near the Dutch coast. The variation in BCFs for terrestrial systems may be even higher as result of the heterogeneous distribution of chemicals in and the variability of physicochemical properties.

The variation in BCF values illustrates the need for methods to estimate the exposure and uptake of chemicals in the field. This is a rather undeveloped

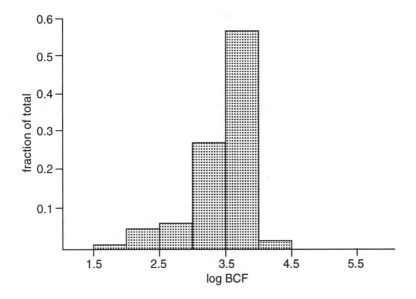

Figure 10.6 Distribution of bioconcentration factors (BCF) for cadmium in mussels (BCF = concentration in mussels/concentration in sea water), on logarithmic scale. The cadmium concentrations in mussels and sea water were measured at seven different locations on the Dutch coast at several occasions over a 10-year period. (From Haenen *et al.*, 1993.)

area in ecotoxicological modelling. Recently, a simple model has been developed to estimate concentrations of chemicals in soil organisms in relation to the spatial variability of the chemical and the size of the organisms (Marinussen and Van der Zee, 1994). The model is based on very simple assumptions, and should be considered to be a first attempt to model chemical uptake in heterogeneously contaminated soils.

The model calculates cadmium uptake by (theoretical) organisms, which have differently sized home ranges, and move randomly in a defined area that is polluted heterogeneously with cadmium. The model shows that the risk of exceeding certain threshold levels in the organisms becomes higher with increasing mobility, i.e. increasing size of the home range, of the organisms. The cadmium concentrations in organisms that do not move (e.g. plants) follow the concentrations in the soil. With increasing mobility the organisms are increasingly able to reach all different cadmium concentrations present at the modelled area and will sooner or later be exposed to high concentrations, i.e. eventually, all organisms with high mobility contain high cadmium concentrations.

10.4 VALIDATION AND LIMITATIONS: THE PERFECT MODEL

10.4.1 THE RELATIONSHIP BETWEEN MODELS AND DATA

In this chapter, the term 'calibration' is used to describe the process of adjusting parameter values and initial values for variables by comparing model calculations with data sets. As the models are used to predict effects of chemicals, calibration is executed using data sets derived from situations without chemicals. 'Validation' is defined as the process of comparing model calculations for effects of chemicals with field data sets. Validation, however, may not be the correct word; one must keep in mind that agreement between model calculations and field data does not necessarily mean that a model is valid. More information about the validity of the model can be gained from situations in which a model is falsified using field data.

A very important application of models in ecological risk assessment is the use for estimation of 'safe levels' in the environment. At present, models for risk assessment are often implemented in environmental protection policies (at least in the Netherlands) before they are validated properly. This is a result of the lack of field data, as well as laboratory data, on effects of chemicals. As a result, models are often 'prematurely' used in environmental policy-making; once a model has been accepted by policy makers, scientists appear to lose their grip on it. To avoid this situation, validation should be considered as a part of the development of models for ecological risk assessment.

Consequently, validation of models for ecological risk assessment with field data is one of the future challenges in ecotoxicology. Validation of models may lead to considerable research efforts, taking into account that an agreement between model predictions and field data does not necessarily mean that the model is valid. For example, at a recent workshop on the behaviour of pesticides in the environment, it was agreed that a satisfactory validation of a model for the abiotic distribution of pesticides needs at least a comparison with 10 different sets of field data.

10.4.2 FOOD-WEB AND ECOSYSTEM MODELS

LERAM and the food-web models are most similar: LERAM calculates effects of chemicals in terms of biomass changes in the compartments and the food-web models calculate effects in terms of ecosystem function expressed as rates of carbon and nitrogen mineralization. CATS calculates accumulation of chemicals; this is primarily relevant for our understanding of the behaviour of chemicals in ecosystems and for risk assessment for groups of organisms at higher trophic levels.

All food-webs in the models are simplified by using functional groups instead of individual species. The advantage of functional groups is that the number of parameters is limited. The disadvantage is, however, that the parameter values for functional groups have to be constructed from the values for individual species, which is often hazardous, especially with respect to effects of chemicals. This problem depends partly on the definition of a 'functional group'. In LERAM and the food-web models, functional groups appear to be relatively homogeneous, mainly because the groups are based on taxonomy, but in the CATS model the functional group 'worms' for example consists of all detritivore and microbivore species, including isopods, oribatid mites, springtails and annelid worms.

The influence of environmental conditions, the growth equations and toxic effects are modelled in very different ways (or not taken into account at all). Table 10.2 shows very clearly that an increasing level of 'realism' in the model is paid for by an increasing ratio between the number of parameters and state variables (compartments) in the model. Considering the results of the validation of LERAM and the food-web models, as opposed to models when considering the level of 'realism', no conclusion can be drawn about which type of model offers the best description of the field situation.

All the food-web and ecosystem models mentioned are essentially ecological models. The incorporation of toxic effects into the models seems to be problematic. The most obvious way to incorporate toxic effect in the models is to adjust parameter values in relation to the presence of the chemical. However, this requires knowledge of dose–effect relationships and other chemical-specific information, which is often not present. In LERAM this problem is avoided by calculating a dimensionless factor that works on all physiological parameters, but it was concluded that this part of the model especially should be improved (Hanratty and Stay, 1994). This also means that the use of food-web and ecosystem models for routine screening of

Table 10.2 The number of parameters, state variables and their ratios in LERAM (Hanratty and Stay, 1994), the food-web model for an arable farming system (De Ruiter *et al.*, 1993b) and the CATS model for a peaty meadow (Traas and Aldenberg, 1992)

	Littoral ecosystem model	Food-web model	CATS model
Parameters	275	95*	83/99
State variables	10	19*	25
Ratio	28	5*	3–4

*(Bio)Mass estimates are considered as state variables; on average, five parameters per model compartment were assumed.

chemicals seems out of the question, because of the high number of chemical specific parameters in these models. Therefore, the application for risk assessment will probably be limited to the 'scientific use' of the models, as a tool both for understanding a system and to indicate research needs.

10.4.3 MODELS FOR SECONDARY POISONING

The models presented for secondary poisoning were developed for different purposes; those of Romijn *et al.* (1993, 1994) were developed as a part of a procedure for the derivation of 'safe' levels in the environment, but because of their simplicity, large numbers of chemicals can be screened. The herring gull model is species- and location-specific, and requires a considerable amount of chemical-specific information. However, the model of Noppert *et al.* (1993) takes an intermediate position with regard to simplicity and specific data requirement.

One of the main problems in using the relatively simple models for secondary poisoning, i.e. Romijn *et al.* (1994), and to a lesser extent Noppert *et al.* (1993), is the scarcity or complete lack of data. NOECs for wildlife species do not generally exist, and values for standard laboratory test species (rat, mouse, guinea pig, fowl, etc.) are used as an alternative. This requires NOECs for laboratory animals to be extrapolated to NOECs for wildlife. Differences in laboratory versus field conditions, metabolic rate, food choice, binding of chemicals to the food, sensitivity to the chemical, exposure routes and exposure time have to be taken into account. Ruys and Pijnenburg (1991) proposed an extrapolation method which calculates NOECs for wild, fish-eating birds from NOECs obtained for laboratory birds, eating cereals. However, this is mainly a correction for the energy content of food, and only partly solves the problem.

10.4.4 THE PERFECT MODEL

While the models discussed are still under development, Hanratty and Stay (1994) refer in their paper on the field validation of LERAM to a description of a 'perfect' model for ecological risk assessment. Such models should be able to predict the probability of hazard for many types of chemicals, during all seasons, in different geographical regions that contain different species and environmental conditions. The models should be able to estimate ecological risk due to multiple doses of one or more chemicals, should extrapolate to doses that have not been tested, and should represent biological and environmental interactions that influence the effect of chemical stressors. Models must incorporate basic ecological theory that can be tested scientifically.

It is obvious that it is much easier to give a description of a 'perfect' model than actually to develop this model. We might even question if we need 'perfect' models for the purposes to which we apply them. Considering the models that have been discussed in this chapter in relation to the 'perfect' model, there is still a very long way to go. It appears that even the incorporation of toxic effects, which may be considered as the essential part of models for ecological risk assessment, is problematic. A 'high level of realism' and 'being able to predict the probability for many types if chemicals' appear to be opposing demands. Although in practice (very) simple models are used for the screening of large numbers of chemicals, realistic models are limited in their general applicability by their high system- and chemical-specific data requirements.

10.5 SUMMARY

Model predictions can be used in ecological risk assessment for chemicals, in addition to data obtained from single-species tests and micro- and mesocosm studies. Models that predict effects of chemicals on ecosystem structure and nutrient cycling (food-web and ecosystem models), as well as those for predicting the risk of secondary poisoning, have been developed recently and appear to be promising tools in ecological risk assessment. Examples of both categories are discussed. The incorporation of toxic effects of chemicals in food-web and ecosystem models appears to be difficult, due to the large number of parameters in the models that may be affected by a chemical. The models for secondary poisoning avoid this problem mainly by using bioconcentration factors, which are, however, very variable. The validation of the models is often limited and considered as one of the major future challenges in ecotoxicology. When looking at descriptions of a 'perfect' model, it is concluded that there is still a long way to go in the development of models for ecological risk assessment. In practice, very simple models are used for the screening of large numbers of chemicals, whereas realistic models are limited in their general applicability by their high (system- and chemical-) specific data requirements.

ACKNOWLEDGEMENTS

I thank Nico van Straalen and peter De Ruiter for their critical comments. Miranda Aldham-Breary corrected the English.

REFERENCES

Aldenberg, T. and Slob, W. (1993) Confidence limits for hazardous concentrations based on logistically distributed NOEC data. *Ecotox. Environ. Safety*, **25**, 48–63.

Bosatta, E. (1982) Acidification and release of nutrient from organic matter – a model analysis. *Oecologia (Berlin)*, **55**, 30–3.

De Ruiter, P.C., Moore, J.C., Zwart, K.B., Bouwman, L.A., Hassink, J., Bloem, J., De Vos, J.A., Marinissen, J.C.Y., Didden, W.A.M., Lebbink, G. and Brussaard L. (1993a) Simulation of nitrogen mineralization in the below-ground food-webs of two winter wheat fields. *J. Appl. Ecol.*, **30**, 95–106.

De Ruiter, P.C., Van Veen, J.A., Moore, J.C., Brussaard, L. and Hunt, H.W. (1993b) Calculation of nitrogen mineralization in soil food-webs. *Plant Soil*, **157**, 263–73.

De Ruiter, P.C., Neutel, A. and Moore, J.C. (1994) Modelling food-webs and nutrient cycling in agro-ecosystems. *Trends Ecol. Evol.*, **9**, 378–83.

Haenen, C.P.L., Van Der Tol-Bakker, M. and Schobben, J.H.M. (1993) *BCF's nader Bekeken. Onderzoek naar Methoden en Variatie*. Rapport DGW-93.031. Ministerie van Verkeer en Waterstaat, RWS/RIKZ, The Netherlands (in Dutch).

Hanratty, M.P. and Stay, F.S. (1994) Field evaluation of the littoral ecosystem risk assessment model's predictions of the effects of chlorpyrifos. *J. Appl. Ecol.*, **31**, 439–53.

Health Council of the Netherlands: Committee on Ecotoxicological Issues (1994) *Ecotoxicology is Well on Course*. Publication no. 1994/13. Health Council of the Netherlands, The Netherlands (in Dutch, with English summary).

Marinussen, M.P.J.C. and Van Der Zee, S.E.A.T.M. (1994) *Spatial Variability, Risk and Extent of Soil Pollution: Conceptual Approach of Estimating the Exposure of Organisms to Soil Contamination*. Groundwater Quality Management, IAHS Publ. No. 220, pp. 245–55.

Mathes, K. and Schulz-Berendt, V.M. (1988) Ecotoxicological risk assessment of chemicals by measurements of nitrification combined with a computer simulation model of the N-cycle. *Tox. Assess.*, **3**, 271–86.

Moore, J.C. and De Ruiter, P.C. (1993) Assessment of disturbance on soil ecosystems. *Vet. Parasitol.*, **48**, 75–85.

Moore, J.C., De Ruiter, P.C. and Hunt H.W. (1993) Soil invertebrate/micro-invertebrate interactions: disproportionate effects of species on food-web structure and function. *Vet. Parasitol.*, **48**, 247–60.

Noppert, F., Dogger, J.W., Balk, F. and Smits, A.J.M. (1993) Secondary poisoning in a terrestrial food chain: a probabilistic approach, in *Integrated Soil and Sediment Research: a Basis for Proper Protection* (eds H.J.P. Eijsackers and T. Hamers), Kluwer Academic Publishers, Dordrecht, pp. 303–7.

Norstrom, R.J., Clark, T.P. and MacDonald, C.R. (1991) *The Herring Gull Bioenergetics and Pharmacodynamics Model*. Compiled spreadsheet ver 1.0. Manual. National Wildlife Research Centre, Canadian Wildlife Service, Environment Canada, Hull, Québec.

Paine, R.T. (1980) Food-webs: linkage, interaction strength and community infrastructure. *J. Anim. Ecol.*, **49**, 667–85.

Paine, R.T. (1992) Food-web analysis through field measurements of per capita interaction strength. *Nature*, **355**, 73–5.

Romijn, C.A.F.M., Luttik, R., Van der Meent, D., Slooff, W. and Canton, J.H. (1993) Presentation of a general algorithm to include effect assessment on secondary poisoning in the derivation of environmental quality criteria. Part 1. Aquatic Food Chains. *Ecotox. Environ. Safety*, **26**, 61–85.

Romijn, C.A.F.M., Luttik, R. and Canton, J.H. (1994) Presentation of a general algorithm to include effect assessment on secondary poisoning in the derivation of environmental quality criteria. 2. Terrestrial Food Chains. *Ecotox. Environ. Safety*, **27**, 107–27.

Ruys, M.M. and Pijnenburg, J. (1991) *Maximaal Toelaatbare Risicowaarden in het Aquatische Milieu van Steltlopers en Zeevogels. Methode en Berekening op Basis van doorvergiftiging*. Intern rapport Rijkswaterstaat DGW/AOCE, The Netherlands (in Dutch).

Slooff, W. (1992) *RIVM Guidance Document – Ecotoxicological Effect Assessment: Deriving Maximum Tolerable Concentrations (MTC) from Single-Species Toxicity Data*. Report No. 719102018, RIVM, The Netherlands.

Traas, Th.P. and Aldenberg, T. (1992) *CATS-1: A Model for Predicting Contaminant Accumulation in a Meadow Ecosystem. The Case of Cadmium*. Report No. 719103001, RIVM, The Netherlands.

Van Wensem, J. (1992) *Isopods and Pollutants in Decomposing Leaf Litter*. PhD Thesis, Vrije Universiteit, Amsterdam.

Yodzis, P. (1988) The indeterminacy of ecological interactions, as perceived by perturbation experiments. *Ecology*, **72**, 1964–72.

11 A *physiologically driven mathematical simulation model as a tool for extension of results from laboratory tests to ecosystem effects*

JØRGEN AAGAARD AXELSEN

11.1 SIMULATION MODELS FOR POPULATION DYNAMICS

One of the great challenges in ecotoxicology is to extrapolate results from exposure–effect experiments in the laboratory to a prediction or understanding of exposure–effect relationships at ecosystem level. A range of tests has been developed or is under development (Kula *et al.*, 1995) for the assessment of exposure – effect relationships in terrestrial systems. Most of these are single-species tests and the interpretation of the results are problematic because they do not include the interactions, which take place in a community consisting of a large number of interacting species under natural climatic conditions.

The most important factors ignored by single-species tests carried out at constant temperature are:

1. Interactions between species
- predator–prey interactions
- competition
2. Temperature effects
- development and activity of poikilotherm organisms
- degradation and evaporation of chemicals.

When these factors are ignored, a good correlation between laboratory test results and field effects cannot be expected.

A tempting way to solve the problem of extending results from laboratory tests to prediction of ecosystem effects, is to develop mathematical simulation models. In such a model it is possible to integrate:

Ecological Risk Assessment of Contaminants in Soil. Edited by Nico M. van Straalen and Hans Løkke. Published in 1997 by Chapman & Hall, London. ISBN 0 412 75900 4

- climatic input;
- dose–response relations on survival, growth and reproduction;
- simulations of chemical fate;
- development of the populations of the most important species of the ecosystem;
- exposure; and
- complicated ecological interactions such as competition and predator–prey relationships.

A simulation model with these capabilities using results from well-tested and reliable laboratory tests as input would be a strong tool in hazard and risk assessment. There are at least four ways to use a model. Firstly, it is possible to make predictions of population densities (including extinction) of the involved species at any concentration of the chemical relative to clean soil. Secondly, it is possible to make predictions of short-term and long-term effects. Thirdly, it is possible to make predictions at different climatic conditions, including the worst case. Fourthly, by Monte Carlo simulations it is possible to carry out risk assessment by allowing random choices of both initial densities of the involved species and climatic input files (different locations and years). When using Monte Carlo simulations it is important to allow a realistic range of input parameters since predictions are dependent on the range of allowed parameter values (Hommen *et al.*, 1993).

To make a reliable model, it is important to have reliable input data, since high-quality input is a prerequisite for producing high-quality output. Further, it is of utmost importance that the model incorporates ecosystem interactions. It should incorporate the mechanisms concerning interactions between species, interactions between the chemicals and the involved species, and the fate of chemicals. Making models which do not take these mechanisms into account is not true modelling, but rather a fitting process, and predictions made with such models will too often fail. Last, but certainly not least, the model should have a very high degree of credibility, which can be obtained by a number of positive tests against independent field data, before being released for use as a tool in hazard and risk assessment.

The core of a simulation model with the abilities mentioned above is a model which can describe the interactions between the species of the ecosystem without the influence of xenobiotics. In this chapter I will focus on a population dynamic model which is built on a few very important driving variables and incorporates the mechanisms of the interaction. Firstly, I will describe a model which can simulate predator–prey interactions at different concentrations of the insecticide dimethoate in a microcosm laboratory test system (Axelsen *et al.*, 1997a), and secondly I will discuss the possibilities of further developing the simulation model in relation to predicting ecological risks in terrestrial systems.

11.2 THE MICROCOSM MODEL

The model works on the population level and has been developed to simulate a test system consisting of the collembolan *Folsomia fimetaria* and the predacious mite *Hypoaspis aculeifer* in 30 g of soil in a plastic microcosm (5.5 cm high × 6.0 cm diameter). The test animals are taken from synchronous cultures 16–19 days old and left in the microcosm for 21 days at 20°C with a surplus of baker's yeast as food for *F. fimetaria*. When dimethoate is added to the microcosm, it is mixed into the soil. The experiment is terminated by extraction of the animals from the soil in a high gradient extractor for 2 days.

The interactions between the mites and Collembola are simulated by a physiologically driven population dynamic model. The model type has been developed by Gutierrez *et al.* (1984, 1987, 1988a) and Graf *et al.* (1990a), and is able to simulate multi-trophic interactions in agricultural ecosystems. The model has been used to simulate the growth and yield of plants (Gutierrez *et al.*, 1988a, 1991, 1994; Graf *et al.*, 1990b), competition between plant species for light, nitrogen and water (Graf *et al.*, 1990c; Graf and Hill, 1992), damage caused by insect pests (Gutierrez *et al.*, 1984, 1987, 1988b; Tamó *et al.*, 1993), and predator–prey interactions (also host–parasitoid interactions; Gutierrez *et al.*, 1987, 1988c; Axelsen, 1994).

The microcosm model is a two-trophic level model which generates a very detailed simulation of the growth of male and female populations of *F. fimetaria* and *H. aculeifer* and the interactions between the two species. In the model *F. fimetaria* was split into 11 stages (eggs, juveniles, eight adult instars and dead individuals), while *H. aculeifer* was split into five stages (eggs, non-feeding larvae, juveniles, adults and dead individuals). Thus, the model describes complex interactions between twenty two prey types (eleven stages of both males and females) and four predator types (only juveniles and adults are predaceous; Figure 11.1)

A simulation runs through the following steps:

1. Read input data and open the four populations.
2. Enter daily loop:
   (a) Read temperature input and convert to physiological time units
   (b) Calculate the food demands of all populations
   (c) Let all populations forage, i.e. obtain their food supply (*F. fimetaria* populations get their demand fulfilled).
   (d) Allocate the supply to the different physiological needs.
   (e) Let all populations grow, age and suffer mortality.
   (f) Print output, which can be number, mass, average weight, numbers killed by predators, food demand, and % satiation of food demand of all instars of all species.
3. Stop after 21 days.

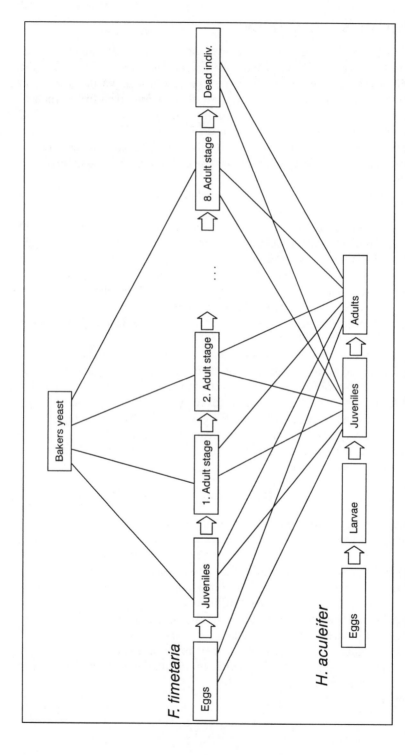

Figure 11.1 The model simulates the complicated predator–prey interactions that takes place when different stages of the animals are present within the microcosm.

The steps (b)–(e) within the daily loop are the corner stones of the simulation for the population development of the species involved (Figure 11.2). These corner stones are described in words below and some important parts in mathematical terms in Boxes 11.1–11.3.

Although the model simulates the interactions at constant temperatures, it has been prepared for simulations at fluctuating temperatures by using a physiological time-scale (degree-days, °D), which should always be used when dealing with poikilothermic organisms.

The demand for food includes the demands for chemical conversion, respiration, egestion, reproduction, growth and reserves. Daily demands for reproduction, respiration and growth are strongly temperature-dependent and are based on a respiration rate, maximum reproduction rate and maximum growth rates, which are all expressed in units of °D^{-1}.

Food (supply) acquisition is simulated by the Gutierrez–Baumgärtner functional response equation (Gutierrez *et al.*, 1981), which makes food acquisition dependent on the amount of prey and the demand of the predator. The functional response equation is used for all combinations of prey and predator, and therefore the total food acquisition of predators is the sum of the acquisitions from predation processes towards all kinds of prey. Similarly, the losses from a prey population are the cumulated losses to all predators. The mathematical formulation of the foraging process is described in Box 11.1.

The allocation of the supply to the different physiological needs followed a metabolic pool model (Graf *et al.*, 1990a), where resources are allocated according to a rule of priority. First priority is excretion which has to be subtracted from the acquired amount of food, second the cost of chemical conversion, third respiration, fourth reproduction, fifth growth and finally

Population

Demand	Supply
Egestion	Prey availability
Chemical conversion	Capture efficiency
Respiration	Prey density
Reproduction	Preference
Growth	Functional response
Reserves	Predator density

Allocation	Growth
1. Egestion	Survival
2. Chemical conversion	Loss to predators
3. Respiration	Migration
4. Reproduction	Ageing
5. Growth	
6. Reserves	

Figure 11.2 Diagram illustrating the contents of the 'corner stones' of the population model.

reserves. Resources are allocated to the different needs in relation to the sup-ply/demand ratio. If supply exceeds demand, the demand is satisfied, and if supply is less than demand, the demand is not satisfied. Thus, if there is a shortage of food, the growth rate or the reproduction rate of a population can-not reach maximum. If a shortage is severe, growth and reproduction will cease, and if the demand for respiration cannot be satisfied the fraction of the population that cannot satisfy the respiratory demand will die.

Ageing is controlled using varying distributed delay (Manetch, 1976; Vansickle, 1977). A distributed delay procedure is a book-keeping device that splits each life stage of an organism into a number of substages (k) and keeps track of the amount/numbers of individuals in each substage. Substage 1 con-tains the youngest part of the population in a life stage, and substage k contains the oldest part. Ageing is done by transferring the content of one substage (i) to the next substage ($i+1$) (Figure 11.3). Transfer from the last substage of a life stage is the input to the first substage of the following life stage. The distributed delay adds variation to the average transit time of a life stage, a variation which can be described by an Erlang density distribution (Manetch, 1976). The dis-tributed delay is described in mathematical terms in Box 11.2.

Growth is performed by multiplying the content in each delay process (the content of each substage) by a growth rate, which depends on the supply/demand ratio for growth, survival rate and loss to predators.

The toxicity of dimethoate was included by a modification of the dose–response equation given by Lacey and Mallett (1991):

$$z = \frac{CS}{1 + e^a\, C^{-a/\ln(LC_{50})}} \qquad (11.1)$$

where z is the survival, CS is the control survival, C is the concentration of the toxin and a is a constant. The parameters were derived from the results of an experiment with the test system described in this paper (Krogh, 1995). Since these parameters were used to describe survival over a 21-day period and the model makes daily calculations, the curve could not be used directly. Therefore, it was necessary to turn a LC_{50} curve with an exposure duration of 21 days into a LC_{50} curve with the duration of 1 day, by fitting an LC_{50} to the

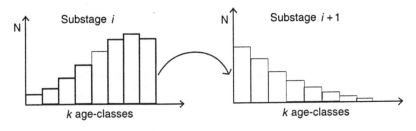

Figure 11.3 Distributed delay splits life stages into k substages. The arrow indicates transfer from one life stage to the next.

Box 11.1 Outline of the simulation model

Food acquisition is described by a Gutierrez–Baumgaertner functional response model (Gutierrez *et al.*, 1981):

$$M^*_{z,v,x,y} = D_{z,v,x,y}\left(1 - \exp\left(\frac{-s_{z,v}M'_{z,v,x,y}}{D_{z,v}}\right)\right) \tag{11.2}$$

where '*x,y*' denotes prey species *x* instar *y*, '*z,v*' denotes predator species *z* instar *v*, $M'_{z,v,x,y}$ is the available biomass of prey species *x* instar *y* to predator species *z* instar *z*, $D_{z,v}$ is the demand of predator *z* instar *v* (g prey per g predator per degree-day) and $s_{z,v}$ is the search rate of predator *z*, instar *v* (per predator per degree-day). The amount of prey acquired by predator species *z* instar *v* ($M^*_{z,v,x,y}$) is in balance with the amount lost from prey species *x* instar *y* to predator species *z* instar *v* ($L_{x,y,z,v}$). Total food acquisition ($M^*_{z,v}$) by predator *z* instar *v*, is the accumulation of acquisitions from all prey species and instars:

$$M^*_{z,v} = \sum_{x=1}^{P}\sum_{y=1}^{I_x} M^*_{x,y,z,v} \tag{11.3}$$

where *x* denotes the prey species, *P* is the number of prey species, *y* is the prey instar, I_x is the number of instars of the prey species *x*, and $M^*_{x,y,z,v}$ instar is the mass of prey instar y consumed by the predator z instar v.

The loss ($L_{x,y}$) from prey *x* instar *y* may be calculated in a very similar way:

$$L_{x,y} = \sum_{z=1}^{P_r}\sum_{v=1}^{I_z} L_{z,v,x,y} \tag{11.4}$$

where P_r is the number of predator species, I_z is the number of instars of predator species *z* and $L_{z,v,x,y}$ is the loss to predator *z* instar *v* from prey species *x* instar *y*. The available amount of prey species *x* instar *y* for predator *z* instar *v*, $M'_{z,v,x,y}$ may be calculated by:

$$M'_{z,v,x,y} = -q_{x,y}\,\gamma_{z,v,x,y}\,M_{x,y} \tag{11.5}$$

where $\gamma_{z,v,x,y}$ is the preference of predator *z* instar *v*, for prey *x* instar *y* and $q_{x,y}$ is a factor which secures that the amount of prey species *x* instar *y* available for the predators does not exceed the amount present in the population, and $M_{x,y}$ is the amount of prey *x* instar *y*. The factor $q_{x,y}$ is given by:

$$q_{x,y} = \frac{M_{x,y}}{\displaystyle\sum_{z=1}^{P_r}\sum_{v=1}^{I_z}\gamma_{z,v,x,y}\,M_{x,y}} = \frac{1}{\displaystyle\sum_{z=1}^{P_r}\sum_{v=1}^{I_z}\gamma_{z,v,x,y}} \qquad q \in [0,1] \tag{11.6}$$

where the symbols have been defined above and *q* must not exceed 1. This means that the different kinds of prey were distributed as possible prey for the different kinds of predators according to relative availability.

Box 11.2 The use of distributed delay to add variation to stage durations

When distributed delay is used to describe insect ageing, each life stage is split into a number of substages, k. The ageing is done by transferring a proportion of the quantity in a substage to the next substage. In mathematical terms the dynamics of a distributed delay is described by:

$$\frac{dQ(t)}{dt} = x(t) - y(t) \tag{11.7}$$

where $Q(t)$ is the quantity (numbers or mass) in the system, $x(t)$ is the input to the substage, and $y(t)$ is the output. If a stage i is divided into k substages, Equation (11.7) expands to a series of Equations (11.8):

$$dQ_1/dt = x(t) - r_1(t)$$
$$dQ_2/dt = x(t) - r_2(t)$$

$$\begin{matrix} \cdot & \cdot & \cdot \\ \cdot & \cdot & \cdot \\ \cdot & \cdot & \cdot \end{matrix} \tag{11.8}$$

$$dQ_k/dt = r_{k-1}(t) - y(t)$$

where Q_i is the quantity in substage $i = 1, 2, 3 \ldots k$, and r_i is the flow from Q_{i-1} to Q_i. The flow through the system is described by:

$$r_i(t) = \frac{k}{\text{DEL}} Q_i(t) \tag{11.9}$$

where DEL is the average duration of the life stage.
The delay process adds variation to the transit time $f(t)$ which is described by an Erlang density function:

$$f(t) = \frac{(k/\text{DEL})^k t^{k-1} \exp(-kt/\text{DEL})}{(k-1)!} \tag{11.10}$$

---

results and keeping the a-value given by Krogh (1995). This is not the ideal way to include toxicity, since it makes the model dependent on temperature and time step. Therefore, the model cannot simulate toxic effects at temperatures other than 20°C and a time step of 1 day. I will return to this point later.

11.3 PROGRAMMING AND INPUT

11.3.1 PROGRAMMING

The model has been programmed in the C++ programming language using the object-oriented modelling approach. The central part of the model is a number of C++ classes, containing the characteristics of a delay process, a population, a chemical and an arthropod community. In C++ it is possible to define classes that inherit the characteristics from a base class and add some new special characteristics. Thus, new species, described by a new class which inherits most properties from a basic 'population class', are easily added to the program. Similarly, the basic characteristics of a chemical have been described in a basic class which can be inherited by classes describing specific chemicals. All species and chemicals are variables in a 'community class' which takes care of the interactions. This program construction makes it easy to add new species and chemicals to the simulation model.

11.3.2 SIMULATION RESULTS

Most simulation results will be presented elsewhere (Axelsen *et al.*, 1997, a,b), but a few examples of the behaviour of the model will be presented here.

The model was used to investigate the dose–response curve of juvenile *H. aculeifer* in a normal microcosm test series compared with tests series where the food demand of *H. aculeifer* was satiated throughout the test (Figure 11.4). The latter is a simulation of a test series where *H. aculeifer* is provided with a surplus of food (Krogh, 1995), i.e. there is no starvation due to a shortage of food at dimethoate concentrations which are harmful to *F. fimetaria* but harmless to *H. aculeifer*. Due to the priority in the allocation of available resources to the different demands of a population, starvation reduces the reproductive output. The simulation results in Figure 11.4 demonstrate an indirect effect of a chemical. The number of juveniles is rather low in microcosm simulations relative to the satiated situation, even without the presence of the dimethoate effect (Figure 11.4(a)), suggesting a shortage of food in the microcosm test. This is supported by Krogh (1995), who observed optimal reproduction when adult mites each consume eight adult Collembola per day at 20°C, which is not possible in standard microcosm tests where initially 15 mites and 100 Collembola interact in 21 days.

When the dose–response curve is shown on a percentage scale, the indirect effect appears as a difference between the microcosm situation and the food surplus situation. The relatively weak effect of higher doses of dimethoate on the mites in the microcosm (Figure 11.4(b)) is due to eggs laid on day 1 of the simulated experiments. In the model some eggs will be laid on day 1, even at high doses of dimethoate, and the eggs were in the simulations insensitive to dimethoate because they are immobile, and the only uptake is direct exposure which was ignored. When the initial dimethoate dose is not very high (1 mg kg^{-1} soil), the concentration has dropped below a detrimental concentration by the time the eggs hatch. Although there is a very limited number of prey available, the model takes the possibility of consuming dead Collembola into account. The number of surviving mite's eggs is not dependent on the food

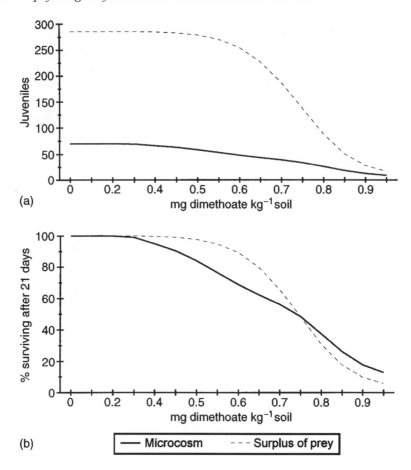

Figure 11.4 The simulated reproduction of *H. aculeifer* versus dimethoate concentration in microcosms, where there is a shortage of food and in a situation where a surplus of prey is provided. (a) Data presented in absolute numbers; (b) data presented as percentage of control. In (b), note the early decline in the microcosm, which is caused by a reduction in available prey, i.e. an indirect effect.

availability, and constitutes a relatively small fraction of the juveniles in the situation with food satiation compared with the microcosm situation.

The model was used to investigate the impact of different food qualities of the prey species by varying the fraction of the prey which was not assimilated by the mites (Figure 11.5). If the prey is of high quality, the predator (*H. aculeifer*) has a high reproductive output and the resulting number of juvenile prey (*F. fimetaria*) is low, and vice versa at low food qualities. These results stress the importance of solid biological information for the output from a simulation model.

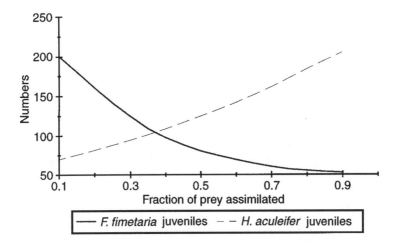

Figure 11.5 Simulated numbers of juvenile *F. fimetaria* and *H. aculeifer* versus quality of *F. fimetaria* as food for *H. aculeifer*.

11.3.3 POSSIBILITIES OF EXTENDING THE MODEL TO FIELD ECOSYSTEM MODELLING

Due to the object-oriented programming approach it is easy to add new species and chemicals to the simulation program, and turn it into a field ecosystem model (Figure 11.6). The procedure taking care of the interactions has been prepared to be able to add any number of species, although computation time will increase with each new species. The use of a physiological time-scale makes it possible for the model to simulate population development at a wide range of temperature regimes, but there are still two problems left, namely modelling of exposure to a chemical, and the degradation of the chemical. Submodels to describe degradation and exposure are beyond the scope of this presentation, but the model outlined here places some demands on an exposure submodel.

11.3.3 INPUT TO THE MODEL

The model is very demanding in terms of basic biological information about the involved species. It is important to know developmental threshold temperatures, duration of life stages in degree-days, maximum growth rate, maximum reproduction rate, survival rate, respiration rate and egestion rate in order to simulate population development. Simulation of trophic interactions and competition in a food-web requires reliable information about who eats whom, predator preference for different kinds of prey, and suitability of the

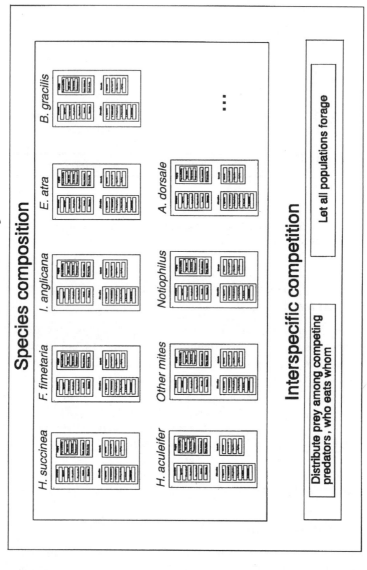

Figure 11.6 Diagram of the arthropod community included in the extended model.

prey for each predator. For species which are regarded as important for agricultural yield and human health, a large part of this information is available, but for most of the species occurring in more natural ecosystems, the information is lacking. This problem is most pronounced when information on biological information is needed on a physiological time-scale.

11.4 THE USE OF PHYSIOLOGICAL TIME

The model needs an exposure submodel which takes both exposure duration and temperature into consideration. For poikilotherm organisms these two parameters are closely coupled, and can be turned into one by using a physiological time-scale, e.g. degree-days (see Box 11.3), which would make exposure–effect equations generally applicable and useful in a general model. For instance, imagine an experiment with an insect species having a threshold temperature of 5°C. The insects are exposed to a chemical for 1 day and the experiment is carried out at two temperatures, 10°C and 20°C. Comparing results from these two temperatures does not make much sense, since the exposure times are 5 and 15 degree-days for the 10°C and 20°C experiments, respectively. If the comparison is done, one may reach the conclusion that the compound is more toxic at higher temperatures, which is true on a human time-scale but not on the time-scale of the insect. If we want to understand the impact of chemicals on insects and other poikilotherm organisms we have to interpret results on their time-scale. In the example above, the insects may have died after 1 day at 20°C and not at 10°C, but after 2 more days the insects kept at 10°C have also been exposed for 15°D and may actually die. This is a plea for exposure–effect experiments to be related, not just to temperature, but to both temperature and time simultaneously using a physiological time-scale in the future.

Further, the degradation of many anthropogenic compounds is temperature-dependent. In a model which uses temperature input, the degradation can be simulated. Thus, the model can keep track of the actual concentration of the chemical and calculate how long the insect has been exposed to the chemical (on the insect time-scale).

11.5 THE RELATIONSHIP BETWEEN MODELS AND EXPERIMENTS

In a review, Barnthouse (1992) concludes that individual-based population dynamic models provide a means for using information on physiological and behavioural effects of toxicants or other relevant properties of individual organisms in risk assessment. Although the model presented in this chapter operates on stage-distributed populations, it also has the capacity to include available information on physiological and behavioural effects, and life-history parameters. Most often the parameters used are average values, but it is possible to take

---

Box 11.3 Degree-day summation

The concept of degree-days seems to be new in ecotoxicology although it has been known for a long time in entomology (Uvarov, 1931). Basically, when using physiological time, the development of a poikilotherm species is regarded as a chemical process, and it is well known that chemical processes are strongly temperature-dependent. In fact, poikilotherm growth is regarded as an enzymatically driven process, and the reaction rate of these processes depends on temperature in an exponential manner, which is modified by changes in enzyme and membrane structural activity. These changes are responsible for linearizing the developmental rate curve (basis for the degree-day concept) and the inactivation seen at low and high temperatures resulting in a sigmoid developmental rate curve (Sharpe and Hu, 1980).

When using degree-days, a linear relationship between temperature and development is assumed. This is not far from reality within the range of the sigmoid developmental rate curve, which can be regarded as linear. The conversion of temperature to degree-days is done according to:

$$\Delta^\circ D = \overline{T}(t) - T_0 \qquad (11.11)$$

where $\Delta^\circ D$ is the time step in degree-days, $T(t)$ is the average temperature of the time step and T_0 is a species specific threshold temperature under which the development can be ignored. Due to the linearity, the degree-day concept fails at low and high temperatures. Therefore, more detailed physiological time-scales have been proposed by Taylor (1981), who describes insect development rate by a normal distribution; Stinnner *et al.* (1974) who use a logistic model; and Sharpe *et al.* (1977), who describe a chemically founded time-scale. For a comparison of these time-scales, see Lamb *et al.* (1984) and Wagner *et al.* (1984).

---

variation into consideration. Variation on the developmental time is already included due to the use of distributed delay to simulate population development, and the importance of variation on other parameters can be included by Monte Carlo simulations where one or more parameters are chosen from frequency distributions. Compared with individual-based models which have rather long computation times (Hommen *et al.*, 1993), this population-based model has the advantage of a relatively short computation time.

Apparently, no models have been suggested for use in terrestrial ecotoxicological risk assessment, which include the physiological demands of the involved organisms. In aquatic ecotoxicological risk assessment Bartell *et al.* (1992) suggest a model that integrates fate and population effects (taking the physiological parameters into consideration) as a possible tool in aquatic risk assessment. Furthermore, a littoral ecosystem model (Hanratty and Stay, 1994) and three other freshwater ecosystem models (Hommen *et al.*, 1993) have been presented. These models were tested against data from mesocosm studies and were found to give good predictions of toxic effects on most of the involved organisms. Consequently, models with modelling concepts similar to the fundamental concepts of the model presented in this chapter have been suggested as tools in aquatic ecotoxicology, which is encouraging for further development of the model which links physiology, population dynamics and ecotoxicology. A lack of combination of these disciplines was recognized by Kooijman *et al.* (1987) as a main obstacle to ecotoxicological progress.

Hanratty and Stay (1994), however, do not regard their model as fully validated and give recommendations for further evaluation, while Hommen *et al.* (1993) point out that the ecological knowledge is insufficient to perform risk studies to be used in the regulatory process. This raises the question of how it should be possible to reach the goal of making a model that is actually able to simulate the ecosystem effects, when it is so demanding in terms of input data? It will only be possible if scientists make well-planned joint efforts to reach this goal. The ideal way of working towards the establishment of a reliable model is to create larger research projects which integrate modelling and experimental research. These two disciplines of science work very well together. Ideally a project should include the following:

1. Construction of a model based on input parameters available in literature. If information on the species in question is not available, information from the nearest relative may be used.
2. Sensitivity analysis reveals which parameters are the most important ones. (During the development of the model important gaps in our knowledge have been identified.)
3. Experiments are planned and carried out in order to obtain reliable inputs on the parameters identified to be most important.
4. Simulations with the model produce predictions and hypotheses of which some can be tested experimentally.
5. Testing of the predictions and hypotheses experimentally.

It may prove necessary to repeat steps 2–5 several times before the model gains sufficient credibility. Step 5 is usually regarded as model validation, which is unfortunate. A model cannot be validated; at best it can avoid 'devalidation' and gain credibility. When a model has gone through step 5 several times without being 'devalidated', it may be regarded as reliable. Stepping through these steps once or several times will demonstrate another important aspect of projects with integrated modelling and experimental work, namely that **modelling constitutes a kind of framework for the experimental effort**. Last but not least, models are very strong 'hypothesis makers', and good hypotheses are very important parts of a research process.

The prospectives of ecosystem modelling in ecotoxicology are intriguing but modelling is very demanding in terms of biological information. Therefore, scientists should plan large joint projects with participants dealing with basic biological information, exposure (including behavioural aspects), degradation of toxins, and modelling, to produce a tool that can extend results from a wide range of laboratory tests to the ecosystem level.

11.6 SUMMARY

A detailed mathematical simulation model combining population dynamics and energetics is presented as a tool for extension of results from laboratory tests to field ecosystem effects, which makes it a possible tool in future risk assessment. The model described in this chapter simulates the predator–prey interaction between a predaceous mite, *Hypoaspis aculeifer*, and the collembolan *Folsomia fimetaria* in a laboratory microcosm test system at constant temperature. Results are presented where the model demonstrates the ability to simulate an indirect effect of the insecticide dimethoate on *H. aculeifer* at sublethal doses through a lethal effect on the prey species, *F. fimetaria*. In order to prepare it for elaboration into a field ecosystem model, the model uses a physiological time-scale (degree-days) and has been prepared to use temperature measurements as input. Furthermore, an object-oriented program structure makes the inclusion of new species easy. The model is very demanding in terms of biological knowledge, which requires a close interaction between development and parameterization of the model and experiments in both laboratory and field. Thus, the model can serve as a framework for the experimental effort.

ACKNOWLEDGEMENTS

This work was supported by the Danish Strategic Environmental Research Programme.

REFERENCES

Axelsen, J.A. (1994) Analysis of host–parasitoid relationships in an agricultural ecosystem. A computer simulation. *Ecol. Modelling*, **73**, 189–203.

Axelsen, J.A., Holst, N., Hamers, T. and Krogh, P.H. (1997a) Simulations of the predator–prey interactions in a two species ecotoxicological test system.

Axelsen, J. A., Holmstrup, M. and Krogh, P.H. (1997b). A simulation study of the impact of synchronisation, temperature and selection of individuals for the results of ecotoxicological tests with the Collembola *Folsomia fimetaria* L. and the predatory mite *Hypoaspis aculeifer* Canestrini. *Pedobiologica*, in press.

Barnthouse, L.W. (1992) The role of models in ecological risk assessment: a 1990's perspective. *Environ. Toxicol. Chem.*, **11**, 1751–60.

Bartell, S.M., Gardner, R.H. and O Neill, R.V. (1992) *Ecological Risk Estimation*, Lewis Publishers, Boca Raton, Florida.

Graf, B. and Hill, J.E. (1992) Modelling the competition for light and nitrogen between rice and *Echinochloa crus-galli*. *Agricult. Syst.*, **40**, 345–59.

Graf, B., Baumgärtner, J. and Gutierrez, A.P. (1990a) Modeling agroecosystem dynamics with the metabolic pool approach. *Mitt. Schweiz. Entomol. Ges.*, **63**, 465–76.

Graf, B., Rakotobe, O., Zahner, P., Dellucchi, V. and Gutierrez, A.P. (1990b) A simulation model for the dynamics of rice growth and development: Part I – The carbon balance. *Agricult. Syst.*, **32**, 341–65.

Graf, B., Gutierrez, A.P., Rakotobe, O., Zahner, P. and Delluchi, V. (1990c) A simulation model for the dynamics of rice growth and development. Part II – the competition with weeds for nitrogen and light. *Agricult. Syst.*, **32**, 367–92.

Gutierrez, A.P., Baumgärtner, J.U. and Hagen, K.S. (1981) A conceptual model for growth, development and reproduction in the ladybird beetle *Hippodamia convergens* G.-M. (Coccinellidae: Coleoptera). *Can. Ent.*, **113**, 21–33.

Gutierrez, A.P., Baumgärtner, J.U. and Summers, C.G. (1984) Multitrophic models of predator–prey energetics. *Can. Ent.*, **116**, 923–63.

Gutierrez, A.P., Schulthess, F., Wilson, L.T., Villacorta, A.M., Ellis, C.K. and Baumgärtner, J.U. (1987) Energy acquisition and allocation in plants and insects: a hypothesis for the possible role of hormones in insect feeding patterns. *Can. Ent.*, **119**, 109–29.

Gutierrez, A.P., Wermelinger, B., Schulthess, F., Baumgärtner, J.U., Herren, H.R., Ellis, C.K. and Yaninek, J.S. (1988a) Analysis of biological control of cassava pests in Africa. I. Simulation of carbon, nitrogen and water dynamics in carbon. *J. Appl. Ecol.*, **25**, 901–20.

Gutierrez, A.P., Neuenschwander, P., Schulthess, F., Herren, H.R., Baumgärtner, J.U., Wermelinger, B., Löhr, B. and Ellis, C.K. (1988b) Analysis of biological control of cassava pests in Africa. II. Cassava mealybug *Phaenococcus manihoti*. *J. Appl. Ecol.*, **25**, 921–40.

Gutierrez, A.P., Yaninek, J.S., Wermelinger, B., Herren, H.R. and Ellis, C.K. (1988c) Analysis of biological control of cassava pests in Africa. III. Cassava green mite *Mononychellus tanajoa*. *J. Appl. Ecol.*, **25**, 941–50.

Gutierrez, A.P, Dos Santos, W.J., Villacorta, A., Pizzamiglio, M.A., Ellis, C.K., Caarvalho, L.H. and Stone N.D. (1991) Modelling the interaction of cotton and the cotton boll weevil. I. A comparison of growth and development of cotton varieties. *J. Appl. Ecol.*, **28**, 371–97.

Gutierrez, A.P., Mariot, E.J., Cure, J.R., Wagner Riddle, C.S., Ellis, C.K., and Villacorta, A.M. (1994) A model of bean (*Phaseolus vulgaris* L.) growth types I–III: factors affecting yield. *Agric. Syst.*, **44**, 35–63.

Hanratty, M.P. and Stay, F.S. (1994). Field evaluation of the littoral ecosystem risk assessment model's predictions of the effects of chlorpyrifos. *J. Appl. Ecol.*, **31**, 439–53.

Hommen, U., Poethke, H.-J., Dülmer, U. and Ratte, H.T. (1993) Simulation models to predict ecological risk of toxins in freshwater systems. *ICES J. Mar. Sci.*, **50**, 337–47.

Kooijman, S.A.L.M., Hanstveit, A.O. and Van der Hoeven, N. (1987) Research on the physiological basis of population dynamics in relation to ecotoxicology. *Wat. Sci. Tech.*, **19**, 21–37.

Krogh, P.H. (1995) Effects of pesticides on the reproduction of *Hypoaspis aculeifer* (Gamasida: Laelapidae) in the laboratory. *Acta Zool. Fenn.*, **196**, 333–7.

Kula, H., Heimbach, U. and Løkke, H. (eds) (1995) Progress Report 1994 of SECOFASE, Third Technical Report. *Development, improvement and standardization of test systems for assessing sublethal effects of chemicals on fauna in the soil ecosystem*. National Environmental Research Institute, Denmark.

Lacey, R.F. and Mallett, M.J. (1991) Further statistical analysis of the EEC ring test of a method for determining the effects of chemicals on the growth-rate of fish. Room Document 3. OECD Ad-Hoc Meeting of Experts on Aquatic Toxicology, at WRC Medmenham, 10–12 December 1991.

Lamb, R.J., Gerber, G.H. and Atkinson, G.F. (1984) Comparison of developmental rate curves Applied to egg hatching data of *Entomoscelis americana* Brown (Coleoptera: Chrysomelidae). *Environ. Entomol.*, **13**, 868–72.

Manetch, T.J. (1976) Time-varying distributed delays and their use in aggregative models of large systems. *IEEE Transactions on systems, man, and cybernetics*, **8**, 547–53.

Sharpe, P.J.H. and Hu, L.C. (1980) Reaction kinetics of nutrition dependent poikilotherm development. *J. Theor. Biol.*, 82, 317–33.

Sharpe, P.J.H., Curry, G.L., DeMichele, D.W. and Cole, C.L. (1977) Distribution model of organism development times. *J. Theor. Biol.*, **66**, 21–38.

Stinner, R.E., Gutierrez, A.P. and Butler, G.D. (1974) An algorithm for temperature-dependent growth rate simulation. *Can. Ent.*, **106**, 519–24.

Tamó, M., Baumgärtner, J. and Gutierrez, A.P. (1993) Analysis of the cowpea agro-ecosystem in West Africa. II Modelling the interactions between cowpea and the bean flower thrips *Megalurothrips sjostedti* (Trybom) (Thysanoptera, Thripidae). *Ecol. Modelling*, **70**, 89–113.

Taylor, F. (1981) Ecology and evolution of physiological time in insects. *Am. Nat.*, **117**, 1–23.

Uvarov, B.P. (1931) Insects and climate. *Trans. Entomol. Soc. London*, **79**, 1–247.

Vansickle, J. (1977) Attrition in distributed delay models. *IEEE Transactions on systems, man, and cybernetics*, **7**, 635–8.

Wagner, T.L., Wu, H-I., Sharpe, P.J.H., Schoolfield, R.M. and Coulson, R.N. (1984) Modelling insect developmental rates: a literature review and application of a biophysical model. *Ann. Entomol. Soc. Am.*, **77**, 208–25.

PART SIX

Ecological approaches: case studies

12 *Extrapolation of laboratory toxicity results to the field: a case study using the OECD artificial soil earthworm toxicity test*

DAVID J. SPURGEON

12.1 EXTRAPOLATING THE RESULTS OF STANDARDIZED TESTS

Six factors that can alter the sensitivity of organisms in standardized tests compared with the field have been described (Van Straalen and Denneman, 1989). These factors may cause problems when attempts are made to use laboratory results to predict the impact of pollutants in the field, since it is not clear if an extrapolation factor should be applied. Clearly, if results from laboratory assays are to be placed in their correct context within the risk assessment procedure, discrepancies between toxicity in test and natural systems must be evaluated and strategies developed to rationalize such variations. To understand the relationship between toxicity in the laboratory and field, it is essential that studies extrapolating the results of standardized tests to polluted sites are undertaken. The work described in this chapter details such a study. Data from a widely used laboratory test procedure (the earthworm artificial soil toxicity test) for the metals cadmium, copper, lead and zinc has been related to effects on earthworms at polluted sites. This allows the capacity of the test to predict effects on populations and individuals in the field to be assessed. Particular attention has been paid to the influence of soil conditions on the availability and toxicity of the tested chemicals.

The earthworm artificial soil test was adopted and standardized by the OECD in 1984 and EEC in 1985. The test was designed to allow toxicity to be determined in a simulated soil medium, as it was anticipated that this would provide replicative results that could be extrapolated directly to the field (Goats and Edwards, 1988; Van Gestel and Van Dis, 1988; Heimbach,

Ecological Risk Assessment of Contaminants in Soil. Edited by Nico M. van Straalen and Hans Løkke. Published in 1997 by Chapman & Hall, London. ISBN 0 412 75900 4

1992). Since its inception, the use of artificial soil in ecotoxicological tests studies increased. New tests have been proposed for species such as the collembolan *Folsomia candida* (ISO, 1994) and the enchytraeid *Enchytraeus albidus* (Römbke, 1989). Consequently, as tests with this medium will form an increasing proportion of the soil invertebrate ecotoxicological database, it is important to understand how the toxicity of different pollutants will be modified in such test compared with effects in the field.

To compare the effects of metals predicted in the laboratory to those at polluted sites, the results of OECD artificial soil tests with the earthworm *Eisenia fetida* have been related to earthworm population survey data and the results of tests conducted with contaminated natural soils. All field work was conducted at sites in the area around Avonmouth in south-west England, a region subject to high metal input, from a primary cadmium, lead and zinc smelter (Coy, 1984; Vale and Harrison, 1994). Emissions released from the factory result in the presence of elevated metal levels at sites up to 20 km downwind from the source (Read *et al.*, 1987; Hopkin, 1989; Jones, 1991; Hopkin and Hames, 1994). In this chapter, an overview of the main parts of the study will be presented. Data from the four pieces of experimental work listed below are discussed.

1. **Assessing the impact of metals on earthworm populations in the field**: earthworm populations were estimated at 22 sites in the region of the smelting works.
2. **Mapping laboratory toxicity data to predict the effects of metals in the field**: artificial soil tests with the metals cadmium, copper, lead and zinc were undertaken. Calculated LC_{50} and EC_{50} values were superimposed on to the levels of these metals measured in soils at Avonmouth. This process permitted the area over which soil metal levels exceed the calculated toxicity values to be ascertained, thus estimating the effects due to each pollutant.
3. **Determination of toxic effects in contaminated field soils**: the toxicities to *Eisenia fetida* of eight soils collected from sites around the smelter were determined. Results are related to the predictions made from the artificial soil tests to allow the agreement between predicted and actual toxicity to be judged.
4. **Assessing the influence of soil factors on zinc availability and toxicity**: the relationships between toxic effects on the survival and cocoon production of *Eisenia fetida* and zinc concentration in two soil fractions and the tissues of the exposed worms were examined. Results have been used to rationalize any differences found for metal toxicity in the artificial soil tests and in naturally contaminated soils.

12.2 ASSESSING THE IMPACT OF METALS ON EARTHWORM POPULATIONS

The effects of pollutants on populations and communities have frequently been ignored in favour of laboratory toxicity studies with individuals

(Bengtsson *et al.*, 1985). An assessment of earthworm distribution should be the starting point for any comparisons of toxicity in laboratory tests and the field. Such studies will also allow the influence of ecological, physiological and behavioural differences on species sensitivity to be assessed and provides an ultimate measure of the impact of pollutants at contaminated sites.

Earthworm populations are known to be reduced by high concentrations of metals in soil (see Bisessar, 1982; Bengtsson *et al.*, 1983; Hunter *et al.*, 1987). An earlier population survey conducted at sites in the Avonmouth area has indicated that some earthworm species are absent from sites 2 km from the factory (Hopkin *et al.*, 1985). However, a larger study was required to provided a more comprehensive estimate of populations effects around the smelter. For this, 22 sites in the area to the north-east of the factory were sampled. The results of this work indicated there is a significant reduction in earthworm abundance at sites less than 1.8 km from the factory, with all worms absent within 1 km (Figure 12.1). Regression of log cadmium, copper, lead and zinc concentration in the soil and earthworm numbers gave significant negative relationships, intimating that elevated metal levels are responsible for the decrease in earthworms numbers close to the smelter.

The effects of metals are clearly more severe for some earthworms than others, since these species are eliminated from sites close to the smelter where others persist (Figure 12.1). Among the relatively sensitive species are representatives of the three functional earthworm groups, i.e. epigeics (*Allolobophora chlorotica*), endogeics (*Aporrectodea rosea*) and anequeic worms (*Aporrectodea caliginosa*) (see Bouché, 1992) also both *r* (*Allolobophora chlorotica*) and *K* (*Aporrectodea rosea*) -selected species (Lee, 1985). Sensitivity cannot therefore be explained by similarities of ecology or life-history. Physiological similarities seem to provide a better explanation. The 'sensitive' species *Allolobophora chlorotica*, *Aporrectodea caliginosa* and *Aporrectodea rosea* all have lower calcium gland activity than the 'tolerant' *Lumbricus terrestris*, *Lumbricus rubellus* and *Lumbricus castaneus* (Piearce, 1972; Morgan and Morris, 1982; Morgan and Morgan, 1988a, 1991). Calcium has an important role in metal sequestration and elimination in earthworms (Prentø, 1979; Andersen and Laursen, 1982; Morgan A.J. *et al.*, 1989; Morgan J.E. *et al.*, 1989). Thus, in 'sensitive' species with low calcium gland activity, a lower rate of metal sequestration and elimination may occur which results in the presence of higher metal concentrations in their most sensitive tissues (Morgan and Morris, 1982; Morgan and Morgan, 1988b, 1991).

12.3 MAPPING LABORATORY TOXICITY DATA TO PREDICT THE EFFECTS OF METALS IN THE FIELD

The toxicity of cadmium, copper, lead and zinc was measured for *Eisenia fetida* using the adaptation of the OECD test proposed by Van Gestel *et al.* (1989).

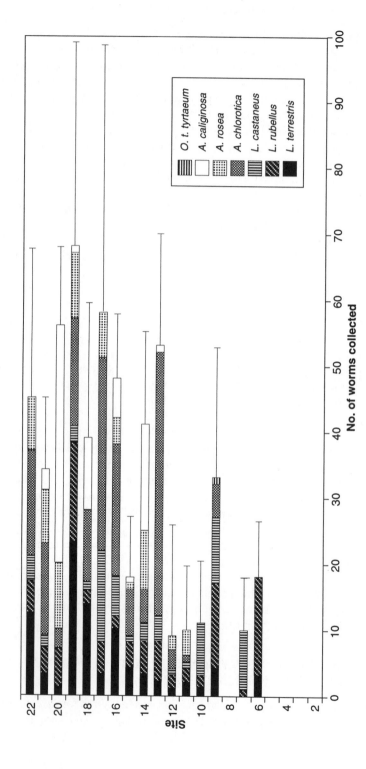

Figure 12.1 Number of worms collected from four 0.25×0.25 m quadrats taken at 22 sites in the Avonmouth area. Error bars indicate the standard deviation of the mean.

Effects on mortality occurred at lower copper concentrations than those for lead and zinc (Table 12.1), which is in good agreement with toxicity values reported in the literature (Ma, 1983; Neuhauser *et al.*, 1985; Bengtsson *et al.*, 1986; Van Gestel and Van Dis, 1988; Van Gestel *et al.*, 1989, 1991, 1992). An LC_{50} value could not be determined for cadmium, since significant mortality was not observed even at the highest test concentration (300 μg Cd g^{-1}). Van Gestel and Van Dis (1988) recorded a 14 day LC_{50} for cadmium of >1000 μg Cd g^{-1} in artificial soil, while Neuhauser *et al.* (1985) calculated a 14 day LC_{50} of 1843 μg Cd g^{-1}. Thus, the relative toxicity of cadmium in terms of mortality is between those for zinc and lead.

The EC_{50}s determined for growth and cocoon production in this study are in close agreement with previously reported values (Ma, 1983; Bengtsson *et al.*, 1986; Van Gestel *et al.*, 1989, 1991, 1992; Spurgeon *et al.*, 1994). Cadmium had the greatest detrimental effect on cocoon production, followed by zinc, copper and lead (Table 12.1).

The metals with the lowest toxicity values are not necessarily those causing the greatest detrimental effects on worm populations at Avonmouth. The impact of each metal will depend on both its toxicity and field soil concentrations. If the LC_{50} and cocoon production EC_{50} values obtained for each element are superimposed on soil metal levels at Avonmouth, the areas over which soil cadmium, copper, lead and zinc concentrations exceeded those found to increase the mortality and reduce the fecundity of *Eisenia fetida* can be defined (for details of the method, see Spurgeon *et al.*, 1994). Extrapolation of current toxicity values in this way predicts almost no effect due to cadmium (Figure 12.2(a)), similar effects for copper and lead, with the cocoon production EC_{50} exceeded over 5.8 km² and 7.2 km² (Figures 12.2(b,c)), while for zinc, the cocoon production EC_{50} is exceeded in soils over 411 km² around the smelter (Figures 12.2(d)). Thus, of the four metals studied, zinc would be anticipated to be have the greatest deleterious effects on earthworms at Avonmouth.

A comparison of the artificial soil test results with the earthworm population survey data indicates a lack of agreement between the predicted and actual impact of metals in the region. Observations of earthworm populations at

Table 12.1 LC_{50} and EC_{50} values (with 95% confidence intervals in parentheses where calculable) for mortality, growth, and cocoon production of *Eisenia fetida* exposed to a geometric series of cadmium, copper, lead and zinc in artificial soil. (Values are based on a mean of four replicates and are given in μg g^{-1})

	Mortality 14-day LC_{50}	**Growth** 21-day EC_{50}	**Cocoon production** 21-day EC_{50}
Cadmium	>300	295 (-)	215 (167–292)
Copper	836 (721–939)	716 (-)	601 (383–5823)
Lead	>10 000	1629 (-)	2249 (-)
Zinc	1078 (789–1149)	>400	357 (-)

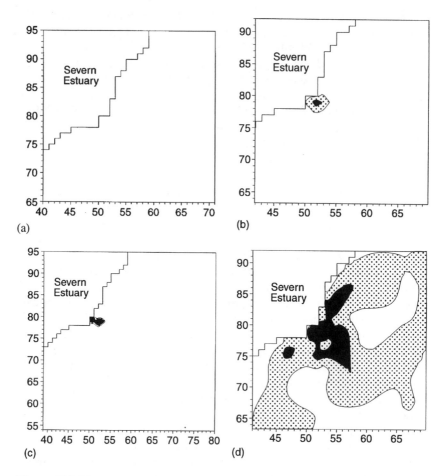

Figure 12.2 Maps showing areas within which estimated toxicity values (LC_{50} and cocoon production EC_{50}) for (a) cadmium, (b) copper, (c) lead and (d) zinc for the earthworm *Eisenia fetida* calculated from a 3-week modified OECD artificial soil test (see Table 12.1 for actual values) are exceeded by total (nitric acid-extractable) metal concentrations in soils at Avonmouth in south-west England. The area over which soil metal levels exceed cocoon production EC_{50} are show in grey, while black areas indicate the area over which the LC_{50} is exceeded. Units on the axes are km. (For details of the method used for drawing maps, see Spurgeon *et al.*, 1994.)

Avonmouth indicated that certain species are eliminated over an area of approximately 20 km² around the factory, while the LC_{50} and cocoon production EC_{50} values for zinc are exceeded in soils over 77 km² and 411 km². Thus, a normal earthworm fauna is present well within the area where soil metal levels exceed both LC_{50} and EC_{50} values for zinc. Earthworm populations are more resilient that would be anticipated from the results of the laboratory test.

Van Straalen and Denneman (1989) elaborated three reasons for the increased resilience of organisms in the field compared with laboratory tests: the development of genetic adaptation, the presence of ecological compensation and regulatory mechanisms and changes in pollutant availability. Additionally for this study, a fourth explanation can be added, that field populations are able to persist despite effects on individual performance, since effects on individuals in laboratory tests are being compared with population effects in the field. Of these four explanations, it has been suggested that differences in availability may be particularly important for mediating the toxicity of chemicals under laboratory and field conditions. Differences in metal availability between soils were therefore examined further in a series of toxicity tests.

12.4 DETERMINATION OF TOXIC EFFECTS IN CONTAMINATED FIELD SOILS

For this experiment, *Eisenia fetida* were exposed to soils from seven sites around Avonmouth and a control site situated on Reading University campus using a procedure adapted from the test protocol proposed by Van Gestel *et al.* (1989). Effects on survival, growth and cocoon production were measured. The pH, organic matter (OM) content and concentrations of cadmium, copper, lead and zinc in each soil are given in Table 12.2.

Table 12.2 Location of the sites in the vicinity of the Avonmouth smelter and the pH, organic matter content and concentration of metals in the soils collected from each site (mean ±SD values given; $n = 6$ for metals, $n = 3$ for % organic matter)

	Ordnance survey grid ref	Distance from smelter (km)	pH (range of 4 replicates)	% Organic matter	Cadmium ($\mu g\ g^{-1}$)	Copper ($\mu g\ g^{-1}$)	Lead ($\mu g\ g^{-1}$)	Zinc ($\mu g\ g^{-1}$)
Site 1	529794	0.5	6.49–6.66	17.2 ±2.8	312.2 ±81.3	2609 ±230	15996 ±2551	32871 ±4860
Site 2	533790	0.5	6.13–6.4	22 ±0.4	129.9 ±22.0	780 ±169	6723 ±1360	7945 ±1540
Site 3	535786	0.8	6.28–6.4	17.8 ±0.6	32.4 ±10.2	159 ±46	842 ±246	19878 ±733
Site 4	532803	1.8	6.6–6.75	27.1 ±0.8	33.5 ±12.7	164 ±69	1245 ±447	2793 ±1178
Site 5	537817	3.0	7.29–7.45	18.5 ±9.8	14.3 ±1.5	108 ±11	930 ±260	1848 ±223
Site 6	552853	5.8	7.25–7.42	12.9 ±1.5	0.9 ±0.5	36 ±15	245 ±93	657 ±252
Site 7	578816	7.0	6.83–6.98	19.7 ±2.6	2.7 ±0.5	42 ±7	290 ±48	925 ±81
Site 8 (Control)	737714	110	5.5–5.6	9.4 ±0.1	0.1 ±0.2	31 ±6	30 ±9	38 ±12

No significant mortality occurred in any of the soils collected from the field, despite the fact that soil from site 1 contained over 30 times as much zinc as the 14-day LC_{50} value determined for this metal in artificial soil (Neuhauser *et al.*, 1985; Van Gestel *et al.*, 1993; Spurgeon *et al.*, 1994). Cocoon production rates were significantly decreased in worms exposed to the four most contaminated soils, while growth was reduced significantly in the two most contaminated. To calculate metal LC_{50} and EC_{50} values for the field soils, two approaches using independent (i.e. assuming no interaction for toxicity between metals) and additive model (following Sprague, 1970) were applied (Table 12.3). Both models were used, since Kraak *et al.* (1993) concluded that the effects of mixtures could not always be predicted from summation of the acute or chronic effects of the individual metals.

To determine which metal within the field soils is reducing the performance of worms around the factory, a simple ratio approach comparing toxicity values calculated for each metal in the field (from the independent toxicity model) (Table 12.3) and the artificial soil test (Table 12.1) was used. Since no mortality was recorded for the field soils, LC_{50} ratios could not be determined; however, comparisons of EC_{50}s for the most sensitive sub-lethal parameter cocoon production could be made. Cocoon production EC_{50} ratios for cadmium, copper, lead and zinc are (field soil : artificial soil) 0.19 : 1, 0.5 : 1, 0.95 : 1 and 10.1 : 1. If the reduction in toxicity in the field soils is similar for all metals, it follows that the highest ratio will be for the most important limiting element. The highest ratios were for zinc; therefore, it is probable that this metal is affecting earthworm reproduction and consequently population viability close to the smelter.

The high comparative toxicity ratio found for zinc suggests that the toxicity of this metal is reduced by an order of magnitude in soils at Avonmouth. The differences in the toxicity of zinc between the laboratory and field can be demonstrated by superimposition of toxicity values on to metal concentrations at Avonmouth. Mapping of artificial soil test results indicate that the

Table 12.3 LC_{50} and EC_{50} values (with 95% confidence intervals in parentheses where calculable) for mortality, growth, and cocoon production of *Eisenia fetida* exposed to soils collected from eight sites in the Avonmouth area (see Table 12.2 for details of the soils from each site). Individual metal values have been calculated using an independent toxicity model, toxic unit values are determined using the model of Sprague (1970). Values are based on a mean of four replicates. All values given in $\mu g\ g^{-1}$

	Mortality 14-day LC_{50}	**Growth** 21-day EC_{50}	**Cocoon production** 21-day EC_{50}
Cadmium	>312	211 (172–264)	40 (-)
Copper	>2609	1763 (1392–2303)	296 (-)
Lead	>15 996	10 830 (8839–13 521)	2131 (-)
Zinc	>32 871	223 71	3605 (-)
Toxic units	>115	22.75 (20.47–26.54)	6.1 (4.53–6.22)

LC_{50} and cocoon production EC_{50} values are exceeded over a large area around the factory (Figure 12.2(d)), while for the field soil the cocoon production EC_{50} is only exceeded over 1 km² (Figure 12.3).

The increased sensitivity to zinc of worms in artificial soil can be explained by differences in the availability of this metal between the two soil types. These differences can be attributed to the physical and chemical characteristics of the respective soils and the experimental protocols used. Although the pHs of the soils were similar (6.1 in the artificial soil, 5.5–7.4 in the field soils), the artificial soil had a lower OM content than any of the field soils used (10% in the artificial soil, 12.9–27.15% in the field soils). Additionally, the kaolin clay used in the artificial soil (which is uncommon in soils in Europe), has a lower cation adsorption capacity than clay minerals such as montmorillonite, illite and vermiculite that occur widely in natural

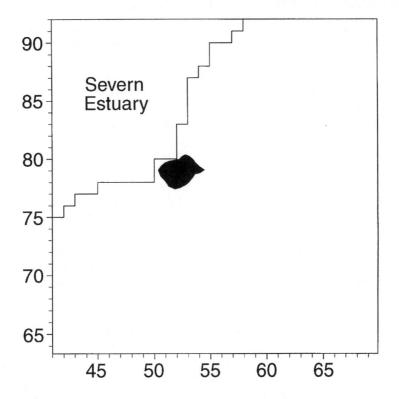

Figure 12.3 Map showing the area within which the cocoon production EC_{50} of 3605 μg Zn g⁻¹ is exceeded by the total (nitric acid-extractable) zinc concentrations in soils at Avonmouth. The EC_{50} was calculated using an independent toxicity model (see section 12.4) from the results of a toxicity test in which *Eisenia fetida* were exposed to soils collected from eight sites located on a gradient from a smelting works. (For details of the method used for drawing maps, see Spurgeon *et al.*, 1994.)

soils (Wild, 1993). Thus, artificial soil has a lower metal adsorption capacity than the field soils, resulting in higher metal availability (see section 12.5.1 for further details).

The form in which the test chemical is added may also be critical in determining the increased toxicity of metals in the artificial soil. Zinc was added to the test soil as a soluble salt $(Zn(NO_3)_2)$. Although a short period (72 h) was allowed for the metal to stabilize, it is unlikely that the sorption kinetics had reached equilibrium after this time (Van Wensem *et al.*, 1994). Thus, in the laboratory test with artificial soil, the worms may be exposed to higher concentrations of unnatural metal species than they would normally encounter in the field.

12.5 ASSESSING THE INFLUENCE OF SOIL FACTORS ON ZINC AVAILABILITY AND TOXICITY

To develop a procedure for coping with differences in toxicity between artificial and field soils, an understanding of the effect of different soil conditions on metal availability is required. To develop such knowledge, a further series of toxicity tests with zinc was conducted to examine the effects of changes in soil conditions on the availability and hence toxicity of zinc.

The bioavailability and hence bioaccumulation and toxicity of a chemical are influenced by physical factors that determine pore-water concentrations. For organic chemicals, bioavailability can be predicted from QSAR relationships between sorption to soil organic matter (K_{om}) or organic carbon (K_{oc}), and lipophilicity (K_{ow}) (Van Gestel and Ma, 1990, 1993; Van Gestel, 1992; Belfroid *et al.*, 1993a,b). However, for metals the situation is more complicated, since toxicity and bioaccumulation are mediated by a number of soil factors such as pH, cation exchange capacity (CEC) and organic matter and clay content (Ma *et al.*, 1983; Ma, 1984, 1989; Van Gestel and Van Dis, 1988; Hopkin *et al.*, 1993). Of these factors, pH and soil OM content have been cited as most important for determining availability to soil organisms.

To examine how changes in pH and soil OM content may alter solubility and hence toxicity, the survival and cocoon production of *Eisenia fetida* was assessed in nine artificial soil tests in which soil pH and OM content were varied. Three levels of OM (5, 10 and 15%) and three pHs (4, 5 and 6) were used in factorial combination. Effects on survival and reproduction have been related subsequently to zinc concentrations in two soil fractions (nitric acid-extractable, NAX, and water-extractable, WX) and in the tissues of the exposed worms (dead worms could not be analysed due to their rapid rates of decay). Total, soluble and worm zinc concentrations were also analysed for the soils and test organisms used for the field toxicity study detailed in section 12.4. Results of this work have been used to explain, in terms of metal solubility and pore-water concentrations, the reduction of metal toxicity in field soils and to establish which metal fraction correlates most strongly to effects on *Eisenia fetida* in a range of soils.

12.5.1 SOIL AND EARTHWORM METAL ANALYSIS

The concentration of soluble zinc increased at low artificial soil pH and OM content, a similar trend to that found for cadmium by Crommentuijn (1994). The increased solubility of zinc at low pH and low OM content can be explained in terms of the cation exchange properties of the medium. The two most important components that bind metals within the soil matrix are OM and clay, since these bear negative charges on the particle surfaces that bind metals from solution (Wild, 1993). Thus in artificial soils, a reduction in the percentage OM content will increase metal solubility, as the total number of negative sites is reduced.

Avonmouth soils have a higher CEC capacity than any of the artificial soils tested. For example, there were only three sites from which more than 1% of total zinc was present in the WX fraction, while 1% was always extracted from the artificial soils. The increase in the binding capacity of the field soils can be attributed directly to their composition. All field soils have a high OM content (Table 12.2); furthermore, this OM is likely to be more degraded and thus have a greater surface area/mass ratio than the *Sphagnum* peat used in the artificial medium. This increase in surface area will increase soil binding capacity, as negative sites occur only at the particle surface. In addition, the clays that are most common in natural soils have a greater binding capacity than the kaolin used for the OECD medium. The CEC of kaolin is 0.02–0.06 mol_c kg^{-1}, compared with values of 0.3 mol_c kg^{-1} and 1.4 mol_c kg^{-1} for the clay minerals illite and vermiculite respective (Wild, 1993). Thus, the binding capacity of vermiculite is over 20-fold that of kaolin.

The charges associated with soil constituents can be either permanent or temporary (broken down by low pH). The clays illite, smectite and vermiculite have permanent charges, while for kaolinite and OM they are temporary (Wild, 1993). Thus, the two components of artificial soil that bind metals both have reduced adsorption at low pHs. In addition to altering the cation exchange properties of a soil, low pH also affects metal binding by competing directly for the available negative sites. Herms and Brümmer (1984) found a positive correlation between pH and metal binding in soils; this was attributed to antagonism between the H^+ and metal cations in solution. At low pH, the H^+ ion concentration is raised, increasing site competition and ultimately resulting in a reduction in metal adsorption. Thus at low pH, a combination of a reduction in the number of negative sites and an increase in competition from H^+ ions reduces zinc binding and increase the soluble concentration.

Log worm zinc levels showed a significant positive relationship with both NAX and WX metal. The strongest correlation in both the artificial and field soils was found for the relationship with soluble zinc (Figure 12.4(a,b)). In all cases, the slope parameter for the regression was below 1, indicating that zinc is regulated by *E. fetida*, probably through the mechanisms already in place for the control of this essential element (Morgan and Morgan, 1988b, 1989). For metal body burdens to be useful for estimating

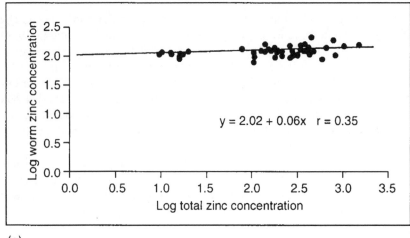

(a)

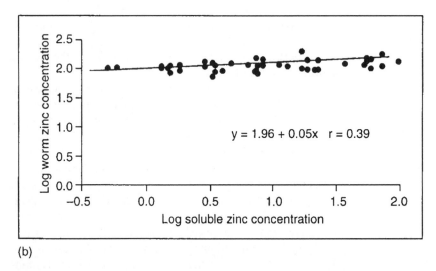

(b)

Figure 12.4 Scatter plots and regression equations of log zinc concentrations in the tissues of *Eisenia fetida* exposed to soils contaminated with zinc in nine toxicity tests in which OM and pH were varied, plotted against (a) log total zinc (nitric acid-extracted) and (b) log soluble zinc (water-extractable) in the respective test soils.

bioavailable concentrations, tissue concentrations should both increase in contaminated soils and reflect changes in solubility. Thus, although worm zinc levels are marginally increased at high soluble zinc concentrations, such values are not a good indicator of bioavailable zinc, as the low level of accumulation makes it difficult to estimate exposure. This is indicated by the low correlations found for the regressions for both soil fractions (Figure 12.4(a,b)).

12.5.2 COMPARISONS OF TOXIC EFFECTS WITH MEASURED ZINC CONCENTRATIONS

High total and water-soluble zinc concentrations decreased *Eisenia fetida* survival. A comparison of the log-likelihood ratios for the logistic regressions indicated that the best fit is obtained with the water-soluble as opposed to the NAX or worm zinc levels. (Figure 12.5(a–c)). Cocoon production (calculated as a percentage of the control) fell as the concentrations of NAX and WX zinc increased (Figure 12.6(a,b)). Analysis of the relationships between log NAX and WX zinc and cocoon production using a logistic model indicated a 'better' fit between cocoon and WX zinc levels than for the NAX data, while regression to earthworm zinc gives a poorer fit than for either soil fraction (Figure 12.6(c)).

The stronger correlation found between toxicity parameters and WX rather than NAX zinc is perhaps unsurprising, since nitric acid extraction will strongly overestimate the concentration of zinc available for the earthworms. If soluble zinc levels, as suggested here, indicate pore water concentrations, these results support the use of the pore water hypothesis for predicting differences in metal toxicity between soil types (see Van Gestel, 1992).

If it is assumed that metal toxicity is exerted via pore water concentrations, the calculation of toxicity values based on soluble zinc levels should give comparable estimates for all the soils tested (Crommentuijn, 1994). For the artificial and the field soil tests used in this survey, the LC_{50}s, and cocoon production EC_{50}s determined from the soluble zinc, were not raised consistently in high pH or OM soils, while values calculated from total metal levels increased. For example, an LC_{50} value of 71.8 μg Zn g^{-1} was calculated from the 5% OM–pH 6.0 test which is almost identical to that calculated in the test with 5% OM–pH 4.0 (67.7 μg Zn g^{-1}). LC_{50}s and EC_{50}s calculated from total zinc concentrations for each toxicity test indicated that metal toxicity was decreased at high soil pH and OM contents. Thus, toxicity values calculated from soluble zinc appear to give a better indication of the effects of this metal in different soils.

The zinc cocoon production EC_{50}s calculated from total zinc concentrations in Avonmouth soils, calculated from the independent toxicity model, gave a higher values than for any of the artificial soil tests (3605 μg Zn g^{-1}). However, if toxicity values for field soils are calculated from soluble rather than total zinc concentrations, the value obtained, 20.9 μg Zn g^{-1}, is within the range of values calculated for the artificial soil tests. Thus, it appears that water-soluble zinc levels can be used to predict effects on earthworm over a range of soil types in which soil conditions and CECs vary.

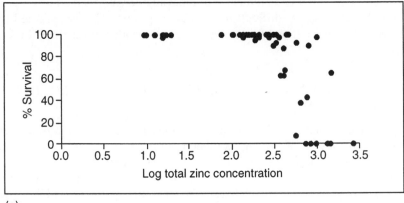

(a)

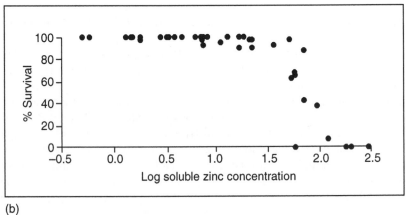

(b)

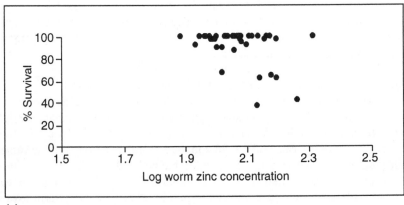

(c)

The poor relationship found between internal zinc and toxicity was somewhat more of a surprise. It is normally anticipated that, for a given species, an internal threshold concentration exists above which toxic effects occur. However, in the current example there is a lack of reciprocity between worm zinc levels and toxicity as a results of the regulation of the metal in the tissues of the exposed worms (Figure 12.4(a,b)). As a consequence of this regulation, toxic effects occur at internal concentrations only marginally above those of unaffected worms, thus making it difficult to define a clear threshold value.

12.6 ASSESSMENT OF HAZARDOUS CONCENTRATIONS

A number of papers have been published recently that propose models for use in establishing 'critical concentrations' for pollutants in ecosystems. Some of these risk-assessment models describe procedures for calculating pollutant concentration above which only 5% of species are adversely affected (= hazardous concentration for 5% of species = 'HC_5' value: Van Straalen and Denneman, 1989; Wagner and Løkke, 1991; Aldenberg and Slob, 1993). The toxicity values used as an input for these models are reproductive NOECs obtained from laboratory tests. In the procedure described by Van Straalen (1993) a model based on an the assumption of a log-logistic distribution of NOEC toxicity values has been used to predict HC_5 values for cadmium, copper and lead in soil.

Much of the toxicity data used were from reproductive toxicity tests with earthworms (see Table 1 in Van Straalen, 1993). However, the work described in this chapter has demonstrated that such laboratory tests may be too sensitive for predicting effects in field soils, as the availability of the pollutant is increased (see sections 12.3 and 12.4). Consequently, using laboratory data for HC_5 calculations, may yield 'critical concentrations' that are over-cautious when compared with the metal concentrations that affect individuals and populations at polluted sites. This is indicated by the fact that there is currently no evidence of deleterious effects on any species caused by the concentrations currently proposed as soil metal HC_5s (Forbes and Forbes, 1993; Hopkin, 1993).

A number of strategies have been proposed to rationalize the differences that exist in the toxicity of pollutants between laboratory test and field soils before the use of data for risk assessment. Van Wensem *et al.* (1994) suggested that the internal threshold concentration (ITC) of exposed organisms could be used as an input to HC_5 models to calculate ecologically relevant soil quality criteria. This technique was found to be useful for cadmium, permitting a 'critical con-

Figure 12.5 (see facing page) Scatter plots of percentage survival at each concentration of *Eisenia fetida* in nine toxicity tests for which soil pH and OM content were varied, plotted against (a) log total soil zinc (nitric acid-extracted), (b) log soluble soil zinc (water-extractable), and (c) log worm tissues zinc at the respective test concentration.

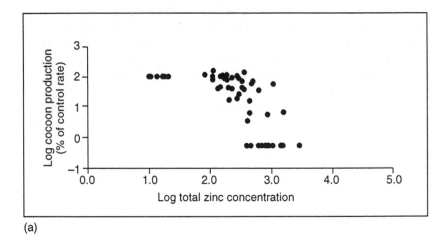

(a)

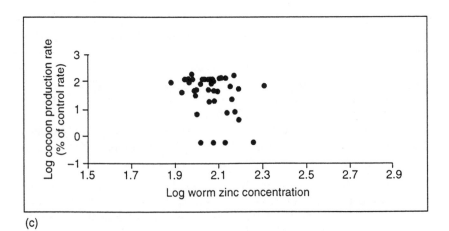

(b)

(c)

centration' of 0.12 μg Cd g^{-1}. However, the results of the work described in section 12.5.1 indicate that for zinc, the use of body burdens to establish internal thresholds may be limited by the regulation of this metal. For such regulated pollutants, the use of water-soluble concentrations for establishing critical concentrations and investigating polluted sites could be considered, since these relate more closely to toxic effects than total pollutant levels (see Figures 12.5(a,b) and 12.6(a,b)). However, if soluble levels are to be used for the calculation of realistic and applicable HC$_5$ values, a major effort will be required to increase the available toxicity data, since at present insufficient information exists to undertake such an assessment for any pollutant.

12.7 SUMMARY

The results of single-species toxicity tests should be validated by extrapolation to contaminated field sites before they can be placed in their correct context for risk assessment. The work described in this chapter details such a study. The LC$_{50}$ and EC$_{50}$ values calculated for cadmium, copper, lead and zinc in artificial soil tests with *Eisenia fetida* were compared with metal concentrations in soils at sites around a smelting works. Mapping toxicity data in this way indicated deleterious effects on earthworms over an area of 400 km^2 around the factory. However, an earthworm population survey in the region only indicated significant effects on total numbers over an area of 20 km^2. Worms were present well within the area where soil concentrations exceed one or more LC$_{50}$ and EC$_{50}$ values, suggesting that OECD test results could not be related directly to the field. A further toxicity assay conducted with contaminated soils collected from the region around the smelting works indicated that metals were 10 times more toxic in artificial soil than in the field. These results suggest that differences in availability are the major reason for the differences found between the test and the population data. Further work studying the impact of availability on toxicity show that, for risk assessment, it may be better to evaluate effects as soil soluble or worm tissue concentrations. For example, comparisons of zinc levels in two soil fractions and in earthworm tissues with effects on worm survival and reproduction indicated that the strongest correlation was for water-soluble metal. Internal concentrations did not correlate well with toxicity, since zinc levels are regulated by earthworms. However, for pollutants such as cadmium that are strongly accumulated, previous work denotes that internal concentration may describe toxic effects with greater accuracy than either the total or soluble soil metal level.

Figure 12.6 (see facing page) Scatter graphs of cocoon production rate at each concentration of *Eisenia fetida* in nine toxicity tests for which soil pH and OM content were varied, plotted against (a) log total soil zinc (nitric acid-extracted), (b) log soluble soil zinc (water-extractable), and (c) log worm tissues zinc at the respective test concentration.

REFERENCES

Aldenberg, T, and Slob, W. (1993) Confidence limits for hazardous concentrations based on logistically distributed NOEC toxicity data. *Ecotox. Environ. Safety*, **25**, 48–63.

Andersen, C. and Laursen, J. (1982) Distribution of heavy metals in *Lumbricus terrestris, Aporrectodea longa* and *A. rosea* measured by atomic absorption and X-ray fluorescence spectrometry. *Pedobiologia*, **24**, 347–56.

Belfroid, A., Seinen, W., Van Gestel, C.A.M. and Hermens, J. (1993a) The acute toxicity of chlorobenzenes for earthworms (*Eisenia andrei*) in different exposure systems. *Chemosphere*, **26**, 2265–77.

Belfroid, A., Van Wezel, A., Sikkenk, M., Van Gestel, C.A.M., Seinen, W. and Hermens, J. (1993b) The toxicokinetic behaviour of chlorobenzenes in earthworms (*Eisenia andrei*) – experiments in water. *Ecotox. Environ. Safety*, **25**, 154–65.

Bengtsson, G., Nordström, S. and Rundgren, S. (1983). Population density and tissue metal concentration of Lumbricids in forest soils near a brass mill. *Environ. Pollut. (Ser. A.)*, **30**, 87–108.

Bengtsson, G., Gunnarsson, T. and Rundgren, S. (1985) Influence of metals on reproduction, mortality and population growth in *Onychiurus armatus* (Collembola). *J. Appl. Ecol.*, **22**, 967–78.

Bengtsson, G., Gunnarsson, T. and Rundgren, S. (1986) Effects of metal pollution on the earthworm *Dendrobaena rubida* (Sav) in acidified soils. *Water Air Soil Pollut.*, **28**, 361–83.

Bisessar, S. (1982) Effect of heavy metals on the micro-organisms in soils near a secondary lead smelter. *Water Air Soil Pollut.*, **17**, 305–8.

Bouché, M.B. (1992) Earthworm species and ecotoxicological studies, in *Ecotoxicology of Earthworms* (eds P.W. Greig-Smith, H. Becker, P.J. Edwards and F. Heimbach), Intercept Ltd, Andover, Hants, UK, pp 20–35.

Coy, C.M. (1984) Control of dust and fumes at a primary zinc and lead smelter. *Chemistry in Britain*, May, 418–20.

Crommentuijn, T. (1994) *Sensitivity of Soil Arthropods to Toxicants*. PhD Thesis, Vrije Universiteit, Amsterdam, The Netherlands.

EEC (1985) EEC Directive 79/831. Annex V. Part C. Methods for the determination of ecotoxicity. Level I. C(II)4: Toxicity for earthworms. Artificial soil test. DG XI/128/82.

Forbes, V.E. and Forbes, T.L. (1993) A critique of the use of distribution-based extrapolation models in ecotoxicology. *Funct. Ecol.*, **7**, 249–54.

Goats, G.C. and Edwards, C.A. (1988) The prediction of field toxicity of chemicals to earthworms by laboratory methods, in *Earthworms in Waste and Environmental Management* (eds C.A. Edwards and E. Neuhauser), SPB Academic Publishing, The Hague, pp. 283–94.

Heimbach, F. (1992) Effects of pesticides on earthworm populations: comparison of results from laboratory and field tests, in *Ecotoxicology of Earthworms* (eds P.W. Greig-Smith, H. Becker, P.J. Edwards and F. Heimbach), Intercept Ltd, Andover, Hants, UK, pp. 100–6.

Herms, U. and Brümmer, G. (1984) Einflussgrössen der Schwermetallöslichkeit und-bindung in Böden. *Z. Pflanzenernaehr. Bodenkd.*, **147**, 400–24.

Hopkin, S.P. (1989) *Ecophysiology of Metals in Terrestrial Invertebrates*, Elsevier Applied Science, London, UK.

Hopkin, S. P. (1993) Ecological implications of '95% protection levels' for metals in soils. *Oikos*, **66**, 137–41.

Hopkin, S.P. and Hames, C.A.C. (1994) Zinc, among a 'cocktail' of metal pollutants, is responsible for the absence of the terrestrial isopod *Porcellio scaber* from the vicinity of a primary smelting works. *Ecotoxicology*, **2**, 68–78.

Hopkin, S.P., Watson K.M., Martin, M.H. and Mould, M.L. (1985) The assimilation of heavy metals by *Lithobius variegatus* and *Glomeris marginata* (Chilopoda; Diplopoda). *Bijdr. Dierkunde*, **55**, 88–94.

Hopkin, S.P., Jones, D.T. and Dietrich, D. (1993) The terrestrial isopod *Porcellio scaber* as a monitor of the bioavailability of metals: towards a global 'woodlouse watch' scheme. *Sci. Total Environ.*, Suppl., Part 1, 357–65.

Hunter, B.A., Johnson, M.S. and Thompson, D.J. (1987) Ecotoxicology of copper and cadmium in a contaminated grassland ecosystem. 2. Invertebrates. *J. Appl. Ecol.*, **24**, 587–99.

ISO (1994) ISO/DRAFT. Soil quality-effects of pollutants on Collembola (*Folsomia candida*): method for the determination of effects on reproduction.

Jones, D.T. (1991) *Biological monitoring of Metal Pollution in Terrestrial Ecosystems*. PhD Thesis, University of Reading, UK.

Kraak, M.H.S., Schoon, H., Peeters, W.H.M. and Van Straalen, N.M. (1993) Chronic ecotoxicity of mixtures of Cu, Zn, and Cd to the zebra mussel *Dreissena polymorpha*. *Ecotox. Environ. Safety*, **25**, 315–27.

Lee, K.E. (1985) *Earthworms: Their Ecology and Relationship with Soil and Land Use*, Academic Press, London, UK.

Ma, W.C. (1983) *Regenwormen als bio-indicators van bodemverontreiniging*. Bodembescherming 15. Staatsuitgeverij, The Hague.

Ma, W.C. (1984) Sub-lethal toxic effects of copper on growth, reproduction and litter breakdown activity in the earthworm *Lumbricus rubellus*, with observations on the influence of temperature and soil pH. *Environ. Pollut. Ser. A*, **33**, 207–19.

Ma, W.C. (1989) Effects of soil pollution with metallic lead pellets on lead bioaccumulation and organ/body weight alterations in small mammals. *Arch. Environ. Contam. Toxicol.*, **18**, 617–22.

Ma, W.C., Edelman, T., Van Beresum, I. and Jans T. (1983) Uptake of cadmium, zinc, lead and copper by earthworms near a zinc-smelting complex: influence of soil pH and organic matter. *Bull. Environ. Contam. Toxicol.*, **30**, 424–7.

Morgan, A.J. and Morris, B. (1982) The accumulation and intracellular compartmentation of cadmium, lead, zinc and calcium in two earthworm species (*Dendrobaena rubida* and *Lumbricus rubellus*) living in highly contaminated soil. *Histochemistry*, **75**, 269–85.

Morgan, A.J., Roos, N., Morgan, J.E. and Winters, C. (1989) The subcellular accumulation of toxic heavy metals: qualitative and quantitative x-ray microanalysis, in *Electron Probe Microanalysis*: *Applications in Biology and Medicine* (eds K. Zierold and H.K. Hagler), Springer-Verlag, London, pp. 59–70.

Morgan, J.E. and Morgan, A.J. (1988a) Calcium–lead interactions involving earthworms. Part 1. The effects of exogenous calcium on lead accumulation by earthworms under field and laboratory conditions. *Environ. Pollut.*, **54**, 41–53.

Morgan, J.E. and Morgan, A.J. (1988b) Earthworms as biological monitors of cadmium, copper, lead and zinc in metalliferous soils. *Environ. Pollut.*, **54**, 123–38.

Morgan, J.E. and Morgan, A.J. (1989) Zinc sequestration by earthworm (Annelida: Oligochaeta) chloragocytes, an in vivo examination using fully quantitative electron probe X-ray micro-analysis. *Histochemistry*, **90**, 405–11.

Morgan, J.E. and Morgan, A.J. (1991) Differences in the accumulated metal concentrations in two epigeic earthworm species (*L. rubellus* and *D. rubidus*) living in contaminated soils. *Bull. Environ. Contam. Toxicol.*, **47**, 296–301.

Morgan, J.E, Norey, C.G., Morgan, A.J. and Kay, J. (1989) A comparison of the cadmium-binding proteins isolated from the posterior alimentary canal of the earthworm *Dendrodrilus rubidus* and *Lumbricus rubellus*. *Comp. Biochem. Physiol.*, **92**, 15–21.

Neuhauser, E.F., Loehr, R.C., Milligan, D.L. and Malecki, M.R. (1985) Toxicity of metals to the earthworm *Eisenia foetida*. *Biol. Fertil. Soils*, **1**, 149–52.

OECD (1984) Guidelines for the testing of chemicals No. 207. Earthworm acute toxicity tests. OECD Adopted 4 April 1984.

Piearce, T.G. (1972) The calcium relations of selected Lumbricidae. *J. Anim. Ecol.*, **41**, 167–88.

Prentø, P. (1979) Metals and phosphate in the chloragosomes of *Lumbricus terrestris* and their possible significance. *Cell Tissue Res.*, **196**, 123–34.

Read, H.J., Wheater, C.P. and Martin, M.H. (1987). Aspects of the ecology of Carabidae (Coleoptera) from woodlands polluted by heavy metals. *Environ. Pollut.*, **48**, 61–76.

Römbke, J. (1989) *Enchytraeus albidus* (Enchytraeid, Oligochaeta) as a test organism in terrestrial laboratory systems. *Arch. Toxicol.*, **13**, 402–5.

Sprague, J.B. (1970) Measurement of pollutant toxicity to fish. II. Utilizing and applying bioassay results. *Water Res.*, **4**, 3–32.

Spurgeon, D.J., Hopkin, S. P. and Jones, D.T. (1994) Effects of cadmium, copper, lead and zinc on growth, reproduction and survival of the earthworm *Eisenia fetida* (Savigny): Assessing the environmental impact of point-source metal contamination in terrestrial ecosystems. *Environ. Pollut. (Ser. A)*, **84**, 123–30.

Vale, J.A. and Harrison, J. (1994) Aerial inputs of pollutants to the Severn estuary. *Biol. J. Linn. Soc.*, **51**, 45–54.

Van Gestel, C.A.M. (1992) The influence of soil characteristics on the toxicity of chemicals for earthworms: a review, in *Ecotoxicology of Earthworms* (eds P.W. Greig-Smith, H. Becker, P.J. Edwards and F. Heimbach), Intercept Ltd, Andover, Hants, UK, pp. 44–54.

Van Gestel, C.A.M. and Ma, W.C. (1990) An approach to quantitative structure–activity relationships (QSARs) in earthworm toxicity studies. *Chemosphere*, **21**, 1023–33.

Van Gestel, C.A.M. and Ma, W.C.(1993) Development of QSARs in soil ecotoxicology – earthworm toxicity and soil sorption of chlorophenols, chlorobenzenes and chloroanilines. *Water Air Soil Pollut.*, **69**, 265–76.

Van Gestel, C.A.M. and Van Dis, W.A. (1988). The influence of soil characteristics on the toxicity of four chemicals to the earthworm *Eisenia andrei* (Oligochaeta). *Biol. Fertil. Soils*, **6**, 262–5.

Van Gestel, C.A.M., Van Dis, W.A., Van Breemen, E.M. and Sparenburg, P.M. (1989). Development of a standardized reproduction toxicity test with the earthworm species *Eisenia andrei* using copper, pentachlorophenol and 2,4-dichloroaniline. *Ecotox. Environ. Safety*, **18**, 305–12.

Van Gestel, C.A.M., Van Dis, W.A., Dirven-Van Breemen, E.M., Sparenburg, P.M. and Baerselman, R. (1991) Influence of cadmium, copper and pentachlorophenol on growth and sexual development of *Eisenia andrei* (Oligochaeta: Annelida). *Biol. Fertil. Soils*, **12**, 117–21.

Van Gestel, C.A.M., Dirven-Van Breemen, E.M., Baerselman, R., Emans, H.J.B., Janssen, J.A.M., Postuma, R. and Van Vliet, P.J.M. (1992) Comparison of sublethal and lethal criteria for nine different chemicals in standardized toxicity tests using the earthworm *Eisenia andrei*. *Ecotox. Environ. Safety*, **23**, 206–20.

Van Gestel, C.A.M., Dirven-Van Breemen, E.M. and Baerselman, R. (1993) Accumulation and elimination of cadmium, copper, chromium and zinc and effects on growth and reproduction in *Eisenia andrei* (Oligochaeta, Annelida). *Sci. Total. Environ.* Suppl., Part 1, 585–97.

Van Straalen, N.M. (1993) Soil and sediment quality criteria derived from invertebrate toxicity data, in *Ecotoxicology of Metals in Invertebrates* (eds R. Dallinger and P.S. Rainbow), Lewis Publishers, Chelsea, pp. 427–41.

Van Straalen, N.M. and Denneman, C.A.J. (1989) Ecotoxicological evaluation of soil quality criteria. *Ecotox. Environ. Safety*, **18**, 241–51.

Van Wensem, J., Vegter, J.J. and Van Straalen, N.M. (1994) Soil quality criteria derived from critical body concentrations of metals in soil invertebrates. *Appl. Soil Ecol.*, **24**, 185–91.

Wagner, C. and Løkke, H. (1991) Estimation of ecotoxicological protection levels from NOEC toxicity data. *Water Res.*, **25**, 1177–86.

Wild, A. (1993) *Soils and the Environment: An Introduction*, Cambridge University Press, Cambridge, UK.

13 Life-table study with the springtail Folsomia candida *(Willem)* exposed to cadmium, chlorpyrifos and triphenyltin hydroxide

TRUDIE CROMMENTUIJN, CONNIE J.A.M. DOODEMAN, ANJA DOORNEKAMP
AND CORNELIS A.M. VAN GESTEL

13.1 LIFE-HISTORY ASSESSMENT IN ECOTOXICOLOGY

The life-history of an organism is its lifetime pattern of growth, differentiation, storage and reproduction (Begon *et al.*, 1990). These characteristics together determine the population growth, and energy spent on each of these may vary depending on many environmental factors. A change in one of the environmental factors will therefore result in a response of the population and hence may change the occurrence of the species (Underwood, 1989).

Toxic substances are known as environmental stresses with the possibility of reducing a population, depending on the exposure level. Effects of toxic substances such as reduced reproduction, growth and survival are often determined under laboratory conditions in standardized toxicity tests (OECD, 1984; ISO, 1994).

When assessing the risk of toxicants for the environment, extrapolation methods are used to estimate safe concentration levels in the environment (Dutch Health Council, 1989). In these methods toxicity data on single species are used as input, each data representing the effect of the toxicant on one of the life-history characteristics of the species (reproduction, growth or survival). Although these methods aim to derive safe levels protecting species, it is, however, not clear if these 'individual level-oriented methods' also protect species at the population level.

In an earlier study we showed that cadmium affects the life-history of soil arthropods in a variety of ways (Crommentuijn *et al.*, 1995). Each species

Ecological Risk Assessment of Contaminants in Soil. Edited by Nico M. van Straalen and Hans Løkke. Published in 1997 by Chapman & Hall, London. ISBN 0 412 75900 4

could be characterized by a combination of the concentrations at which sublethal and lethal effects occur, showing a species-dependent influence on the different life-history characteristics.

To gain insight into effects of substances at the population level, the most appropriate way to describe effects on life-history characteristics is the life-table approach. In this procedure, observations on survival and fertility as a function of age are used to estimate the intrinsic rate of natural increase of a population. This approach has been applied in several ecotoxicological studies by Daniels and Allan (1981), Allan and Daniels (1982), Gentile *et al.* (1982), Day and Kaushik (1987),Van Leeuwen *et al.* (1987) and Van Straalen *et al.* (1989). At the same time, population models have emphasized the importance of life-history characteristics and their interactions (Snell, 1978; Kooijman and Metz, 1984; Nisbet *et al.*, 1989; Sibly and Calow, 1989).

In this study we examine the effects of cadmium (Cd), chlorpyrifos (CPF) and triphenyltin hydroxide (TPT-OH) on different life-history characteristics of the parthenogenetic species *Folsomia candida* (Willem) (Collembola; Isotomidae). Age-specific life-cycle observations were made to evaluate these effects. The intrinsic rate of natural increase was determined and compared with the effects on the different life-history characteristics. In this way we aim to analyse the interactions between life-history components and how these interactions affect the capacity for population increase.

13.2 EXPERIMENTS WITH CONTAMINATED FOOD

13.2.1 ANIMALS

A parthenogenetically reproducing culture of *Folsomia candida* (Willem) (Collembola; Isotomidae) was maintained in the laboratory at $18 \pm 1°C$ at a light : dark cycle of 12 : 12 h, in culture pots with bottoms of plaster of Paris mixed with charcoal. The culture was set up in 1986 with animals which originated from a pine stand of the Roggebotzand Forest, the Netherlands. The experiments were started with synchronized animals of the same age and weight which were obtained by allowing adult animals to lay eggs in a container on a bottom of plaster of Paris mixed with charcoal. After 2 days the adult animals were removed. The eggs present in the containers hatched approximately 2 weeks later. When these offspring were 1 day old the experiment was started. The food used in the culture and in the synchronization period was baker's yeast (Dr Oetker).

13.2.2 SUBSTANCES

The substances used were cadmium ($CdCl_2 \cdot 2.5H_2O$, BDH, AnalaR), chlorpyrifos (CPF) (Pestanal, Riedel-de Haen) and triphenyltin hydroxide (TPT-OH) (Aldrich).

13.2.3 PREPARATION AND CONTAMINATION OF THE FOOD

F. candida was fed with baker's yeast dispensed on filter paper discs (diameter 0.75 cm). Five stock solutions of $CdCl_2 \cdot 2.5H_2O$ in deionized water were prepared for the cadmium experiment. Every week new food was prepared by adding 1.5 ml stock solution to 0.5 g dry yeast mixing the suspension on a vortex and putting it on filter paper discs, 25 μl per disc.

Stock solutions in acetone were prepared for the CPF and TPT-OH experiment. One day before using the food, 300 μl stock solution was added to 0.5 g dry yeast and left overnight in a fume cupboard to let the acetone evaporate. The next day 1.5 ml deionized water was added and the suspension was mixed on a vortex and put on filter paper discs, 25 μl per disc.

13.2.4 EXPERIMENTAL DESIGN

Five treatments and a control were applied for each substance to be tested. The food was prepared freshly each week and old food was removed. The nominal concentrations for cadmium and CPF were 139, 300, 646, 1392 and 3000 μg g⁻¹ dry food and 0.93, 2.00, 4.31, 9.28 and 20 μg g⁻¹ dry food, respectively. In the TPT-OH experiment the nominal concentrations were 139, 300, 646, 1392, and 3000 μg g⁻¹ dry food. The choice of the concentration range for cadmium was based on effects found in an earlier study (Crommentuijn *et al.*, 1995). For the CPF and TPT-OH experiments this choice was based on range-finding experiments (unpublished results) from which it was concluded that sublethal effects could be expected at the three highest concentrations from the selected range. An extra acetone control was prepared for the CPF and TPT-OH experiments.

The experiment was started with synchronized animals of 1 day old (mean fresh weight 18 ± 4 μg). The animals were kept individually in perspex containers (diameter 1 cm) with a bottom of plaster of Paris mixed with charcoal. For each treatment 30 replicates were prepared, so in total 180 for Cd, 210 for CPF and 210 for TPT-OH. For each substance tested the replicates were placed in a random order on a wet plate of plaster of Paris, on the bottom of a box (100 × 50 × 30 cm). Each box contained about 70 replicates and three of these boxes were used per substance. The experiments were conducted in a climate room at an average temperature of 20°C (ambient fluctuation range maximal 2°C), 75% relative air humidity and a light : dark cycle of 12 : 12 h.

Fresh weight was determined from day 16 until day 80, once weekly, for 15 individuals of each treatment. Animals were weighed individually on a microbalance (Sartorius, Super micro, S4). Individuals were transferred carefully into a capsule with a small opening, to prevent desiccation of the animals during the measurements. Each week the surviving individuals were counted. Since *F. candida* is parthenogenetic, eggs appeared in each replicate container. The number of eggs produced was counted for 15 individuals per treatment and all the eggs laid were removed weekly for all the individuals. The age at which the first eggs were produced was recorded for all individuals.

13.2.5 CALCULATIONS AND STATISTICS

Effects on weight and number of eggs produced due to the acetone treatment were evaluated using a *t*-test for each time interval. Effects on survival due to the acetone control were tested using a Pearson chi-square test.

The increase of body weight was described using the Von Bertalanffy growth model for each treatment tested (Kooijman, 1988):

$$W(t) = \left\{ W_\infty^{1/3} - (W_\infty^{1/3} - W_0^{1/3}) \exp(-\gamma t) \right\}^3$$

where: $W(t)$ = body weight at day t (μg fresh weight); W_∞ = final weight (μg fresh weight); W_0 = initial weight (μg fresh weight); γ = specific growth rate (day^{-1}); and t = time (days).

Differences between the estimated Von Bertalanffy growth curves were analysed using a generalized likelihood ratio test.

Survival data at the different time intervals were fitted to a Weibull distribution (Christensen and Nyholm, 1984) in which the proportion of survival (p) is described by:

$$p = \exp\left\{ -\ln 2 \left(\frac{t}{LT_{50}} \right)^k \right\}$$

where: p = proportion of individuals surviving; LT_{50} = median life span (days); t = time (days); and k = a dimensionless exponent determining the shape of the survival curve.

Differences between the estimated Weibull distributions were analysed using a generalized likelihood ratio test.

To evaluate for possible differences in the length of the juvenile period and the total number of eggs in 120 days between control and treatments, a Kruskall–Wallis multiple comparisons test was used. Life-table data were used to determine the intrinsic rate of natural increase (r) with a computer program listed in Stearns (1992) using the Euler–Lotka equation (Lotka, 1913):

$$\sum_{x=0}^{120} l_x m_x e^{-rx} = 1$$

where: l_x = proportion of individuals surviving to age x; m_x = number of eggs per surviving female between age x and $x+1$; and r = intrinsic rate of natural increase (day^{-1}).

All the statistical tests mentioned and the curve-fitting procedures using the Von Bertalanffy growth model and the Weibull distribution were performed using the SYSTAT 5.1. statistics package for MS-DOS computers. The computer program used for the estimation of the intrinsic rate of natural increase was written in APL (A Programming Language, version 5.0).

13.3 EFFECTS ON LIFE-HISTORY CHARACTERISTICS

13.3.1 EFFECTS ON GROWTH

No effects on growth due to the use of acetone as a solvent were found at each of the time intervals considered ($P > 0.05$). Growth curves shown in Figures 13.2 and Figure 13.3 for the control are therefore based on the lumped data of the control and the acetone control.

Cadmium (Cd) and triphenyltin hydroxide (TPT-OH) reduced the specific growth rate g as well as the final weight W_∞ of *Folsomia candida* individuals ($P < 0.05$) (Figures 13.1 and 13.3; Table 13.1). Springtails exposed to CPF experienced neither effects on the specific growth rate nor on the final weight ($P > 0.05$) (Figure 13.2; Table 13.1). There were also slight differences between the specific growth rate for the three control treatments ($P < 0.05$).

Table 13.1 Effect of different treatments (1 to 5) of cadmium (Cd), chlorpyrifos (CPF) and triphenyltin hydroxide (TPT-OH) on body growth of *Folsomia candida*. Shown are the estimated specific growth rate (γ day^{-1}) and final fresh weight ($W_\infty^{1/3}$ in $\mu g^{1/3}$ and W_∞ in μg)

Sub-stance	Parameter	Control	Control acetone	Treatment 1	2	3	4	5
Cd	γ	0.051		0.051	0.046	0.045	0.054	0.042*
		(0.003)		(0.004)	(0.003)	(0.002)	(0.006)	(0.004)
	$W_\infty^{1/3}$	6.736		6.493	6.516	6.388	5.892	4.843*
		(0.056)		(0.065)	(0.067)	(0.059)	(0.077)	(0.063)
	W_∞	306		274	277	261	205	114
CPF	γ	0.045	0.043	0.039	0.034	0.037	0.039	0.043
		(0.001)	(0.002)	(0.001)	(0.001)	(0.002)	(0.002)	(0.002)
	$W_\infty^{1/3}$	6.699	6.634	6.768	6.721	6.651	6.533	6.488
		(0.047)	(0.055)	(0.037)	(0.046)	(0.049)	(0.058)	(0.047)
	W_∞	301	292	310	304	294	279	273
TPT-OH	γ	0.065	0.062	0.045	0.048	0.048	0.056*	0.044*
		(0.002)	(0.008)	(0.003)	(0.003)	(0.003)	(0.004)	(0.004)
	$W_\infty^{1/3}$	6.448	6.548	6.399	6.218	5.701	4.988*	4.550*
		(0.051)	(0.080)	(0.062)	(0.053)	(0.038)	(0.037)	(0.043)
	W_∞	268	281	262	240	185	124	94

Note: standard errors of each estimate are given between parentheses.
For concentrations of treatments 1–5, see section 13.2.4.
*Significantly different from lumped data from control and acetone control ($P < 0.05$).

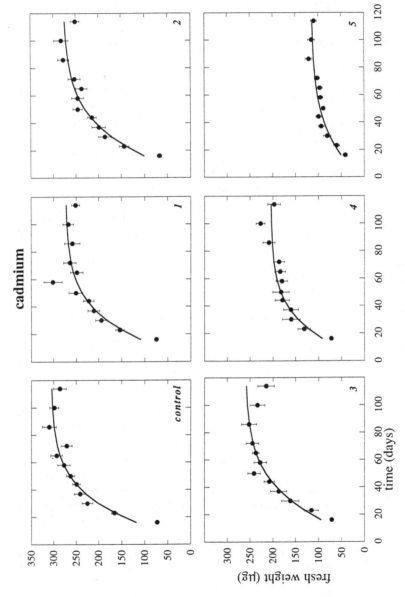

Figure 13.1 Effect of cadmium on animal growth. Measured weight with standard error (bars) and estimated weight using the Von Bertalanffy growth model (—) of *Folsomia candida* at different time intervals. The graph numbers correspond to increasing concentrations of cadmium (μg g⁻¹) in the dry food: graph 1, 139; graph 2, 300; graph 3, 646; graph 4, 1392; graph 5, 3000.

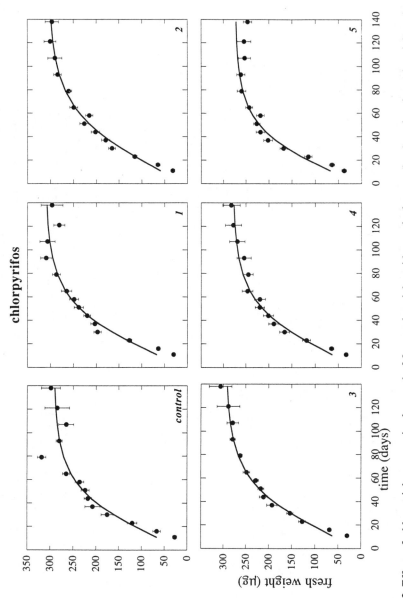

Figure 13.2 Effect of chlorpyriphos on animal growth. Measured weight with standard error (bars) and estimated weight using the Von Bertalanffy growth model (—) of *Folsomia candida* at different time intervals. The graph numbers correspond to increasing concentrations of chlorpyrifos (μg g^{-1}) in the dry food: graph 1, 0.93; graph 2, 2.00; graph 3, 4.31; graph 4, 9.28; and graph 5, 20.

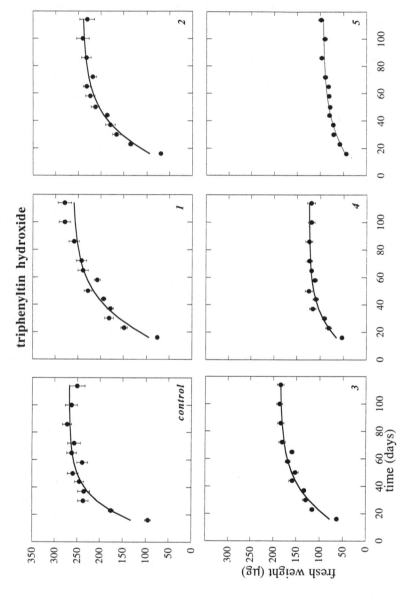

Figure 13.3 Effect of triphenyltin hydroxide on animal growth. Measured weight with standard error (bars) and estimated weight using the Von Bertalanffy growth model (—) of *Folsomia candida* at different time intervals. The graph numbers correspond to increasing concentrations of triphenyltin hydroxide ($\mu g\ g^{-1}$) in the dry food: graph 1, 139; graph 2, 300; graph 3, 646; graph 4, 1392; and graph 5, 3000.

13.3.2 EFFECTS ON REPRODUCTION AND SURVIVAL

No significant effects on the total number of eggs and survival due to the use of acetone as a solvent were found ($P > 0.05$). Data presented in Figures 13.5 and 13.6 for the controls are therefore the grouped data of the control and the acetone control.

The length of the juvenile period increased with increasing concentrations for individuals exposed to Cd, CPF and TPT-OH (Table 13.2). In the Cd experiment the juvenile period was affected only at the highest treatment. Juvenile period was affected at the two highest treatments in the CPF experiment and at the three highest treatments in the TPT-OH experiment.

Number of eggs produced was the highest during the first 40 days in all the three experiments at all treatments and decreased after some time (Figures 13.4–13.6). The total number of eggs produced in 120 days decreased with increasing concentrations for all the substances tested. Significant effects were found at the fifth treatment in the Cd experiment and from the third treatment onwards in the CPF experiment and the TPT-OH experiment (Table 13.2).

The median lifespan increased with increasing Cd and TPT-OH concentrations and decreased with increasing CPF concentrations.

13.3.3 RELATIONSHIP BETWEEN LIFE-HISTORY CHARACTERISTICS AND EFFECTS ON POPULATION GROWTH

There was a very strong negative correlation between the estimated final weight and the estimated median lifespan in the Cd and TPT-OH experiment (Figure 13.7). LT_{50} values increased linearly with decreasing weights. Significant positive correlations were found between estimated final weight and total number of eggs produced in the Cd and TPT-OH experiment (Figure 13.8). No significant correlations between estimated final weight and the estimated median lifespan and between estimated final weight and total number of eggs produced were found in the CPF experiment.

The intrinsic rate of natural increase decreased with increasing concentrations of Cd, CPF and TPT-OH. In the Cd experiments the intrinsic rate of natural increase declined only slightly, whereas the strongest effect was found in the CPF experiment (Table 13.2).

13.4 CONCLUSIONS ON THE LIFE-CYCLE APPROACH

This study demonstrated substance-specific effects of toxicants on the different life-history characteristics of *Folsomia candida* exposed to Cd, CPF and TPT-OH. Cd and TPT-OH affected growth of the springtails while CPF did not. Reproduction was affected by all the three substances tested. A significant correlation was found between final weight and total number of eggs produced in the Cd and TPT-OH experiment, while this was not the case for springtails exposed to CPF. We conclude therefore that Cd and TPT-OH act

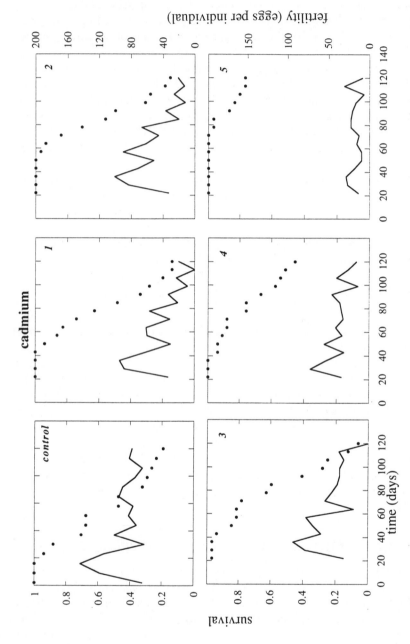

Figure 13.4 Effects of cadmium on fertility and survival of *Folsomia candida*. The counted number of eggs produced per individual (—) and the proportion of animals surviving (bars) at the different time intervals are shown. The graph numbers correspond to increasing concentrations of cadmium ($\mu g\ g^{-1}$) in the dry food: graph 1, 139; graph 2, 300; graph 3, 646; graph 4, 1392; and graph 5, 3000.

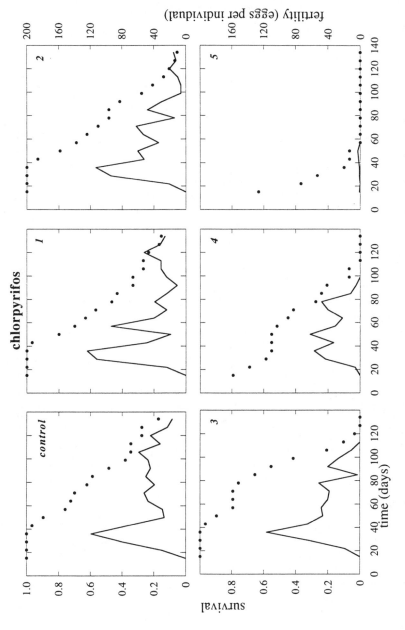

Figure 13.5 Effects of chlorpyrifos on fertility and survival of *Folsomia candida*. The counted number of eggs produced per individual (—) and the proportion of animals surviving (bars) at the different time intervals are shown. The graph numbers correspond to increasing concentrations of chlorpyrifos ($\mu g\ g^{-1}$) in the dry food: graph 1, 0.93; graph 2, 2.00; graph 3, 4.31; graph 4, 9.28; and graph 5, 20.

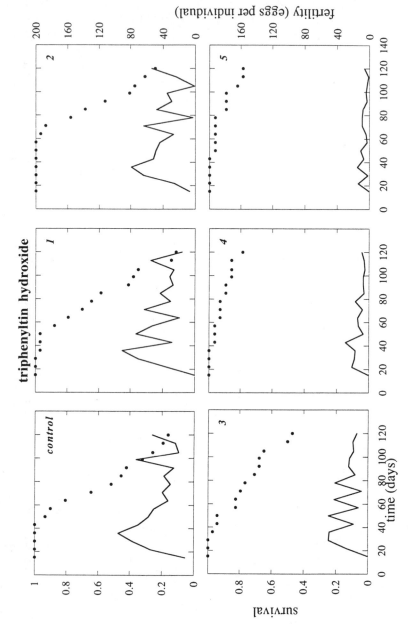

Figure 13.6 Effects of triphenyltin hydroxide on fertility and survival of *Folsomia candida*. The counted number of eggs produced per individual (—) and the proportion of animals surviving (•) at the different time intervals are shown. The graph numbers correspond to increasing concentrations of triphenyltin hydroxide (μg g^{-1}) in the dry food: graph 1, 139; graph 2, 300; graph 3, 646; graph 4, 1392; and graph 5, 3000.

Table 13.2 Effect of different treatments (1 to 5) of cadmium (Cd), chlorpyrifos (CPF) and triphenyltin hydroxide (TPT-OH) on life-table characteristics of *Folsomia candida*. Shown are the juvenile period (jp in days), total number of eggs in 120 days (eggs), estimated median survival time (LT_{50} in days) and intrinsic rate of natural increase (r in day^{-1})

Sub-stance	Parameter	Control	Acetone control	Treatment 1	2	3	4	5
Cd	jp	18.6		19.0	19.1	19.6	19.0	21.7*
		(0.3)		(0.3)	(0.3)	(0.4)	(0.3)	(0.4)
	eggs	646		496	599	550	469	255*
		(65)		(61)	(63)	(43)	(46)	(23)
	LT_{50}	81.6		85.6	91.9	87.0	112**	147**
		(1.3)		(0.8)	(0.9)	(0.9)	(1.6)	(5.1)
	r	0.191		0.189	0.189	0.183	0.185	0.147
CPF	jp	18.7	19.4	19.0	20.0	20.0	23.8*	33.3*
		(0.4)	(0.4)	(0.4)	(0.7)	(0.5)	(1.0)	(7.1)
	eggs	742	837	616	544	454*	288*	9*
		(73)	(98)	(62)	(63)	(55)	(85)	(6)
	LT_{50}	93.1	81.6	81.3	79.0	90.6	48.4	18.2
		(1.1)	(0.8)	(1.1)	(0.7)	(0.8)	(1.6)	(0.2)
	r	0.187	0.183	0.189	0.184	0.177	0.137	0.011
TPT-OH	jp	16.1	16.5	16.9	17.1	17.8*	19.3*	21.9*
		(0.3)	(0.4)	(0.4)	(0.2)	(0.2)	(0.3)	(0.8)
	eggs	525	602	576	515	373*	185*	97*
		(41)	(37)	(43)	(30)	(26)	(20)	(12)
	LT_{50}	77.0	85.0	89.0	98.0**	119**	199**	162**
		(0.9)	(1.3)	(1.0)	(1.3)	(2.8)	(12)	(6.6)
	r	0.214	0.220	0.187	0.197	0.174	0.155	0.131

Note: Standard errors of each estimate are given between parentheses.
For concentrations in treatments 1–5, see section 13.2.4.
*Significantly different from control ($P < 0.05$) using Kruskall–Wallis test.
**Significantly different from control using generalized likelihood test.

mainly on growth and through this indirectly on reproduction while CPF directly influences reproduction.

The indirect effect of Cd on reproduction and the direct effect on growth was also found in studies using the springtails *Orchesella cincta* (Van Straalen *et al.*, 1989) and *Onychiurus armatus* (Bengtsson *et al.*, 1985). Indeed, we derived the same conclusion in an earlier study in which *F. candida* was exposed to cadmium in artificial soil for 28 days (Crommentuijn *et al.*, 1993).

Differences in effects may be due to the different modes of action of the toxicants used: Cd, a heavy metal with a general reactivity; CPF, an

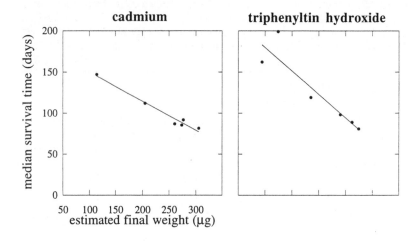

Figure 13.7 Relationship between estimated final weight and estimated median lifespan of *Folsomia candida* exposed to cadmium and triphenyltin hydroxide in the food.

organophosphorus insecticide acting by cholinesterase inhibition; and TPT-OH, an organotin fungicide, acting by inhibition of the oxidative phosphorylation.

Substance-specific effects were also found in other studies. In life-table studies with the aquatic species *Daphnia magna*, Van Leeuwen *et al.* (1987) found that reproduction was inhibited by bromide and 2,4-dichloroaniline; cadmium, bichromate and metavanadate mainly affected growth while bro-

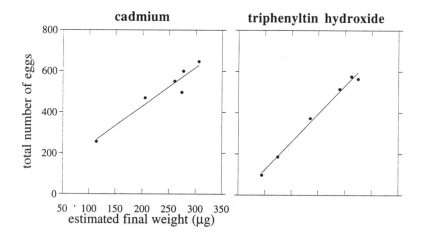

Figure 13.8 Relationship between estimated final weight and total number of eggs produced by *Folsomia candida* exposed to cadmium and triphenyltin hydroxide in the food.

mide affected both growth and reproduction. Fenvalerate affected reproduction of *D. magna* in a study performed by Day and Kaushik (1987). Vranken *et al.* (1991) found differences in toxicity ranks based on fecundity, mortality and development rate for the marine nematode *Monhystera disjuncta* when exposed to different metals. These examples also indicate a substance-specific influence on the different life-history characteristics.

This substance-specific influence is, however, not the same for all species. Following exposure to Cd, the mite *Platynothrus peltifer* experienced no effects on growth, but reproduction was affected directly (Van Straalen *et al.*, 1989). Species-specific differences in sensitivity were also found when evaluating effects on life-history characteristics of different species exposed to the same material, as shown in an earlier study (Crommentuijn *et al.*, 1995).

Besides the substance- and species-specific effects on the different life-history characteristics, interactions between effects also exist. In our study of springtails, the lifespan was increased when exposed to Cd and TPT-OH, but reduced following exposure to CPF. The effects of Cd and TPT-OH on the number of eggs produced were partly compensated by a longer lifespan. In a study with *Daphnia magna* exposed to Cd an increased lifespan of the individuals was also found which corresponded with a decrease in reproduction (Klüttgen and Ratte, 1994). No such trade-off existed in individuals exposed to CPF.

Folsomia candida appears to compensate the effects on body growth with an increase in lifespan, which makes the intrinsic population growth *r* less sensitive than expected. When reproduction is directly affected (as with CPF), no compensation seems to be possible and *r* is severely reduced. These observations are in line with the theory developed by Sibly and Calow (1989) who argued that the effects of toxicants on the fitness of a species depend on the trade-off between mortality and growth.

The interactions found between body growth, egg production and survival demonstrate that it is impossible to predict population level effects from only one of these characteristics, a conclusion also reached by Nisbet *et al.* (1989). The combination of effects on the different life-history characteristics determine the population growth, which is a species- and substance specific interaction. Which of the life-history characteristics has a larger effect on *r* cannot be stated categorically (Snell, 1978). The relative effects of the different life-history characteristics on *r* are also determined by the values of the variables (Kooijman and Metz, 1984).

In conclusion, it can be stated that there is a need to apply a life-cycle approach in order to better understand the subtle species- and substance-specific interaction of different life-history characteristics and the consequences of such interaction at the population level. It is not clear if the single species-individual level-oriented approach will fully protect species in the field. The integration of ecological concepts in life-history theory in combination with toxicological aspects will improve the understanding of the effects expected in the field and will help the development of a more ecologically oriented risk assessment procedures.

13.5 SUMMARY

Chronic life cycle experiments were performed using the collembolan species *Folsomia candida* (Willem) to evaluate the effects of toxicants at the population level. Springtails were exposed to cadmium (Cd), chlorpyrifos (CPF) and triphenyltin hydroxide (TPT-OH) at each of five concentrations, the test compounds being mixed with the food. Effects on weight increase, number of eggs produced and survival were examined for 120 days and the intrinsic rate of natural increase (r) was estimated from the age-specific vital rates using the Euler–Lotka equation. Cd and TPT-OH primarily reduced body growth while CPF mainly affected number of eggs produced. The reduction in body growth observed in springtails exposed to Cd and TPT-OH was compensated by an increase in lifespan. The intrinsic rate of natural increase was reduced by all three toxicants. This effect was most severe in the case of CPF due to the lack of a compensatory mechanism. These results show the importance of evaluating the effects on different life-history characteristics to understand the effects at the population level.

ACKNOWLEDGEMENTS

The authors are indebted to T.C. Van Brummelen and N.M. Van Straalen for comments on the manuscript. Mrs M. Aldham-Breary improved the English. The research was supported by the Netherlands Integrated Soil Research Programme (NISRP), project nò. 5002 and the Ministry of Housing, Physical Planning and Environment, project no. 361284.

REFERENCES

Allan, J.D. and Daniels, R.E. (1982) Life-table evaluation of chronic exposure of *Eurytomora affinis* (Copepoda) to kepone. *Mar. Biol. (Berlin)*, **66**, 179–84.

Begon, M., Harper, J.L. and Townsend, C.R. (1990) *Ecology, Individuals, Populations and Communities*, Blackwell Scientific Publications, Oxford.

Bengtsson, G., Gunnarsson, T. and Rundgren, S. (1985) Influence of metals on reproduction, mortality and population growth of *Onychiurus armatus* (Collembola). *J. Appl. Ecol.*, **22**, 967–78.

Christensen, E.R. and Nyholm, N. (1984) Ecotoxicological assays with algae: Weibull dose–response curves. *Environ. Sci. Technol.*, **18**, 713–18.

Crommentuijn, T., Brils, J. and Van Straalen, N.M. (1993) Influence of cadmium on life-history characteristics of *Folsomia candida* (Willem) in an artificial soil substrate. *Ecotoxicol. Environ. Safety*, **26**, 216–27.

Crommentuijn, T., Doodeman, C.J.A.M., Doornekamp, A., Rademaker, M.C.J., Van Der Pol, J.J.C. and Van Gestel, C.A.M. (1995) Sublethal Sensitivity Index as an ecotoxicity parameter measuring energy allocation under toxicant stress.

Application to cadmium in soil arthropods. *Ecotox. Environ. Safety*, **31**, 192–200.

Daniels, R.E. and Allan, J.D. (1981) Life-table evaluation of chronic exposure to a pesticide. *Can. J. Fish. Aquat. Sci.*, **38**, 485–94.

Day, K. and Kaushik, N.K. (1987) An assessment of the chronic toxicity of the synthetic pyrethroid, fenvalerate to *Daphnia galeata mendotea*, using life-tables. *Environ. Pollut*, **44**, 13–26.

Dutch Health Council (1989) *Assessing the Risk of Toxic Chemicals for Ecosystems*. Report no. **28/E**, Health Council, s-Gravenhage, The Netherlands.

Gentile, J.H., Gentile, S.M., Hairston, N.G., Jr and Sullivan, B.K. (1982) The use of life-tables for evaluating the chronic toxicity of pollutants to *Mysidopsis bahia*. *Hydrobiologia*, **93**, 179–87.

ISO (1994) ISO/DRAFT. Soil Quality – *Effects of Soil Pollutants on Collembola (Folsomia candida): Method for Determination of Effects on Reproduction.*

Klüttgen, B. and Ratte, H.T. (1994) Effects of different food doses on cadmium toxicity to *Daphnia magna*. *Environ. Toxicol. Chem.*, **13**, 1619–27.

Kooijman, S.A.L.M. (1988) The Von Bertalanffy growth rate as a function of physiological parameters: a comparative analysis, in *Mathematical Ecology* (eds T.G. Hallam, L.J. Gross and S.A. Levin), World Scientific, Singapore, pp. 3–45.

Kooijman, S.A.L.M. and Metz, J.A.J. (1984) On the dynamics of chemically stressed populations: the deduction of population consequences from effects on individuals. *Ecotox. Environ. Safety*, **8**, 254–74.

Lotka, A.J. (1913) A natural population norm. *J. Wash. Acad. Sci.*, **3**, 241–8, 289–93.

Nisbet, R.M., Gurney, W.S.C., Murdoch, W.W. and McCauley, E. (1989) Structured population models: a tool for linking effects at individual and population level. *Biol. J. Linnean Soc.*, **37**, 79–99.

OECD (1984) *Guideline for Testing of Chemicals. no. 207. Earthworm Acute Toxicity Tests.* Adopted 4 April 1984.

Sibly, R.M. and Calow, P. (1989) A life-cycle theory of responses to stress. *Biol. J. Linnean. Soc.*, **37**, 101–16.

Snell, T.W. (1978) Fecundity, developmental time and population growth rate. *Oecologia (Berlin)*, **32**, 119–25.

Stearns, S.C. (1992) *The Evolution of Life Histories*, Oxford University Press, Oxford.

Underwood, A.J. (1989) The analysis of stress in natural populations. *Biol. J. Linnean Soc.*, **37**, 51–78.

Van Leeuwen, C.J., Niebeek, G. and Rijkeboer, M. (1987) Effects of chemical stress on the population dynamics of *Daphnia magna*: a comparison of two test procedures. *Ecotox. Environ. Safety*, **14**, 1–11.

Van Straalen, N.M., Schobben, J.H.M., De Goede, R.G.M. (1989) Population consequences of cadmium toxicity in soil microarthropods. *Ecotox. Environ. Safety*, **17**, 190–204.

Vranken, G., Van Derhaeghen, R. and Heip, C. (1991) Effects of pollutants on life-history parameters of the marine nematode *Monhystera disjuncta*. *ICES J. Mar. Sci.*, **48**, 325–34.

14 Reaction norms for life-history traits as the basis for the evaluation of critical effect levels of toxicants

JAN E. KAMMENGA, GERARD W. KORTHALS, TOM BONGERS AND JAAP BAKKER

14.1 THE QUESTION OF SENSITIVITY

A main objective of ecotoxicology is to assess the effect of chemicals on ecosystems and to evaluate their potential ecological risk (Moriarty, 1983). Consequently, many theoretical and experimental explorations have been made to provide an adequate basis for the effect assessment of toxicants at the ecosystem level. Much attention is paid to the development of statistical methodologies to extrapolate the results obtained from laboratory toxicity tests to the field situation (Van Straalen and Denneman, 1989; Wagner and Løkke, 1991; Aldenberg and Slob, 1993). Basically these methods estimate safe environmental concentrations of hazardous compounds from distribution models of critical effect levels obtained from single-species toxicity tests for different organisms.

In general, toxicity tests are used to estimate critical effect levels (EC_x or LOEC) from concentration–response relationships for single life-cycle variables such as mortality, growth, reproduction or breeding success. Numerous results concerning these tests have been presented for studies using both aquatic and terrestrial organisms (Jørgensen et al., 1991), and currently a large part of ecotoxicological research is focused on the development of standardized tests for a wide range of terrestrial and aquatic species (Maltby and Calow, 1989; Van Gestel and Van Straalen, 1994).

A prevailing view in many of these studies is that the estimation of critical effect levels for sensitive life-history traits are highly relevant for the ecological effect assessment of toxicants. This premise stems from classic toxicological studies which aim for sensitive indicators in test organisms to evaluate the biological hazard of chemical compounds to humans. For instance,

Ecological Risk Assessment of Contaminants in Soil. Edited by Nico M. van Straalen and Hans Løkke. Published in 1997 by Chapman & Hall, London. ISBN 0 412 75900 4

Nagel *et al.* (1991) implicitly focused on early-life stages in zebrafish because of their sensitivity to toxicants. Thompson *et al.* (1991) stipulated specifically that somatic growth was more sensitive than mortality when evaluating the impact of hydrogen sulphide on sea urchins. Also Burgess and Morrison (1994) pointed out that sublethal growth experiments greatly increased the sensitivity compared with tests based on mortality. Indeed, the rationale of applying the most sensitive life-stage as a suitable parameter is often adopted (McKim, 1985; Norberg-King, 1989; Marchini *et al.*, 1993), and it was mentioned that 'it is important that the most sensitive stages of the life-history are identified if the results are to be ecologically relevant and of use in predicting effects of pollutants in the field' (Green *et al.*, 1986).

At present, standardized toxicity tests – which are explicitly based on the sensitivity concept – are performed with earthworms and daphnids and have been adopted by the Organization for Economic Cooperation and Development (OECD) and the European Union (EU) (Anonymous, 1984, 1985). Although these studies and approaches may be useful for the ranking of the potential toxicity of chemicals, we will illustrate that the general premise of sensitive variables being ecologically relevant is not supported by life-history theory.

This chapter explores the importance of critical effect levels on various traits for two different life-history strategies. The evaluation is based on the reaction norm for different traits which mirrors the range of toxicant concentrations to which life-history traits are exposed. In addition, the implications of this approach for risk assessment procedures will be outlined.

14.2 REACTION NORMS FOR LIFE-HISTORY TRAITS TO TOXICANTS

Life-history theory states that the tendency of natural selection is to maximize fitness by optimizing different components of the life cycle (Fisher, 1958). Following this theorem Charlesworth (1980) and Kozlowski (1993) showed that phenotypic fitness is defined by the intrinsic rate of natural increase which is the root r of the following Euler–Lotka equation:

$$1 = \sum_{t=0}^{\infty} e^{rt} l_t(p, E) n_t(p, E) \qquad (14.1)$$

where t is age, $l_t(p,E)$ is survivorship during time t of phenotype p in environment E; and $n_t(p,E)$ is the number of female offspring per time unit at age t of phenotype p in environment E (Sibly and Calow, 1983; Smith, 1991).

To maintain maximum fitness in a changing and less favourable or contaminated environment many species are able to adapt life cycle phenotypes within one generation, a phenomenon which is called phenotypic plasticity (Stearns, 1983). Plasticity refers to differences in life-history components within populations which do not originate from genetic differentiation (Stearns, 1992). For example, phenotypic plasticity has been found by Dangerfield and Hassall (1992) in breeding phenology for the woodlouse *Armadillidium vulgare*. They observed a range in life-history phenotypes due to spatial as well as temporal variation, and argued that plasticity can be appropriate in the attempt to maximize fitness in a changing environment. Another example is a study by Stibor (1992) who reported that life-history shifts in cladocerans can be induced by external factors such as chemical stimuli released by a predator.

The range of potential phenotypes that a single genotype can develop, if exposed to a specified range of ambient conditions, is called the reaction norm (Woltereck, 1909). At present, reaction norms have been found to temperature for eye-size genotypes in *Drosophila* and to altitude for plant height (Griffiths *et al.*, 1993). The findings of Stearns (1983) have led to the definition of the reaction norm for life-history traits such as age and size at maturity (Stearns and Koella, 1986). However, the classic toxicological dose–response relationship can also be conceived of as a reaction norm if obtained for one genotype. The concentration range of the toxicant represents the ambient environment, and the variation in phenotypes is represented by the range of effect levels of different life-history traits.

A fundamental aspect of plasticity is that life-history traits can be flexible over a certain range without being disadvantageous to the species in question, i.e. without any significant fitness reduction. These findings have serious consequences for the evaluation of effect levels which will be exemplified by means of a life cycle analysis for two different iteroparous strategies.

14.2.1 ANNUAL ITEROPAROUS LIFE CYCLE WITH EQUAL JUVENILE AND ADULT PERIODS

Consider a sexually reproducing organism with homozygous alleles and an annual iteroparous life cycle. This means that each genotype has the same life cycle and that the adults survive after reproduction and live on to next year for the following breeding season (e.g. some invertebrates). Following the approach of Sibly (1989) we further assume that the period between breeding seasons is equal to the length of the juvenile period (Figure 14.1). Hence Equation (14.1) becomes (Sibly, 1989):

$$\frac{1}{2}nS_j e^{-rt_j} + S_a e^{-rt_j} = 1 \qquad (14.2)$$

where n = the age-specific reproduction rate, S_j = survival from birth to the end of the juvenile period, r = fitness, t_j = the juvenile period and S_a = survival during the reproductive period. The hypothetical reaction norm Δx for the genotype is defined as the toxicant-induced change in a life cycle variable

x, from 0 (no change) to 1 (100% change). The reaction norm can be related to changes in fitness (Δr) by performing a sensitivity analysis of Equation (14.2) using multiple iteration processes. The software package MathCad 5.0 (Mathsoft Inc. USA) was used for this purpose. Different maps can now be constructed illustrating the relative sensitivity of fitness to Δx.

(a) Map of Δr versus ΔS_a

Assuming $S_j = 0.63$ and $t_j = 30.6$, which are realistic values obtained from life cycle experiments with nematodes, Figure 14.2(a) shows the relationship between changes in fitness (Δr) and the reaction norm for S_a (ΔS_a) for $n = 4$, $n = 6$ and $n = 30$. It is illustrated that ΔS_a influences Δr most strongly when n is low. However, when n is high the effect is negligible. This implies that species with a high reproduction rate are less vulnerable to stress-induced reduction in adult survival. On the contrary, species with low reproduction rates are most susceptible to impairment of adult survival. The length of the juvenile period did not have any effect on the relationship between ΔS_a and Δr.

Figure 14.2(b) shows the same map; however, in this case $S_j = 0.05$, 0.5 and 1.0 for $n = 4$ and $t_j = 30.6$. It shows that ΔS_a has a very strong effect on Δr when S_j is low. The influence on Δr will never be negligible because even at maximum juvenile survival ($S_j = 1.0$) there is a significant decrease in Δr. The map illustrates that organisms with low S_j are extremely vulnerable to stress-induced reductions in S_a. Even when S_j is 1, a change in S_a will have detrimental consequences for fitness.

(b) Map of Δr versus Δn, ΔS_j and Δt_j

Figure 14.2(c) shows the reaction norm for the length of the juvenile period (Δt_j), juvenile survival (ΔS_j) and reproduction rate (Δn) in relation to changes in fitness. It is illustrated that Δt_j influences Δr differently compared with ΔS_a (Figure 14.2(c)) when $n = 4$, $S_j = 0.63$, $t_j = 30.6$. However changing n, S_j or S_a does not affect this relationship, indicating that the influence of t_j on fitness is not determined by these variables. Also the relationship between Δn or ΔS_j, and fitness appears to be very rigid and non-sensitive to changes in the other variables.

These findings may have far-reaching consequences for the evaluation of critical effect levels of toxicants. For example, a 50% effect (EC_{50}) in t_j or n

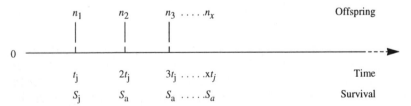

Figure 14.1 Annual iteroparous life cycle with equal juvenile and adult periods.

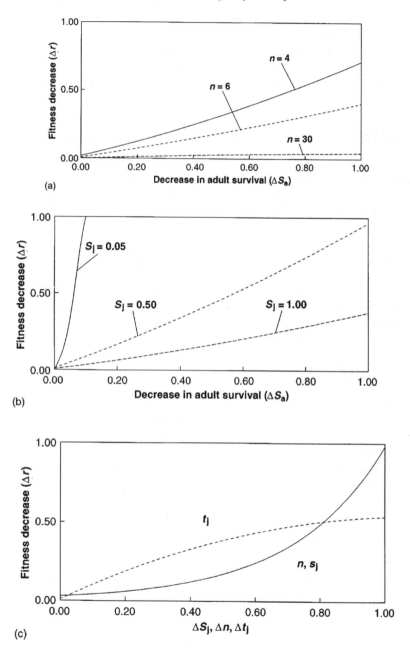

Figure 14.2 Graphs showing the changes in fitness, Δr, as a function of changes in adult survival, S_a, juvenile survival, S_j, rate of reproduction, n, and juvenile period, t_j, for an annual iteroparous life cycle with a juvenile period equal to the adult reproductive period (Equation (14.2)). (a) Δr as a function of ΔS_a for three values of n; (b) Δr as a function of ΔS_a for three values of S_j; (c) Δr as a function of ΔS_j, Δn, and Δt.

has less influence on fitness (Figure 14.2(c)) than 10% (EC_{10}) reduction in S_a if $S_j = 0.05$ (Figure 14.2(b)). On the other hand, a 50% reduction in S_a when $S_j = 1$ (Figure 14.2(b)) has a smaller effect on fitness than even a 20% effect on t_j (Figure 14.2(c)). Also, a 50% reduction of S_a when $n = 30$ (Figure 14.2(a)) has less impact on fitness than a 10% increase in t_j (Figure 14.2(c)).

14.2.2 ANNUAL ITEROPAROUS LIFE CYCLE WITH UNEQUAL JUVENILE AND ADULT PERIODS AND TIME-DEPENDENT MORTALITY

Different results are obtained by focusing on the same iteroparous life cycle except that t_j is not equal to t_a and mortality is time-dependent following an exponential decrease, $S_j = e^{-\mu t_j}$, $S_a = e^{-\tau t_a}$ (e.g. some birds and small vertebrates). Assuming a difference between t_j and t_a and time-dependent mortality (Figure 14.3), then Equation (14.1) can be rewritten as (Sibly, 1989):

$$\frac{1}{2}n\exp(-\mu t_j - rt_j) + \exp(-\tau t_a - rt_a) = 1 \qquad (14.3)$$

where n = age-specific reproduction rate, μ = juvenile mortality rate from birth to the end of the juvenile period, r = fitness, t_j = juvenile period, τ = adult mortality rate, and t_a = adult period. The reaction norm can be related to Δr by performing a sensitivity analysis equivalent to Equation (14.2).

(a) Map of Δr versus Δn

Taking $t_j = 30$, $\tau = 0.02$, $t_a = 60$, and $n = 100..1$ (where 100 is set to 0 and 1 is set to 1), values which are commonly found for some small vertebrates, Figure 14.4(a) shows the relationship between Δr and the reaction norm Δn (0–1) for $\mu = 0.01$, $\mu = 0.05$, and $\mu = 0.08$. Although impairment of reproduction rate is detrimental in all three cases, it appears that the effect of stress-induced reductions in n on fitness is most important in species with high juvenile mortality rates.

(b) Map of Δr versus $\Delta \tau$

Figure 14.4(b) shows the map of changes in the adult mortality rate, $\Delta \tau$, for $t_j = 30$, $\tau = 0.02$, $t_a = 60$, $n = 3$, $n = 5$ and $n = 100$. It is illustrated that $\Delta \tau$

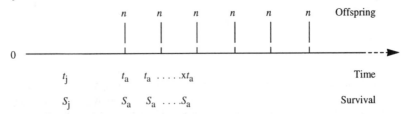

Figure 14.3 Iteroparous life cycle where the juvenile period, t_j, is not equal to the adult period, t_a, and time-dependent mortality.

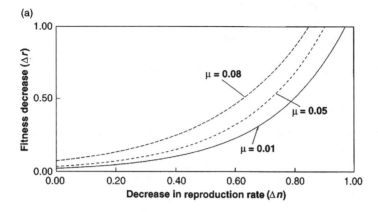

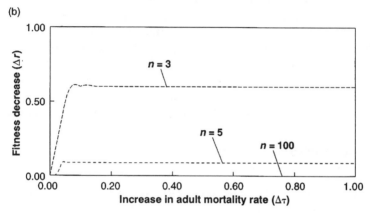

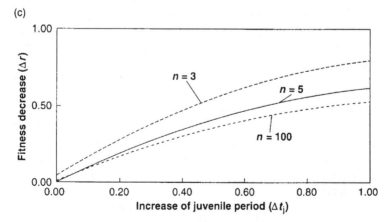

Figure 14.4 Graphs showing the changes in fitness, Δr, as a function of changes in (a) reproduction rate, n; (b) adult mortality rate, τ; and (c) juvenile period, t_j, for different values of juvenile mortality rate, m, and reproduction rate, n. The graphs refer to an annual iteroparous life cycle with unequal juvenile and adult periods (Equation (14.3)).

influences fitness strongly from 0 to 0.1 when $n = 5$ and $n = 3$, whereas further impairment does not affect fitness. Also, it appears that at low n the effect on Δr is very strong compared with high n. At $n = 100$ there is no detrimental effect on fitness. This implies that increased adult mortality rate is only important in species with low reproductive output.

(c) Map of Δr versus Δt_j

Figure 14.4(c) illustrates the relationship between Δt_j (here the juvenile period, t_j, increases from 30, set to 0, to 60, set to 1), and Δr for $\mu = 0.01$, $\tau = 0.02$, $n = 3$, $n = 5$, and $n = 100$. The shape of the curve is the same as for t_j in Figure 14.2(c); however, in this case it depends on n. At low n fitness is much more influenced by Δt_j than at high n, indicating that species with low reproduction rates are more vulnerable to impairment of t_j.

It was indicated that a 50% effect (EC_{50}) in τ when $n = 5$ (Figure 14.4(b)) has a smaller impact on fitness than a 20% effect (EC_{20}) in t_j (Figure 14.4(c)). Also a 50% reduction in n (Figure 14.4(a)) has a smaller effect on fitness than 15% increase in τ (Figure 14.4(b)) when $n = 3$. The conclusion may be that the influence of toxic stress on fitness is not determined by the effect on sensitive traits, but depends on the reaction norm for traits, and the life-history strategy which comprises the relationship between all variables and fitness.

14.3 RISK ASSESSMENT BASED ON REACTION NORMS AND LIFE-HISTORY STRATEGIES

This chapter illustrates that the impact of a toxicant on fitness is determined by: (i) the reaction norm for various life-history traits to chemical stress; and (ii) the life-history strategy. Reaction norms define the variation of life cycle traits to a range of environmental conditions for one genotype. In the field, populations consist of many different genotypes, each having a specific reaction norm to certain stress factors. Using different genotypes of *Daphnia magna*, relatively large differences in reaction norms were observed for stress tolerance to cadmium (Baird *et al.*, 1990). Also, an extensive altitudinal distribution of *Pennisetum setaceum* was noted which could be explained by phenotypic plasticity for leaf photosynthesis to temperature stress (Williams and Black, 1993). Determination of reaction norms requires that different individuals of the same genotype have to be reared in different environments. Therefore, experimental research in this field requires that numerous eggs or juveniles with identical genotypes are produced. At present it is only possible for a few organisms, which are mainly parthenogenetic invertebrates, to obtain a large number of identical genotypes and hence to determine reaction norms for traits to a range of toxicant concentrations (Kammenga, 1994).

In this chapter the reaction norm was hypothesized for various traits from 0 to 100% variation. Experimental data confirm this large variation in phenotypic plasticity for a range of species. A change of 40% was found in plasticity of age at first reproduction in freshwater cladocerans (Stibor, 1992), and in size of reproductive females in woodlice (Dangerfield and Hassall, 1992).

At present, much effort is made in risk-assessment procedures to verify the results obtained from laboratory toxicity tests to the field situation. However, the verification of the test results requires a fundamental understanding of the underlying mechanisms at the individual level. Life-history theory states that in the attempt to maximize fitness, natural selection favours traits which have a strong impact on fitness. Hence, impairment of these traits may give rise to enhanced vulnerability of species to environmental contamination which eventually may lead to extinction. Analysis of the relationship between reaction norms and fitness provides the key to identify these traits for different strategies.

The relationship between the reaction norm and fitness depends on the life-history strategy and determines the ultimate effect of a toxicant on fitness. It was shown in this chapter that for two different life cycles each variable contributes differently to fitness which in many cases depended on the value of the other variables. Strong effects on certain parameters did not always influence fitness to the same extent because reaction norms and fitness are not linearly related. Different maps were drawn relating changes in fitness to the reaction norm of life-history traits for each life cycle strategy. Annual iteroparous species with equal juvenile and adult periods appeared to be vulnerable to stress-induced increase of adult mortality when the juvenile survival or reproduction rate is low. Within the range investigated here, the influence of the juvenile period on fitness was not affected by the values of the other variables. However, species with an iteroparous life cycle and with unequal juvenile and adult periods appeared to be susceptible to increased adult mortality rates when reproduction rate is low, and in this case the relationship between the juvenile period and fitness depended on reproduction rate.

It was reported by other authors that, in general, the length of the juvenile period has the strongest impact on fitness (Lewontin, 1965; Sibly and Calow, 1986). Our results imply that for annual iteroparous organisms with equal juvenile and adult periods the adult survival may have a stronger impact on fitness depending on the value of the juvenile survival. Also in case of the iteroparous life cycle with unequal juvenile and adult periods the influence of adult mortality rate can be stronger at low reproductive output compared with the length of the juvenile period.

This chapter clearly outlines the necessity of novel concepts in the ecological risk assessment of toxicants which unify both ecological and toxicological methodologies. A further step towards a meaningful risk assessment procedure would be to define life-history patterns which can be divided into different classes of vulnerability to toxicants.

14.4 CONCLUSION

The objective of this chapter was to show that critical effect levels of toxicants need to be evaluated on the basis of the reaction norm for life-history traits. By definition the reaction norm is the range of potential life-history phenotypes that a single genotype can develop if exposed to a specified range of toxicant concentrations. Application of the concept to different life-history strategies illustrated that the greatest impact toxicants can have on fitness was not correlated with the critical effect level of the most sensitive traits. In contrast, it was revealed that fitness impairment could be determined by less-sensitive traits depending on the life-history strategy and the reaction norm. These findings imply that the impact of toxicants on fitness is non-intuitive and that current procedures for the risk assessment of toxicants which are based on single and sensitive life cycle variables need to be revisited by life-history theory.

14.5 SUMMARY

A method is proposed to evaluate critical effect levels of toxicants, such as the EC_{50} or LOEC, by unifying life-history theory and toxicological concepts. The method is based on the reaction norm for life-history traits which is defined as the range of potential phenotypes that a single genotype can develop if exposed to a specified range of toxicant concentrations. For two different iteroparous strategies it is illustrated that the impact of toxicants on fitness is not correlated with the critical effect level of the most sensitive traits. In contrast, it was revealed that fitness impairment could be determined by less-sensitive traits depending on the life-history strategy and the reaction norm. These findings denounce current procedures for the risk assessment of toxicants which are based on single and sensitive life-cycle variables. Moreover, these procedures need to be revisited by life-history theory to ensure a proper evaluation of the potential ecological hazards of toxicants.

REFERENCES

Aldenberg, T. and Slob, W. (1993) Confidence limits for hazardous concentrations based on logistically distributed NOEC toxicity data. *Ecotox. Environ. Safety*, **25**, 48–63.

Anonymous (1984) *Guidelines for Testing of Chemicals*. Section 2: Effects on biotic systems - 202 *Daphnia spp*. Acute immobilization test and reproduction test. Organization for Economic Cooperation and Development (OECD). ISBN 64-12221-4.

Anonymous (1985) EEC Directive 79/831. Annex V. Part C. *Methods for the Determination of Ecotoxicity*. Level I. C(II) 4: Toxicity for earthworms. Artificial soil test. DG XI/128/82.

Baird, D.J., Barber, I. and Calow P. (1990) Clonal variation in general responses of *Daphnia magna* Straus to toxic stress. I. Chronic life-history effects. *Funct. Ecol.*, **4**, 339–407.

Burgess, R.M. and Morrison, G.E. (1994) A short-exposure, sublethal, sediment toxicity test using the marine bivalve *Mulinia lateralis*: statistical design and comparative sensitivity. *Environ. Toxicol. Chem.*, **13**, 571–80.

Charlesworth, B. (1980) *Evolution in Age-structured Populations*, Cambridge University Press, Cambridge.

Dangerfield, J.M. and Hassall, M. (1992) Phenotypic variation in the breeding phenology of the woodlouse *Armadillidium vulgare*. *Oecologia*, **89**, 140–6.

Fisher, R.A. (1958) *The Genetical Theory of Natural Selection*, Dover Publications, New York.

Green, D.W.J., Williams, K.A. and Pascoe, D. (1986) The acute and chronic toxicity of cadmium to different life history stages of the freshwater crustacean *Asellus aquaticus* (L). *Arch. Environ. Contam. Toxicol.*, **15**, 465–71.

Griffiths, A.J.F., Miller, J.H., Suzuki, D.T. and Lewontin, R.C. (1993) *An Introduction to Genetic Analysis*, Freeman and Co., New York.

Jørgensen, S.E., Nielsen, S.N. and Jørgensen, L.A. (1991) *Handbook of Ecological Parameters and Ecotoxicology*, Elsevier Scientific Publishers, Amsterdam.

Kammenga, J.E. (1994) *Phenotypic plasticity and fitness consequences in nematodes exposed to toxicants.* PhD Thesis, Agricultural University, Wageningen, The Netherlands

Kozlowski, J. (1993) Measuring fitness in life-history studies. *Trends Ecol. Evol.*, **8**, 84–5.

Lewontin, R.C. (1965) Selection for colonizing ability, in *The Genetics of Colonizing Species* (eds H.G. Baker and L.G. Stebbings), Academic Press, New York, pp. 77–94.

Maltby, L. and Calow, P. (1989) The application of bioassays in the resolution of environmental problems; past, present and future. *Hydrobiologia*, **188/189**, 65–76.

Marchini, S., Hoglund, M.D., Broderius, S.J. and Tosato, M.L. (1993) Comparison of the susceptibility of daphnids and fish to benzene derivatives. *Sci. Tot. Environ.* Suppl., 799–808.

McKim, J.M. (1985) Early life stage toxicity test, in *Fundamentals of Aquatic Toxicology* (eds G.M. Rand and S.R. Petrocelli), Hemisphere, Washington, DC, pp. 58–94.

Moriarty, F. (1983) *Ecotoxicology, the Study of Pollutants in Ecosystems*, Academic Press, London.

Nagel, R., Bresch, H., Caspers, N., Hansen, P.D., Markert, M., Munk, R., Scholz, N. and Ter Höfte B.B. (1991) Effect of 3,4-dichloroaniline on the early life stages of the zebrafish (*Brachidanio rerio*): results of a comparative laboratory study. *Ecotox. Environ. Safety*, **21**, 157–64.

Norberg-King, T.J. (1989) An evaluation of the fathead minnow 7-d subchronic test for estimating chronic toxicity. *Environ. Toxicol. Chem.*, **8**, 1075–89.

Sibly, R. (1989) What evolution maximizes. *Funct. Ecol.*, **3**, 129–35.

Sibly, R. and Calow P. (1983) An integrated approach to life-cycle evolution using selective landscapes. *J. Theor. Biol.*, **102**, 527–47.

Sibly, R. and Calow, P. (1986) Why breeding earlier is always worthwhile. *J. Theor. Biol.*, **123**, 311–19.

Smith, R.H. (1991) Genetic and phenotypic aspects of life-history evolution in animals. *Adv. Ecol. Res.*, **21**, 63–120.

Stearns, S.C. (1983) The evolution of life-history traits in mosquitofish since their introduction to Hawaii in 1905: rates of evolution, heritabilities, and developmental plasticities. *Amer. Zool.*, **23**, 65–76.

Stearns, S.C. (1992) *The Evolution of Life-histories*, Oxford University Press, Oxford.

Stearns, S.C. and Koella, J.C. (1986) The evolution of phenotypic plasticity in life-history traits: predictions of reaction norms for age and size at maturity. *Evolution*, **40**, 893–913.

Stibor, H. (1992) Predator induced life-history shifts in a freshwater cladoceran. *Oecologia*, **92**, 162–5.

Thompson, B., Bay, S., Greenstein, D. and Laughlin, J. (1991) Sublethal effects of hydrogen sulfide in sediments on the urchin *Lytechinus pictus. Mar. Environ. Res.*, **31**, 309–21.

Van Gestel, C.A.M. and Van Straalen, N.M. (1994) Ecotoxicological test systems for terrestrial invertebrates, in *Ecotoxicology of Soil Organisms* (eds M.H. Donker, H. Eijsackers and F. Heimbach), CRC Press, Lewis Publishers, Boca Raton, FL, pp. 205–51.

Van Straalen, N.M. and Denneman, C.A.J. (1989) Ecotoxicological evaluation of soil quality criteria. *Ecotox. Environ. Safety*, **18**, 241–51.

Wagner, C. and Løkke, H. (1991) Estimation of ecotoxicological protection levels from NOEC toxicity data. *Water Res.*, **25**, 1237–42.

Williams, D.G. and Black, R.A. (1993) Phenotypic variation in contrasting temperature environments: growth and photosynthesis in *Pennisetum setaceum* from different altitudes on Hawaii. *Funct. Ecol.*, **7**, 623–33.

Woltereck, R. (1909) Weitere experimentale Untersuchungen über Artveränderung, speziell über das Wesen quantitativer Artenunterschiede bei Dapniden. *Ver. Deutsch. Zool. Gesell.*, **1909**, 110–72.

15 Estimating fitness costs of pollution in iteroparous invertebrates

RYSZARD LASKOWSKI

Life is the art of drawing sufficient conclusions from insufficient premises.

Samuel Butler

15.1 SEMELPAROUS AND ITEROPAROUS INVERTEBRATES

The most important goal of ecotoxicological studies should be to detect what effect a particular toxicant can have on average fitness in a population exposed to it. More detailed studies can provide data on fitness effects in particular phenotypes. If one accepts expected lifetime fertility R_0 as a measure of fitness in laboratory experiments (i.e. in a stationary environment), the task is relatively simple for semelparous invertebrates. In particular, organisms with a short life cycle are suitable for this type of experiment, as they can produce good estimates of R_0 quickly. It is not surprising that these are probably the best-studied organisms in ecotoxicology at present (Daniels and Allan, 1981; Walsh, 1983; Bengtsson *et al.*, 1985; Van Leeuwen *et al.*, 1986; Lawrence *et al.*, 1989; Van Straalen *et al.*, 1989; Chandler, 1990; Kidd, 1991; Snell *et al.*, 1991; Crommentuijn *et al.*, 1993; Fernandez-Caselderrey *et al.*, 1993).

Many invertebrates of paramount importance in the functioning of terrestrial ecosystems are iteroparous. They range from herbivores (e.g. snails and slugs) through carnivores (e.g. centipedes) to detritivores (e.g. earthworms, diplopods). From an ecotoxicological point of view, all these groups deserve special attention for at least two reasons: (i) their well-documented importance in ecosystems (Albert, 1976, 1983a; Jennings and Barkham, 1979; Rożen, 1989); and (ii) the fact that many of them can accumulate substantial amounts of pollutants (Coughtrey and Martin, 1977; Hopkin and Martin,

Ecological Risk Assessment of Contaminants in Soil. Edited by Nico M. van Straalen and Hans Løkke. Published in 1997 by Chapman & Hall, London. ISBN 0 412 75900 4

1983; Bengtsson *et al.*, 1986; Jones, 1991; Laskowski and Maryański, 1993; see also Hopkin, 1989). The latter may have two important consequences: first, these animals can be more endangered by pollution than other short-lived organisms; second, they may be a critical pathway for pollutants along trophic chains (Van Straalen and Ernst, 1991).

Knowledge of the ecotoxicology of iteroparous invertebrates is scant. The main problem is that it is very difficult to estimate the effect of pollutants on the fitness of these organisms in an ecotoxicological experiment. The organisms are perennial, so their whole life cycle takes several years and it is impossible to construct their life-tables during standard short-term laboratory experiments. Moreover, even during the life-span of a single generation the animals are usually exposed to a highly variable environment that is difficult to simulate.

For variable environments the finite rate of increase λ will be used as the general measure of fitness. Nevertheless, any measure of fitness that is chosen will depend on three parameters: birth rate, mortality and somatic growth. The first two have a direct effect on λ, while a decrease in the rate of somatic growth may lead to retardation of maturity and/or decreased size at the age of reproduction. This in turn may decrease the birth rate. All three of these parameters can be affected by toxicants and thus are important in ecotoxicological studies.

This chapter concentrates on problems in estimating and interpreting the effects of toxicants on these parameters for iteroparous invertebrates. The ability to assess these effects for long-lived iteroparous invertebrates would greatly improve prediction of the effects of pollutants on ecosystems. The aim of this study is to propose some solutions to enable ecotoxicologists to study pollution effects on the population parameters of iteroparous invertebrates in a way that will not involve years of experiments. Two examples will be covered. The first is a herbivore, *Helix aspersa* Müller (Gastropoda), and the second a carnivore, *Lithobius mutabilis* L. Koch (Myriapoda). Both species are iteroparous but have different life-history strategies.

15.2 MATRIX MODELLING APPROACH

15.2.1 CONSTRUCTING THE INITIAL PROJECTION MATRIX

As mentioned above, the main problem in studying the ecotoxicology of iteroparous organisms is that it is impossible to provide any direct measure of changes in fitness under the impact of pollutants without spending years on each species. Because it is difficult to measure all the population parameters throughout the life of a cohort, we have to measure a few selected life stages at a time. Then we can estimate the effect of measured changes in these parameters on overall fitness. Of course, these estimates probably will never be as accurate as they can be for short-lived, fast-reproducing animals. Also, they rely to a much greater extent upon statistics and mathematical models.

To relate the estimates to the real world, data on the life-histories of unaffected populations are indispensable. Unfortunately, good life-history data are rarely available, for the same reason that good ecotoxicological information is missing. A good starting point is to attempt to construct a reliable life-table for an unaffected population using data from the literature. Later the calculated population parameters can be adjusted to meet the assumptions (e.g. population stability). These data will be used afterwards as a kind of baseline to which all the changes in life-history parameters caused by the pollutant will be related and used to construct the projection matrix for calculating the population effects of particular treatments.

If the available data allow, one should distinguish between 'birth-flow' and 'birth-pulse' populations (Caswell, 1989) during parametrization of the matrix model. This distinction leads to different methods of calculating entries to the basic projection matrix: survival probabilities (P_i) and fertilities (F_i). Working with invertebrates from temperate climate zones, in the majority of cases we will be dealing with birth-pulse populations, that is, organisms which reproduce during a short breeding season (e.g. snails). Nevertheless, in some cases reproduction can be spread almost continuously over the year (e.g. centipedes; Albert, 1983b), and birth-flow methods may appear more appropriate.

For the purpose of this chapter, however, it will be sufficient to define P for birth-pulse populations counted after reproduction (postbreeding census; for more details see Caswell, 1989). In this case, probability of survival P_i for age class i is written as

$$P_i = \frac{l_i}{l_{i-1}} \tag{15.1}$$

where l_i = probability of surviving from birth to age i. For simplicity, we shall define fertility F_i in age class i as the number of live offspring per female aged i. The reader should be aware, however, that the exact method for calculating values of F_i for a particular population will depend on whether the population is birth-flow or birth-pulse type, on when the census takes place and on what time interval is used for projection. All these methods are reviewed and described in detail by Caswell (1989).

Having done this, a regular Leslie matrix may be constructed in which fertilities F_i appear in the first row and probabilities of survival P_i in the subdiagonal:

$$
\begin{matrix}
F_1 & F_2 & F_3 & \cdots & F_i \\
P_1 & 0 & 0 & \cdots & 0 \\
0 & P_2 & 0 & \cdots & 0 \\
\cdot\cdot & \cdot\cdot & \cdot\cdot & \cdots & \cdot\cdot \\
0 & P & \cdot\cdot & P_{i-1} & 0
\end{matrix}
$$

Together with information about the actual age structure of the population, this matrix permits inferences about the dynamics of the population.

According to the 'strong ergodic theorem' (Cohen, 1979), the long-term dynamics of the population is described completely by the stable age structure of the population and the dominant eigenvalue λ of the Leslie matrix. The dominant eigenvalue λ, being the population growth rate, is related to the intrinsic rate of increase such that $\lambda = e^r$. If $\lambda > 1$, the population increases in number; $\lambda < 1$ means a population decline.

Having parameterized the projection matrix, we may calculate λ for the unaffected population. It will usually differ from 1, and for the purpose of this study it is more convenient to have a stable population for future comparison with toxicant-affected populations. Because we are not really interested in the dynamics of any particular population but are rather attempting to create an artificial population as the baseline for the experimental results, the calculated parameters have to be adjusted to yield $\lambda = 1$. This can be done numerically by introducing small changes in population parameters and calculating a new λ each time. The procedure is finished when $\lambda = 1$. Thus, even if the available life-history data are not very precise, this should not seriously bias the final conclusions. In any event, as in all ecotoxicological studies, the results will be based upon comparison with the control treatment.

15.2.2 SENSITIVITY ANALYSIS

The next step preferably would be an initial sensitivity analysis in order to focus the studies on the parameters and/or life stages that are most important to the final outcome: fitness as measured by λ. The simplest method is to change each entry in the projection matrix by the same fraction, that is, to decrease P_i by ΔP_i and F_i by ΔF_i and to calculate a new λ after each change. Then, for each age class we may calculate the relative sensitivity for both survival probabilities ($\Delta\lambda/\Delta P_i$) and fertilities ($\Delta\lambda/\Delta F_i$). Examples of such sensitivity analyses are provided in the next sections of this chapter. Caswell (1989) discusses various methods of sensitivity analysis in more detail; see also Chapter 14 by J.E. Kammenga *et al.* in this volume.

15.2.3 ESTIMATING POLLUTION EFFECTS

We cannot collect all the data required to construct the full life-table of a pollution-treated population, so all the changes in life-history traits have to be related to the values found in unaffected organisms. Thus, rather than using direct measures of fertility or growth rates, their ratio to the rates observed in control animals will be used. Let us denote the finite rate of increase of the untreated population by $\lambda^{(1)}$ (as for the purpose of such a study we deliberately produce the projection matrix of $\lambda = 1$) and the rate of increase of the population under treatment T by $\lambda^{(T)}$. The respective transition matrices will be $\mathbf{M}^{(1)}$ and $\mathbf{M}^{(T)}$ with fertility entries $F_i^{(1)}$ and $F_i^{(T)}$, and probabilities of survival $P_i^{(1)}$ and $P_i^{(T)}$. The overall effect of treatment T on $\lambda^{(T)}$ can be written as:

$$\lambda^{(T)} = \lambda^{(1)}\delta^{(T)} \tag{15.2}$$

where $\delta^{(T)}$ is the effect of treatment T, that is, the ratio of the growth rate in the treated population to the growth rate in the untreated population. The effect $\delta^{(T)}$ summarizes all of the effects on particular matrix entries $F_i^{(T)}$ and $P_i^{(T)}$. Even if the real values for the whole life-table of the population under treatment are not known, the values of $\delta_i^{(F)}$ and $\delta_i^{(P)}$ will be measured, at least for the most sensitive matrix entries. In fact they will not be measured directly but relative to the control population, as:

$$\delta_i^{(F)} = \frac{F_i^{(E)}}{F_i^{(C)}} \tag{15.3}$$

where $F_i^{(C)}$ and $F_i^{(E)}$ are the fertilities in the control and experimentally treated populations. By analogy,

$$\delta_i^{(P)} = \frac{P_i^{(E)}}{P_i^{(C)}} \tag{15.4}$$

These estimates of $\delta_i^{(F)}$ and $\delta_i^{(P)}$ will enable us to further estimate the values of the population parameters in a 'real' population using the previously constructed projection matrix for the unaffected population ($\mathbf{M}^{(1)}$):

$$F_i^{(T)} = F_i^{(1)}\delta_i^{(F)} \tag{15.5}$$

and

$$P_i^{(T)} = P_i^{(1)}\delta_i^{(P)} \tag{15.6}$$

Having estimated the entries for $\mathbf{M}^{(T)}$, we may now calculate $\lambda^{(T)}$ and determine the population dynamics scenarios under pollution stress. Combined with regression dose–response analysis, this can be quite a powerful method, as it makes it possible to estimate population effects for any dose or concentration of pollutant whose effect on survival and/or fertility can be read from the dose–response curve.

15.3 EXAMPLE 1: THE SNAIL *HELIX ASPERSA*

Snails and slugs are a common component of terrestrial invertebrate fauna. According to Russell-Hunter (1983), molluscs come close to annelid worms (including earthworms) in bioenergetic importance. Thus, they are a crucial element in nutrient turnover, at least in some terrestrial ecosystems (Godan, 1983). On the other hand, they are known to have a relatively high potential for accumulation of heavy metals. For example, Coughtrey and Martin (1977) found c. 50 μg Cd, 400 μg Zn and 90 μg Cu g^{-1} dry mass of snails. Extremely high concentrations were found in *H. aspersa* by Jones (1991) in his studies in the neighbourhood of the Avonmouth smelter (UK): >70 μg Cd, >500 μg Cu, < 300 μg Pb and >900 μg Zn g^{-1} snail tissue (see also Hopkin, 1989). Both of these characteristics are good reasons to learn more about the effects of pollutants on snails.

Some larger snails may live for 6–7 years or even more (Cain, 1983; Godan, 1983). Besides the usual problems in studying the ecotoxicology of long-lived iteroparous invertebrates, with snails there is also a second, very specific problem: their ability to aestivate under unfavourable environmental conditions. This makes them particularly difficult subjects for ecotoxicological studies, because all these studies are based on creating unfavourable conditions. It is not surprising that, although there is relatively abundant literature on the toxicology of pollutants in snails (Beeby and Richmond, 1989; Berger and Dallinger, 1989; Simkiss and Watkins, 1990; Grenville and Morgan, 1991; Dallinger *et al.*, 1993), there are no data on population effects.

Thus, to estimate the effect of any pollutant/toxicant on fitness in snails, some indirect measures have to be used, as explained above. First we have to decide which parameters are crucial and which can be considered negligible from the point of view of their impact on λ (see Chapter 14 by J.E. Kammenga *et al.* in this volume).

Data on the life history of *H. aspersa* collected from various sources (Peake, 1978; Daguzan, 1981; Godan, 1983; Tompa, 1984) allow us to construct a projection matrix to be used in further analyses. The survival and fertility values are adjusted to result in a stable population (see Box 15.1). Sensitivity was analysed numerically by changing (decreasing) each entry in the matrix by a constant percentage (here, 10%). Pollutants may act on survival alone in the first two age classes, because according to our projection matrix (Box 15.1) snails do not reproduce for the first 2 years (though in many large snails such as *H. aspersa* or *H. pomatia* there may be substantial variation in the reproductive age due to natural environmental factors such as droughts (Cain, 1983; Russell-Hunter and Buckley, 1983). In fact, the survival probabilities for the first two classes have the highest sensitivities of all the matrix entries (Figure 15.1). The sensitivity of survival in subsequent age classes seems negligible. Although the sensitivity values for fertility ($\Delta\lambda/\Delta F$) are never as high as those for the probability of survival ($\Delta\lambda/\Delta P$) in age-classes 0 and 1, it is fertility that is clearly more important in class 2 (Figure 15.1).

In snails, if excessive aestivation is caused by pollution rather than natural environmental factors, this will probably lead to delayed maturity which will not be compensated by prolonged life-span; hence reproduction may not occur for one or more years. If we test this, keeping the other matrix entries constant, it appears that this is by far the most important possibility. When one year is lost for reproduction, with other entries unchanged, λ decreases by 0.279. Decreasing all entries by 10% at the same time (but without losing a year) lowers λ by only 0.1. The analysis suggests that one should focus on

Box 15.1 Age-structured (years) projection matrix for *Helix aspersa*, showing age-specific fertilities in the first row and age-specific survival probabilities in the subdiagonal; parameters adjusted to result in a stable population ($\lambda = 1$)

0	0	75	75	75	75
0.0505	0	0	0	0	0
0	0.2	0	0	0	0
0	0	0.25	0	0	0
0	0	0	0.25	0	0
0	0	0	0	0.15	0

survival probabilities for age class 0 and/or 1, and fertility for age class 2. Additionally, and perhaps above all, the growth rate of juveniles should be measured to assess whether maturity can be delayed by pollution.

R. Laskowski and S. Hopkin (unpublished data) have shown that *H. aspersa* fed on a diet highly contaminated with a mixture of Zn, Cu, Pb and Cd eat much less food than do the controls. The animals had prolonged aestivation and did not grow. No mortality increase due to pollution was observed. Fertility also seemed quite insensitive over a broad range of concentrations. For nominal experimental concentrations ranging from $0.2 + 2 + 20 + 20$ µg g^{-1} Cd + Cu + Zn + Pb up to $5 + 50 + 500 + 500$ µg g^{-1}, no significant decrease in fertility was observed. However, above the latter level, fertility declined

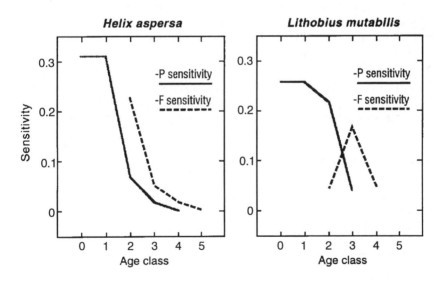

Figure 15.1 The sensitivity of finite rate of increase (l) to changes in age-specific survival probabilities (*P*) and fertilities (*F*) in populations of snails (*Helix aspersa*) and centipedes (*Lithobius mutabilis*).

quickly, and at the highest concentrations (125 + 1250 + 12 500 + 12 500 μg g⁻¹ Cd + Cu + Zn + Pb) the animals ceased to reproduce (Table 15.1). Linear regressions were used to calculate the effects of treatments relative to the controls for fertility $\delta^{(F)}$ (as the number of animals hatched per one parent) at particular heavy metal concentrations.

As calculated from the regressions, combinations of Cd + Cu + Zn + Pb at nominal experimental concentrations of 10 + 100 + 1000 + 1000 μg g⁻¹ caused a barely noticeable decrease in fertility, –2%. However, the consumption rate in juveniles decreased by –38% and the growth rate by approximately –10%, meaning that the animals did not grow at all (final mass/initial mass ≈ 1). These results suggest that at this level of pollution the decreased consumption rate, possibly leading to the loss of one year in reproduction, can be a major cause of population decline ($\lambda = 0.718$; Figure 15.2, line C). If the animals did not lose a year in reproduction, the population probably would not suffer significantly, as even after 50 years its number would remain high and λ would equal 0.994 (see Figure 15.2, line A). During this time, some level of adaptation to a polluted environment could probably evolve (Beeby and Richmond, 1987; Grenville and Morgan, 1991).

Only a doubling of these concentrations, that is, 20 + 200 + 2000 + 2000 μg g⁻¹ Cd + Cu + Zn + Pb, causes a sharp decline in fertility (–31%). At this level of pollution the decrease in fertility has a clear effect on average fitness in the population: $\lambda = 0.895$ (Figure 15.2, line B). Of course, the consumption and growth rates in juveniles decrease further (–55% and –12% respectively) and the loss of one year in reproduction seems even more probable. The latter scenario is illustrated by line D in Figure 15.2. For this level of pollution both scenarios lead to fast extinction of the population.

These results (R. Laskowski and S. Hopkin, unpublished data) suggest that, in the case of iteroparous snails such as *H. aspersa*, the main effect of heavy metal pollution on their fitness may be through delayed maturation due to decreased consumption and prolonged aestivation. At very high concentrations of heavy metals, a sharp decrease in fertility may be crucial. All of these

Table 15.1 Effects of food contamination with a mixture of Cd, Cu, Zn and Pb on consumption and fertility of *Helix aspersa*

Nominal concentration of Cd+Cu+Zn+Pb (mg kg⁻¹)	Consumption (mg day⁻¹ g⁻¹)		Fertility (eggs per snail)
	Juveniles	Adults	
0+0+0+0 (control)	4.65	8.83	32.0
0.2+2+20+20	3.31	8.22	27.5
1+10+100+100	5.17	8.00	33.9
5+50+500+500	4.11	8.88	46.2
25+250+250+250	1.62	3.56	12.9
125+125+12 500+12 500	0.20	0.78	0.00

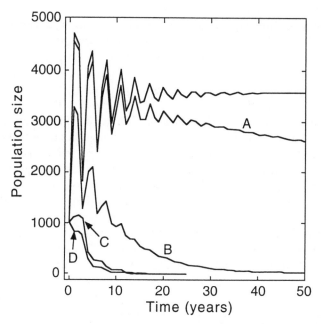

Figure 15.2 The four scenarios for the dynamics of pollution-suppressed populations of *Helix aspersa* as compared with the untreated population (undisturbed). A, food contaminated with 10 μg g⁻¹ Cd + 100 μg g⁻¹ Cu + 1000 μg g⁻¹ Pb + 1000 μg g⁻¹ Zn, without loss of a year in reproduction. B, food contaminated with 20 μg g⁻¹ Cd + 200 μg g⁻¹ Cu + 2000 μg g⁻¹ Pb + 2000 μg g⁻¹ Zn, without loss of a year in reproduction. C and D, as A and B respectively, but with one year lost in reproduction. Initial high variation in population numbers results from the obviously unstable initial age structure.

possibilities are illustrated in Figure 15.2 as probable scenarios for the dynamics of a population of *H. aspersa* exposed to high levels of heavy metal pollution.

15.4 EXAMPLE 2: THE CENTIPEDE *LITHOBIUS MUTABILIS*

In many forest ecosystems, chilopods are the most important carnivorous arthropods. As estimated by Albert (1983a), two species only, *Lithobius mutabilis* and *L. curtipes*, account for 17–27% of total assimilation by predatory arthropods in German beech forests. These two species were clearly dominant in the forests studied by Albert (1976). The average yearly size of the *L. mutabilis* population was estimated as 41 individuals m⁻² (113 mg dry weight m⁻²), and of *L. curtipes* as 32 individuals m⁻² (39 mg dry weight m⁻²). These figures are very similar to those for biomass of spiders and carnivorous beetles in the same stand (Albert, 1976). Poser (1988) found the average yearly biomass of all centipedes in a beech forest to be as high as 265 mg dry weight m^{-2}. Due to their high and constant density throughout the year and wide prey spectrum, Wignarajah and Phillipson (1977) suggested that centipedes are relatively more important than other woodland invertebrates as predators. Because they are relatively high in the trophic chain and also perennial, they certainly deserve close attention from ecotoxicologists.

According to Albert (1983b) there are five larval (L0–L4) and up to nine adult stages during the development of *L. mutabilis*. The animals usually reach sexual maturity no earlier than the fifth adult stage. Oviposition takes place throughout the year, probably with one main peak during the spring and a smaller one during the autumn. The reproductive potential is low; the female produces only one egg over several days. The maximum number of eggs produced yearly by one *L. mutabilis* female was observed to be 42, but medium-sized females lay 22 eggs on average per year. The largest and oldest individuals usually do not exceed 10. Albert (1983b) points out that the real number is probably somewhat higher as the eggs are extremely difficult to find. Egg development time is highly variable, between 27 and 269 days (Albert, 1983b). The long oviposition period together with the variability in egg development time mean that there can be a lag of 1.5 years between the appearance of the first and last newly hatched larvae from one mating (Albert, 1983b). Most juveniles hatch in the spring. Some animals enter the first adult stadium before their first winter. Some overwinter at the last larval stage (L4) or even at L3 and enter the first adult stage in spring. To reach sexual maturity, *L. mutabilis* needs nearly 2 years. Sexually mature individuals can live for about 2–3 more years. The total life span of *L. mutabilis* is estimated at 5 or 6 years (Albert, 1983b).

As there are no good data on egg survival in the literature, only general statements that 'mortality is probably very high', we may try to estimate it in order to construct the full life-table. On the basis of data from Albert (1983b), the calculated average duration of egg development is c. 4.5 months. During the first 4.75 months of life (stages L0–L3), approximately 75% of the newly hatched animals die. This figure can be taken as a rough estimate of egg mortality during their 4.5 month development. Using these estimates and assuming that the untreated population persists in a steady state (i.e. $\lambda = 1$), we may construct the projection matrix given in Box 15.2.

Sensitivity analysis yields results somewhat different from those obtained for *H. aspersa* (Figure 15.1). Generally the sensitivities of age-specific survival probabilities are more evenly distributed among age classes, being the highest for the first three. The most sensitive age-specific fertility is for age class 3, with sensitivities for both earlier and later classes notably lower. Thus, in the case of *L. mutabilis* one should probably focus on pollution effects on survival in age classes between 0 and 2 and fertility in age class 3.

As no aestivation was found in centipedes and it is known that they may remain active even during winter (Lewis, 1981; Tursman *et al.*, 1994), one may assume that losing the first year of reproduction (as in the case of snails)

Box 15.2 Age-structured (years) projection matrix for *Lithobius mutabilis*, showing age-specific fertilities in the first row and age-specific survival probabilities in the subdiagonal; parameters adjusted to result in a stable population ($\lambda = 1$)

0	0	5	22	13
0.049	0	0	0	0
0	0.69	0	0	0
0	0	0.88	0	0
0	0	0	0.456	0

is not a very likely possibility. Even if it happened, the impact on λ would be substantially smaller than a simultaneous 10% decrease in survival probability and fertility (–0.038 versus –0.1; compare with estimates for *H. aspersa*).

Although the distributions of sensitivities between age classes are different for the centipede and the snail (Figure 15.1), the overall effects on fertility or probability of survival are similar. The decrease in λ due to a 10% reduction of fertilities in all age classes is –0.031 for snails, and –0.026 for centipedes; the decreases due to reduction of survival probabilities are –0.071 and –0.076, respectively.

No data are available at present to calculate the 'real' effects of heavy metal pollution on the finite rate of increase in populations of centipedes. Hopefully this will be done in future using the method described above for snails. If the experimenter is able to determine the dose–response relationships, and if the data in the literature allow reliable projection matrices to be constructed, the method described here can be easily adapted for other iteroparous species whose detailed ecotoxicological life-history data are very difficult, too costly or impossible to obtain in the laboratory.

15.5 NEW PROSPECTS FOR RISK ASSESSMENT

The examples in this chapter illustrate the importance of sensitivity analysis for planning ecotoxicological experiments. In various species, different life-history parameters are more sensitive to the effects of pollutants and thus should be primary in particular ecotoxicological studies. As pointed out by J.E. Kammenga (Chapter 14, this volume), neglecting this information may lead to serious misinterpretation of the experimental results in terms of the actual effects of toxicants on population levels.

The scenarios projected for the snail *H. aspersa*, using experimental data on heavy metal toxicity, revealed that a combination of metals at the concentrations present in the soil and plants in areas exposed to industrial pollution may lead to fast extinction of its populations, particularly if reproduction is delayed by metal toxicity or unpalatability of food. Without estimation of pollution effects on particular life-history parameters and use of matrix algebra, such effects are impossible to assess in standard short-term ecotoxicological experiments. Matrix population models have proved useful for estimating the ecotoxicological effects of persistent pollutants for iteroparous long-lived animals such as snails. While the life-table approach for short-lived and fast-reproducing soil invertebrates has been increasingly used in recent years (Van Straalen *et al.*, 1989; T. Crommentuijn *et al.*, Chapter 13, this volume), its use in the ecotoxicology of perennial iteroparous species is non-existent. The approach presented here opens new prospects for risk assessment, making it possible to incorporate a large group of long-lived invertebrates largely neglected thus far. As many animals from this group are important elements in the functioning of the soil/litter subsystem (see the introduction to this chapter), this approach should significantly increase the predictability of pollution effects at levels higher than single populations.

Both examples in this chapter were based on a simple age-classified matrix model, but the same method can be used for size- or stage-classified models (Caswell, 1989). The confidence limits for λ can also be estimated. Unfortunately, at present there is no analytical method for calculating confidence intervals for λ. This problem may be solved to some extent with computer-intensive methods such as jack-knife and bootstrap.

15.6 SUMMARY

Although many iteroparous invertebrates have crucial roles in the functioning of terrestrial ecosystems, knowledge of their responses to chemicals is sparse. There are difficulties in estimating the fitness costs of pollution in iteroparous animals in standard short-term ecotoxicological experiments. This chapter describes how these fitness costs may be estimated using the data in the literature on life-histories and a matrix modelling approach to experimental data. Though not as precise as direct measurements of fitness parameters (feasible for populations of short-lived, fast-reproducing animals), these estimates may allow population dynamics scenarios under various pollution stresses to be examined. Examples are given for the iteroparous snail *Helix aspersa* Müll and the centipede *Lithobius mutabilis* L. Koch. Both are used to illustrate how sensitivity analysis may be used in planning an experiment. For the snails the most sensitive life-history parameters are survival probabilities for age-classes 0 and 1, fecundity for age-class 2, and retardation of sexual maturity. For the centipedes the most important parameters are survival probabilities for age-classes 0–2 and fertility for age-class 3. Scenarios of the population dynamics of snails under pollution stress were projected from the experimental data. The scenarios show that Cd + Cu + Zn + Pb at concentrations of 10 + 100 + 1000 + 1000 μg g^{-1} in dry mass food may decrease λ by approximately 0.6% if no delay in reproduction occurs. However, if one year is lost in reproduction, the decrease in λ is approximately 28%, leading to rapid extinction of the population.

ACKNOWLEDGEMENTS

I wish to thank all members of the Ecotoxicology Group at the University of Reading, UK, for their stimulating discussions, and Drs S. Hopkin and R. Sibly for their critical comments on earlier versions of the manuscript. The comments of the anonymous reviewers and the Editors are also greatly appreciated. M. Jacobs helped edit the text. Funding for this work was provided by the EU Programme Environment 1990–1994 (PECO Programme contract no. ERB-CIPD-CT93-0059). Financial support from the Leverhulme Trust is also gratefully acknowledged.

REFERENCES

Albert, A.M. (1976) Biomasse von Chilopoden in einem Buchenaltbestand des Solling. *Verhand. Ges. Ökologie*, Göttingen, 93–101.

Albert, A.M. (1983a) Energy budgets for populations of long-lived arthropod predators (Chilopoda: Lithobiidae) in an old beech forest. *Oecologia (Berlin)*, **56**, 292–305.

Albert, A.M. (1983b) Life cycle of Lithobiidae – with a discussion of the *r*- and *K*-selection theory. *Oecologia (Berlin)*, **56**, 272–9.

Beeby, A. and Richmond, L. (1987) Adaptation by an urban population of the snail *Helix aspersa* to diet contaminated with lead. *Environ. Pollut.*, **46**, 73–82.

Beeby, A. and Richmond, L. (1989) The shell as a site of lead deposition in *Helix aspersa*. *Arch. Environ. Contam. Toxicol.*, **18**, 623–8.

Bengtsson, G., Gunnarsson, T. and Rundgren, S. (1985) Influence of metals on reproduction, mortality and population growth in *Onychiurus armatus* (Collembola). *J. Appl. Ecol.*, **22**, 967–78.

Bengtsson, G., Gunnarsson, T. and Rundgren, S. (1986) Effects of metal pollution on the earthworm *Dendrobaena rubida* (Sav.) in acidified soils. *Water Air Soil Pollut.*, **28**, 361–83.

Berger, B. and Dallinger, R. (1989) Accumulation of cadmium and copper by the terrestrial snail *Arianta arbustorum* L.: kinetics and budgets. *Oecologia*, **79**, 60–5.

Cain, A.J. (1983) Ecology and ecogenetics of terrestrial molluscan populations, in *The Mollusca, Vol. 6: Ecology* (ed. W. D. Russell-Hunter), Academic Press, London, pp. 597–647.

Caswell, H. (1989) *Matrix Population Models*, Sinauer Associates, Inc. Publishers, Sunderland, Massachusetts.

Chandler, G.T. (1990) Effects of sediment-bound residues of the pyrethroid insecticide fenvalerate on survival and reproduction of meiobenthic copepods. *Mar. Environ. Res.*, **29**, 65–76.

Cohen, J.E. (1979) Ergodic theorems in demography. *Bull. Amer. Math. Soc.*, **1**, 275–95.

Coughtrey, P.J. and Martin, M.H. (1977) The uptake of lead, zinc, cadmium and copper by the pulmonate mollusc, *Helix aspersa* Müller, and its relevance to the monitoring of heavy metal contamination of the environment. *Oecologia (Berlin)*, **27**, 65–74.

Crommentuijn, T., Brils, J., and Van Straalen, N.M. (1993) Influence of cadmium on life-history characteristics of *Folsomia candida* (Willem) in an artificial soil substrate. *Ecotox. Environ. Safety*, **26**, 216–27.

Daguzan, J. (1981) Contribution à l'élevage de l'escargot Petit-Gris: *Helix aspersa* Muller. I. Reproduction et éclosion des juenes, en bâtiment et en conditions thermohygrometriques controlées. *Ann. Zootech.*, **30**, 249–72.

Dallinger, R., Berger, B. and Gruber, A. (1993) Quantitative aspects of zinc and cadmium binding in *Helix aspersa*: differences between an essential and a nonessential trace element, in *Ecotoxicology of Metals in Invertebrates* (eds R. Dallinger and P.S. Rainbow), Lewis Publishers, Boca Raton, pp. 315–32.

Daniels, R.E. and Allan, J.D. (1981) Life-table evaluation of chronic exposure to a pesticide. *Can. J. Fish. Aquat. Sci.*, **38**, 485–94.

Fernandez-Caselderrey, A., Ferrando, M.D. and Andreu-Moliner, E. (1993) Effects of endosulfan on survival, growth and reproduction of *Daphnia magna*. *Comp. Biochem Physiol.*, **106C**, 437–41.

Godan, D. (1983) *Pest Slugs and Snails: Biology and Control*, Springer-Verlag, Berlin.

Greville, R.W. and Morgan, A.J. (1991) A comparison of (Pb, Cd and Zn) accumulation in terrestrial slugs maintained in microcosms: evidence for metal tolerance. *Environ. Poll.*, **74**, 115–27.

Hopkin, S.P. (1989) *Ecophysiology of Metals in Terrestrial Invertebrates*, Elsevier Applied Science Publishers Ltd, London.

Hopkin, S.P. and Martin, M.H. (1983) Heavy metals in the centipede *Lithobius variegatus* (Chilopoda). *Environ. Pollut. (B)*, **6**, 309–18.

Jennings, T.J. and Barkham, J.P. (1979) Litter decomposition by slugs in mixed deciduous woodland. *Holarctic Ecol.*, **2**, 21–9.

Jones, D. (1991) *Biological Monitoring of Metal Pollution in Terrestrial Ecosystems*. PhD Thesis, Department of Pure and Applied Zoology, University of Reading, UK.

Kidd, N.A.C. (1991) The implication of air pollution for conifer aphid population dynamics: a simulation analysis. *J. Appl. Entomol.*, **111**, 166–71.

Laskowski, R. and Maryarski, M. (1993) Heavy metals in epigeic fauna: trophic-level and physiological hypotheses. *Bull. Environ. Contam. Toxicol.*, **50**, 232–40.

Lawrence, S.G., Holoka, M.H. and Hamilton, R.D. (1989) Effects of cadmium on microbial food chain, *Chlamydomonas reinhardii* and *Tetrahymena vorax*. *Sci. Total Environ.*, **87/88**, 381–95.

Lewis, J.G.E. (1981) *The Biology of Centipedes*, Cambridge University Press, Cambridge.

Peake, J. (1978) Distribution and ecology of Stylommatophora, in *Pulmonates, Vol. 2A: Systematics, Evolution and Ecology* (eds V. Fretter and J. Peake), Academic Press, London-New York-San Francisco, pp. 429–526.

Poser, T. (1988) Chilopoden als Prädatoren in einem Laubwald. *Pedobiologia*, **31**, 261–82.

Rożen, A. (1989) The annual cycle in populations of earthworms (Lumbricidae, Oligochaeta) in three types of oak–hornbeam of the Niepolomicka Forest. II. Dynamics of population numbers, biomass and age structure. *Pedobiologia*, **31**, 169–78.

Russell-Hunter, W.D. (1983) Overview: planetary distribution of and ecological constraints upon the Mollusca, in *The Mollusca, Vol. 6: Ecology* (ed. W.D. Russell-Hunter), Academic Press, London, pp. 1–27.

Russell-Hunter, W.D. and Buckley, D.E. (1983) Actuarial bioenergetics of nonmarine molluscan productivity, in *The Mollusca, Vol. 6: Ecology* (ed. W.D. Russell-Hunter), Academic Press, London, pp. 463–503.

Simkiss, K. and Watkins, B. (1990) The influence of gut microorganisms on zinc uptake in *Helix aspersa*. *Environ. Pollut.*, **66**, 263–71.

Snell, T.W., Moffat, B.D., Janssen, C. and Persoone, G. (1991) Acute toxicity tests using rotifers III. Effects of temperature, strain and exposure time on the sensitivity of *Brachionus plicatilis*. *Environ. Toxicol. Water Qual.*, **6**, 63–76.

Tompa, A.S. (1984) Land snails (Stylommatophora), in *The Mollusca, Vol. 7: Reproduction* (ed. A.S. Tompa), Academic Press, London, pp. 47–140.

Tursman, D., Duman, J.G. and Knight, C.A. (1994) Freeze tolerance adaptations in the centipede, *Lithobius forficatus*. *Exp. Zool*, **268**, 347–53.

Van Leeuwen, C.J., Rijkeboer, M. and Niebeek, G. (1986) Population dynamics of *Daphnia magna* as modified by chronic bromide stress. *Hydrobiologia*, **133**, 277–85.

Van Straalen, N.M: and Ernst, W.H.O. (1991) Metal biomagnification may endanger species in critical pathways. *Oikos*, **62**, 255–6.

Van Straalen, N.M., Schobben, J.H.M. and De Goede, R.G.M. (1989) Population consequences of cadmium toxicity in soil microarthropods. *Ecotox. Environ. Safety*, **17**, 190–204.

Walsh, G.E. (1983) Cell death and inhibition of population growth of marine unicellular algae by pesticide. *Aquat. Toxicol.*, **3**, 209–14.

Wignarajah, S. and Phillipson, J. (1977) Numbers and biomass of centipedes (Lithobiomorpha: Chilopoda) in a *Betula–Alnus* woodland in N.E. England. *Oecologia (Berlin)*, **31**, 55–66.

PART SEVEN

Recommendations

16 Soil ecotoxicology: still new ways to explore or just paving the road?

HERMAN EIJSACKERS
WITH CONTRIBUTIONS FROM
S. RUNDGREN, M.H. DONKER, R. SCHULIN, J. WILES AND M. ROSSEL

16.1 MAJOR CHALLENGES

This chapter presents recommendations of a workshop on Ecological Principles for Risk Assessment of Contaminants in Soil, Papendal Arnhem, The Netherlands, 2–5 October, 1994, where the authors presented the draft chapters of this book. The discussions focused on the following three major challenges for the further development of ecotoxicological risk assessment methods for soil contamination:

1. How to provide preventive (provisional) limits for the many thousands of compounds present and further dispersed into the environment.
2. How to tackle the complex composition of soils and the resulting complexities due to availability, both physicochemical and biological, immediate as well as delayed.
3. How to assess the possible impacts on many thousands of species and integrate these to higher levels of biological organization.

These three challenges formed the heart of the workshop. It has generally been accepted that there is a major shortage of data on possible adverse effects of chemical compounds on biological life in soil. Soil is an extremely complicated substrate which does not allow for easy, linear availability relationships. To assess risks in soil properly, a more integrated approach is needed. This was also recognized during a workshop in Silkeborg, Denmark, in January 1992, which formed the start of a research network aiming at Soil Ecotoxicological Risk Assessment Systems (SERAS). The workshop covered:

Ecological Risk Assessment of Contaminants in Soil. Edited by Nico M. van Straalen and Hans Løkke. Published in 1997 by Chapman & Hall, London. ISBN 0 412 75900 4

- soil ecology (natural fluctuations in soil processes and populations of soil organisms);
- chemicals in the environment (their distribution and availability to organisms of chemicals reaching the soil);
- assessment of hazards and risks (potential adverse effects on soil processes and soil organisms caused by chemicals in the soil environment, structural and functional characteristics);
- choice of test species (criteria on which to select organisms appropriate for study);
- field studies (the need for and design of field monitoring and experiments to determine chemical effects); and
- agenda for further research and policy-support activities.

Recommendation 1. Further activities for ecotoxicological risk assessment of soil contaminants, whether policy-supporting or research-oriented, should be structured according to an integrated approach paying attention to preventive comparative assessment procedures, availability assessment and ecological integration. Workshops should be planned accordingly, not only to give overviews, but also to present vision papers.

The Papendal workshop had a special structure. It was built up theme-wise from lower to higher levels of ecological organization: starting with test systems at the species or process level, moving up one level to communities, and the structure–function relationship with, as a next step, the systems and landscape level, and ending at the modelling level. Each session consisted of an overview paper and a vision paper, followed by a number of case studies and incorporating ample discussion time. This 'triple' approach was especially effective, with an overview to set the scene, a series of new and sometimes visionary ideas, and case studies forming an ideal setting for some very rewarding discussions. All speakers were requested to send in extended summaries well before the workshop, and to prepare manuscripts for inclusion in this book.

16.2 GENERALIZATION ACROSS SPECIES AND COMPOUNDS

At present, research is mainly focused on test development at the individual, species, and population level. Research activities are concentrated on further optimization of already developed tests, further calibration of tests, but also (still!) on the development of new tests. Only limited effort is put into a proper ecological validation of these tests, although during the workshop it was clearly recognized that this was one of the most important gaps in our knowledge.

As a further complication, the test species which are used most frequently tend to have been selected for practical experimental reasons, for example, ease of handling, culturing, knowledge of their biology and life cycle, rather

than for ecological criteria, such as their functional importance in soil process-es, their potential as 'keystone' or 'indicator' species or ecosystem engineers or their importance for food-web stability or ecosystem process functioning (see J.C. Moore and P.C. de Ruiter, Chapter 7, this volume). Projects to study and select suites of test species in this way are currently under way.

From a study on Quantitative Species Sensitivity Relationships (QSSR) for a large dataset of aquatic tests it became apparent that, for the species groups studied, the variability in sensitivity between species is far less than the vari-ability in toxicity between the compounds with a general mode of action. This may provide an argument to reduce activities in exchange for a broadening of our already considerable range of test species.

Recommendation 2. Given the observed limited variability in sensitivity between species, it is recommended to investigate which groups are sim-ilar, and which groups deviate, in this way focusing the development of further test systems on proper ecological (field) validation and calibration of tests.

To further restrict the number of necessary tests, the application of inter-polation and extrapolation of QSAR-procedures has proven to be most useful. Especially for non-ionic narcotics, the responses of various test species fall within a rather narrow range which leaves room for the use of a 'mean toxic-ity with a reasonable variability range'. For a number of other compound groups this approach is less successful, with large ranges in variability and greatly differing ranges for chemically related compounds.

Recommendation 3. Future activities should concentrate more on devi-ating or outlying compounds, instead of further supporting the already well-fitted QSAR-relationships for narcotic compounds. For these last compounds inter- and extrapolating QSAR-relationships should be applied.

16.3 IMPROVING EXPOSURE ASSESSMENT

One generally acknowledged problem is the complexity of soils. This com-plexity is present at different levels. Scaling down from the diversity in soil types, via the differences between the various layers within a soil profile, to the complexity between different soil constituents in a soil aggregate, and ending with the complexity of soil minerals and clay particles, we seem to meet a fractal sequence. Due to this variability it is not possible to represent exposure concentrations as being generally valid and uniform. This is espe-cially relevant as most soils do not show the homogeneous composition of the top layer of an arable soil. Contaminants are nearly always distributed in an irregular pattern, in relation to specific soil features, and detailed studies on the actual distribution must be carried out.

Recommendation 4. To assess the exposure of soil organisms to soil pollution in a proper way, it is necessary to study the distribution of contaminants in detail, and to deduce by which route which soil organism is most likely to be exposed, depending on its behaviour and uptake mechanisms.

The deficiency of terrestrial toxicity data is presently compensated by obtaining data from aquatic tests, thus providing a far larger database. It is assumed that, for the majority of compounds and the majority of soil organisms, uptake through the soil water phase is the main route. For a great number of compounds and organisms this is effectively a proper assumption. However, further research is needed with those compounds for which this approach is invalid, and for those non-soft-bodied soil organisms which take up compounds by other routes.

The so-called equilibrium-partitioning approach has been used so far like a closed – or at least only partly opened – box (see C.A.M. van Gestel, Chapter 2, this volume). The various uptake mechanisms should therefore be studied in combination with toxicokinetics in order to place the various mechanisms into a more general valid framework, including those mechanisms involved in uptake, internal transport, storage and excretion.

Recommendation 5. Inter-species differences in sensitivity should be further studied within the context of toxicokinetics, more specifically by grouping organisms with similar biochemical–physiological mechanisms to handle contaminants.

16.4 INTERPRETING BIOCHEMICAL EFFECTS

Over the past few years, there has been a repeated and widespread call for biomarkers, as these should provide general indications (markers) of adverse effects. Although general indications may overcome the great variety in test species, and generally occurring biochemical–physiological mechanisms ensure similarity between species, there is also a great danger in the application of biomarkers. It bears the hidden implications of 'panacea' when biomarker responses are merely coupled to ecotoxicological or toxicological end-points, thereby suggesting a causal relationship. However, this has not been proven at all; indeed, sometimes not even a proper correlation is shown.

Recommendation 6. General biochemical–physiological end-points might be useful as a first and fast indication of adverse impacts. Given the general character of the end-point, a great variety of species might become affected. However, further research is needed on the precise causes of the phenomenon and the potential effects. Proper ecological interpretation of the data plays a central role in this context.

Another, perhaps even more valid, down-scaling process concerns genetic impacts. Activities in this direction have recently begun (L. Posthuma,

Chapter 5, this volume). Impacts on this level influence the population as such, but also indicate long-term developments. Changes in fitness and competitiveness in the longer term may have major implications for species competition and resulting selection processes. This kind of genetic 'erosion' does not only work out on the direct competitiveness of the organisms. Indirect long-term implications are, for example a reduced effectiveness of the moisture regime in isopods under heavy metal stress. The trade-off between adaptation and functioning can also affect processes executed by organisms. For example, microorganisms adapted to heavy metals show a reduced ability to degrade complex organic substrates. The PICT approach provides a powerful and useful tool in this direction (see L. Posthuma, Chapter 5, this volume). Especially with microorganisms and microbial processes it may link diversity of functions with ecosystem processes. The number of soil microorganisms able to carry out (microbial) functions related to the major mineral cycles, can be expressed as functional diversity (see J. Dighton, Chapter 3, this volume). The impact of contaminant stress on functional diversity, the adaptation in functional diversity and the cost by which it is achieved (trade-off), link biochemical–physiological processes at the sub-population level directly to the ecosystem level.

> **Recommendation 7**. More attention to adaptation reactions is needed, not only to assess the potential of an 'adapted' system to maintain basic ecosystem functions in a contaminated area, but also to assess ecologically the long-term implications of soil contamination. More effort must be put into genetic impacts, and tools must be developed to assess the genetic diversity of systems under stress.

16.5 BIOTIC INTERACTIONS ARE THE KEY ELEMENTS OF SOIL COMMUNITIES

Just as important as down-scaling – perhaps even more so – is up-scaling from the population via the community to the ecosystem level. Biotic interactions are the key elements that typify communities and distinguish communities from populations. Interactions have received attention in mesocosm studies, especially in aquatic ecotoxicity. However, within soil ecotoxicology these types of study are rather scarce, except for some classical examples of predator–prey studies and a few on the impact of contaminants on species interactions. Soil–plant interactions have also received only limited attention within ecotoxicology. An interaction of special interest is mycorrhiza, the intimate species-specific symbiosis of a plant and a microorganism which connects the plant as primary food source with the nutrient cycles in the soil.

Facing large-scale processes like acidification and cold-traps for organochlorines, we have to go beyond the ecosystem level and assess contaminant effects on biogeochemical cycles. Within environmental policy the notion of 'life-support systems' can be interpreted in this way as 'the system which provides the abiotic mineral and nutrient cycles plus the biotic interactions, as necessary substrate for optimal development of biological diversity'.

In relation to biogeochemical cycles, 'critical loads' and 'environmental sustainable supplies' are also relevant. So far, these notions have been applied in relation to supra-national problems and essential nutrients such as nitrogen, phosphate, potassium and aluminium. Ecotoxicological impacts have entered this discussion only recently, and a sound ecological theory for these concepts is to be developed urgently. Critical loads are effectively physicochemical variables which describe the balance between input and output: immission into the soil and emission from it by leaching to deep groundwater or surface water, or volatilization into the air. With nitrate, microbial processes such as nitrification and ammonification are involved. The research into biogeochemical cycles, and the policy implications derived from it, have developed separately from ecotoxicology, although the basic biotic processes studied are the same. Therefore, the concept of critical loads should be extended to include the assessment of ecotoxicological effect.

Recommendation 8. Biotic interactions as the key element in communities deserve more attention, especially in relation to nutrient conversion processes. Plant–soil interactions and mycorrhiza should be paid special attention because they link plants as primary energy source with soil processes. In relation to biogeochemical processes, policy concepts such as critical loads, life-support systems and environmental sustainable supplies require a sound ecotoxicological evaluation, interpretation and operationalization.

The systems approach and modelling in general provide a systematic framework, not only to assess individual effects, but also to evaluate their relevant importance. Hence, it provides a suitable basis for comparative analysis and evaluation of risks. By putting the role of the various components in a common and proper perspective, this approach helps to ask the right questions. Within ecotoxicology there has been a recent development and application of ecosystem and food-chain models to assess the impact of contaminants.

In the Netherlands, a series of food-chain models have been developed varying from rather simple models (NOEC over BCF) to more stochastic and complex approaches combining bioconcentration factors (BCFs) with metabolic characteristics of the predator and bioavailable fraction of the contaminant in the soil. System biomass transfer and population dynamics have also been included (see J. van Wensem, Chapter 10, this volume).

Another promising approach is seen in the development of ecosystem models which calculate the consequences of the disappearance of a particular functional or taxonomic group, due to its elevated sensitivity for a particular contaminant, for the total C- and N-cycling of the system (see J.C. Moore and P.C. de Ruiter, Chapter 7, this volume).

Further developments are seen at the supra-population level: dispersal of species in relation to spatial distribution of contaminants, ecological interactions based on energy-driven physiological mechanisms and life-history models – see Chapter 8 (P.C. Jepson) and Chapter 11 (J.A. Axelsen) of this volume. The systems approach provides the necessary structure for applying basic thermodynamic theory, in particular conservation laws for setting up mass balances. This is also a sound basis for linking risk receptors in different environmental compartments. For globally dispersed contaminants this type of intercompartment fate model is of primary importance.

Recommendation 9. There is still a serious gap between scientific models which are useful to guide research and models to be used as practical tools for regulatory purposes, i.e. setting standards and ranking chemicals. Most research models are too complex and require too much input data which in practice are not available or affordable. Cutting corners in a sound and comprehensive way will be a major challenge for this field in the forthcoming period.

16.6 RE-EVALUATING BIODIVERSITY

Models, however useful they may be, are only one necessary instrument which can be used to determine a proper ecotoxicological assessment. Different management practices cannot be assessed individually in artificial isolation from each other. For example, application of pesticides is always combined with other management practices and related to specific soil and landscape features, partly resulting from these practices (see P.C. Jepson, Chapter 8, this volume). Impacts of diffuse chemical compound inputs must be combined with other stressors such as acidifying or eutrophicating compounds in a multistress approach. All these different stressors come together at the landscape level, a situation which may be called a 'toxished', in analogy to a watershed.

In order to assess total toxic impact in an area, all contaminant inputs must be explored, adding for example the various pesticides applied to different crops which have been grown on a particular lot over several years. Subsequently, the exposure of species populations must be assessed over several years, combining the higher stress in a particular area due to a specific pesticide being applied with the lower stress in neighbouring non-treated areas. This must be further combined with the mobility of the species and their dispersal potential in the landscape under assessment.

Within this context of larger spatial scales and longer time frames, biodiversity re-enters the scene – not as the somewhat sterile species lists but as a value in its own right. Diversity should be re-evaluated in the same way that it was evaluated by the classical plant sociologists – by noting in the field the dominating, regular, sparse and rare species. Plain field biology and the carrying out of inventories may provide observations on rare species which indicate subtle changes in the ecological sustainability of that system.

Putting this concept into a longer time-frame, where natural development through adaptation and succession plays a role, one might advocate an overall mapping of genetic diversity, for example in the form of a general index or as more specific mapping of deficient alleles in a population.

Although these concepts may project far into the future by comparison with today's ecotoxicological field – which is dominated by plain practical assessment procedures – we may at least assess their potential implications ultimately to develop a sound suite of 'ecological principles for risk assessment of contaminants in soil'.

Recommendation 10. To give a proper and complete assessment of contaminant stresses these must be placed into the framework of the total set of stressors in a particular area. In an anthropogenically managed landscape, effects have to be assessed over a longer time period, including spatial shifts in application or immission of toxicants. In addition to population life-histories and dispersal, biodiversity in combination with genetic diversity should be studied.

Index